MODERN CARTOGRAPHY

VOLUME TWO

Related Titles of Interest

Books

ANSON & ORMELING
Basic Cartography (for Students and Technicians) Volume 1, 2nd edition

BONHAM-CARTER
Geographic Information Systems for Geoscientists: Modelling with GIS

BOHME
Inventory of World Topographic Mapping (3 volumes)

DENEGRE
Thematic Mapping from Satellite Imagery

TAYLOR
Geographic Information Systems (The Microcomputer and Modern Cartography)

Journals

Computers & Geosciences

Computers & Graphics

Full details of all Pergamon/Elsevier Science Ltd publications/free specimen copy of any Pergamon/Elsevier Science Ltd journal available on request from your nearest Elsevier office.

Visualization in Modern Cartography

Edited by

ALAN M. MACEACHREN
Professor of Geography
The Pennsylvania State University, PA, USA

and

D. R. FRASER TAYLOR
Professor of Geography and International Affairs,
Carleton University, Ottawa, Canada

PERGAMON

U.K.	Elsevier Science Ltd, The Boulevard, Langford Lane, Kidlington, Oxford, OX5 1GB, U.K.
U.S.A.	Elsevier Science Inc., 660 White Plains Road, Tarrytown, New York 10591-5153, U.S.A.
JAPAN	Elsevier Science Japan, Tsunashima Building Annex, 3-20-12 Yushima, Bunkyo-ku, Tokyo 113, Japan

First edition 1994

Library of Congress Cataloging-in-Publication Data

Visualization in modern cartography/edited by Alan M. MacEachren and D.R. Fraser Taylor. — 1st ed.
p. cm. — (Modern cartography ; v. 2)

Includes index.
1. Cartography. 2. Visualization. I. MacEachren, Alan, M., 1952–.
II. Taylor, D.R.F. (David Ruxton Fraser), 1937–.
III. Series.
GA108.7.V58 1994
526—dc20 94-19075

British Library Cataloguing in Publication Data

A catalogue record for this book is available from the British Library

ISBN 0-08-042416-3 Hardcover
ISBN 0-08-042415-5 Flexicover

Printed in Great Britain by Galliard (Printers) Ltd, Great Yarmouth

Preface

VISUALIZATION, our organizing theme, is both an old and a new concept for cartography. This book is replete with new (and reconstituted) problems requiring research and it highlights several potential directions that an evolving cartographic practice might take (e.g. related to design of multimedia products, production and application of visualization systems, interface design, etc.). We hope that the chapters to follow will stimulate fellow cartographers to build on our rich history in exploring the exciting possibilities facilitated by recent developments in scientific visualization and multimedia.

Collaboration on this book began in December 1992 when Fraser Taylor proposed that we co-edit a book dealing with the topic of visualization in cartography. Due to various commitments, real planning did not happen until the May 1993 International Cartographic Association meeting in Cologne, Germany, where we had a chance to discuss the outline for the book and begin the process of inviting potential authors. A draft outline was prepared during a delightful boat cruise down the Rhine at the meeting's end.

Those invited to contribute to the book represent a range of cartographic visualization activities happening around the world. An effort was made to provide a balanced view of issues and activities in visualization that went beyond what was happening in individual countries or individual disciplines. Although the mix of authors is biased toward geography (the discipline of the co-editors), views from researchers in other mapping sciences as well as computer science and psychology are represented.

Our finished product represents the efforts of many people, asked to work on short deadlines. In addition to preparing their own chapters, each lead author was asked to comment on one or more drafts of other chapters in the book. The goal here was both to insure a high quality product and to foster the sharing of ideas about visualization before chapters took their final form. In addition, each chapter (except the first by MacEachren and the last by Taylor) was reviewed by both editors, and by an anonymous referee. We appreciate the patience and perseverance of all of the contributors (who were asked to deal with critiques of their work from at least four sources). We also thank all of the referees for their candid and insightful comments.

ALAN M. MACEACHREN
D. R. FRASER TAYLOR

Front Cover

Cynthia Brewer's aspect/slope color scheme allows terrain visualizaton through relief shading while simultaneously categorizing the surface into explicit aspect and slope classes. Aspect categories are mapped with hues, slope categories with saturation, and near-flat slopes with gray for all aspects. Lightness sequences are built into both the aspect and slope color progressions to approximate relief shading. The map was developed from a database of digitized topography of Hungry Valley State Vehicular Recreation Area (north of Los Angeles, California) provided by the Steven and Mary Birch Foundation Center for Earth Systems Analysis Research in the Department of Geography at San Diego State University. The aspect/slope scheme is an elaboration of Moellering and Kimerling's MKS-ASPECT scheme and is described in Brewer and Marlow's AutoCarto 11 proceedings paper referenced in Chapter 7.

Contents

List of Figures

List of Tables

Introducing Geographic Visualization (GVIS)

CHAPTER 1

Visualization in Modern Cartography: Setting the Agenda

ALAN M. MACEACHREN*

Department of Geography

302 Walker, The Pennsylvania State University

University Park, PA 16802, USA

Introduction

The title of this book, *Visualization in Modern Cartography,* demands some explanation. What cartographers mean by visualization and how we respond to visualization developments outside cartography are critical to what cartography can become as it moves to the 21st century. Cartography is at a crossroads, balancing precariously between links with geography and links with other "mapping sciences", between past traditions and the "threat" of geographical information systems (GIS) to replace cartography as we know it, between the demands to be proficient technologists who can build mapping systems and competing demands to draw upon our cognitive expertise to evaluate whether the system we build will work....and I could go on.

David Rhind, the Director of the Ordnance Survey in the UK, signaled the demise of the paper map in his Cologne International Cartographic Association (ICA) address (Rhind 1993); a profound event for the discipline if (or should I now say "when") it comes to fruition. Over the past five or six years, we have witnessed a dramatic ascension of visualization as an acceptable method of scientific practice. This development has been mirrored by an explosive advance in multimedia technology that promises to deliver interactive visual/audio products to the public. Map-making firms are already involved in the production of animated maps for CD-ROM encyclopedias. In January 1994 I read in the local newspaper that

*e-mail: NYB@PSYVM.PSU.EDU

Blockbuster Video (the largest video rental company in the US) is beginning to rent CD-ROMs as well as videos, and a recent newspaper had two articles on US efforts to built an "information highway" based on fiber optic cables. At least one spatial information provider is marketing a travel information product that runs on personal digital assistants (PDAs) such as the Apple Newton; if your PDA has a fax modem, it will even fax your reservation to a selected restaurant.

In this rapidly evolving scientific/business climate, it seems essential to consider the implications of maps as dynamic interactive spatial information tools (in contrast with their more traditional role as static storage devices for spatial data). This move to interactivity is a key theme that threads through the chapters that follow. While interactivity is not the central issue in every chapter, each chapter considers issues that arise when interactive access to spatial information (often by non-cartographers) is possible.

As noted in the opening paragraph, to produce a book about visualization in modern cartography requires some attention to just what is meant when we use the term "visualization". The section that follows presents my own (continually evolving) thoughts on this question. An earlier draft was circulated to all authors for comments (in August 1993). I invited responses and suggested that those who either agreed or disagreed with the perspective might use my essay as an anchor (or target) for their own views. I received several insightful replies (some of which are recounted in the endnotes to this chapter).

(Cartography)[3]

Visualization through mapping has long been treated as a fundamental geographic method. This point of view is illustrated by Philbrick (Philbrick 1953: 11) who submitted that "....not only is a picture worth a thousand words but the interpretation of phenomena geographically depends upon *visualization* by means of maps" (my emphasis). Philbrick's perspective suggests that cartographic visualization deals with maps as geographic research or spatial analysis tools. DiBiase's more recent research-sequence characterization of cartographic visualization matches this view (DiBiase 1990); see also (MacEachren and collaborators 1992) (Fig. 1.1).

Borrowing from the literature of both scientific visualization and exploratory data analysis (EDA), DiBiase (1990) proposed a framework for thinking about geographic visualization (GVIS) in the context of scientific research (with particular attention to earth science applications).[1] His framework emphasizes the role of maps in a research sequence. It defines map-based scientific visualization as including all aspects of map use in science, from initial data exploration and hypothesis formulation through to the final presentation of results. Emphasis is on re-establishing links between cartography and geography (as well as the earth sciences in general) and on the role of maps at the exploratory end of the research process. A key distinction made is that between maps to foster *private visual thinking* early in the research process and those to facilitate *public visual communication* of research results. Following this approach, visualization is not a new aspect of cartography, but a renewed way of looking at one application of cartography (as a research tool) that balances attention between visual communication (where cartographers have put much of their energy during the past two or three decades) and visual thinking (to which geographic cartographers of the first half of the century devoted considerable attention).

John Ganter and I (MacEachren and Ganter 1990, Fig. 7: 79) also incorporated a public–private distinction in our discussion of how "cartographic" visualization tools might be applied. To this distinction we added one between visualization tools intended to facilitate scientific research and those geared toward architectural and engineering applications (e.g. design of a building, highway, industrial park, or golf course). As DiBiase did, we linked the private use of visualization tools to exploring options and developing an approach to a problem. Our presentation and DiBiase's both suggest that representation options are reduced in number (ultimately to a single view) as the public end of the visualization tool use continuum is approached.

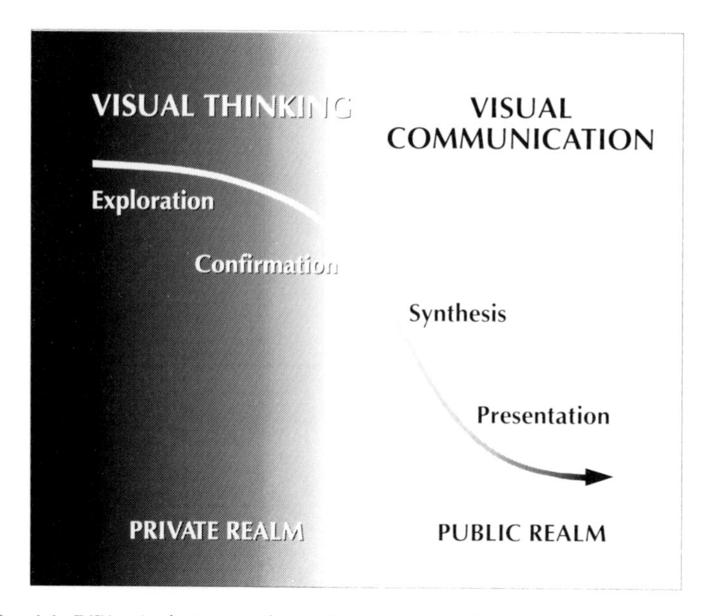

Fig. 1.1. DiBiase's depiction of visualization as a tool of scientific research. The curve depicts the research sequence through which different kinds of graphic depictions play a role. At the exploratory end of the sequence, maps and other graphics act as reasoning tools (i.e. facilitators of visual thinking). At the presentation end of the sequence, the visual representations serve primarily a communication function to a wider audience. Reproduced with permission from DiBiase (1990), *Earth and Mineral Sciences, Bulletin of the College of Earth and Mineral Sciences*, The Pennsylvania State University.

A complementary perspective on visualization as it relates to cartography has been offered by Taylor (1991) who focuses, not on how visualization tools are used or who uses them, but on the place of visualization within various cartographic research approaches of the past few decades. Taylor portrays visualization as occupying center stage, as the meeting ground of research on cartographic cognition, communication, and formalism (with "formalism" being used to suggest the strict adherence to rule structures required when computer technologies are applied) (Fig. 1.2). He calls visualization "a field of computer graphics" that attempts to address both "analytical" and "communication" issues of visual representation. By implication, then, visualization (for cartography) becomes the application of computer mapping to analytical and communication issues of map representation. Taylor stresses, however, that attention to computer formalism has dominated the discipline at the expense of cognitive and communication issues. He contends that research in all three areas is required to support successful cartographic visualization.[2]

A primary difference between Taylor's perspective on visualization and that offered by DiBiase (1990) or the one Ganter and I originally proposed (MacEachren and Ganter 1990) is in the emphasis placed on technology supporting visualization versus uses of visualization. Taylor links visualization directly to computer graphic technology but does not restrict visualization tool use to particular kinds of

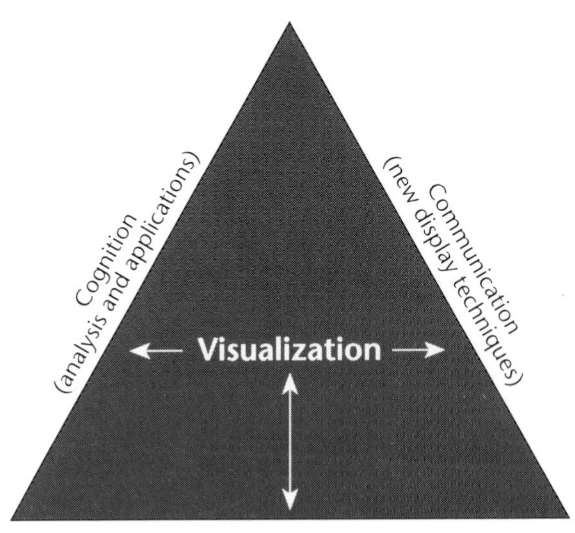

FIG. 1.2. Taylor's representation of visualization as the amalgamation of approaches to cartography associated with cognition, communication, and the formalism of computer technologies. Reproduced with permission from Taylor (1991) *Geographic Information Systems: The Microcomputer and Modern Cartography*, Pergamon Press.

application (i.e. scientific research). Both DiBiase and MacEachren and Ganter de-emphasize the technology producing the visualizations and concentrate on the kinds of uses to which they are put. All authors, however, seem to agree that visualization includes both an analysis/visual thinking component and a communication/presentation component and suggest (or at least imply) that communication is a subcomponent of visualization.

One problem with the views on visualization considered thus far is the appropriation of communication under the umbrella of visualization. If visualization includes both visual thinking and visual communication we might pose the question of what it does not include (and some have done so). Is "cartographic visualization" simply a new name for cartography? Saying that visualization involves computer graphics does not help much. It simply equates visualization with computer cartography. While Taylor drew attention to the links between visualization and computer graphics, Monmonier and I took this one step farther to place the emphasis on changes in computer technology that have made real-time interaction possible. We suggest not only a technological difference in tools for representation, but a "fundamental" difference in the nature of how analysts interact with those representations:

> The computer facilitates direct depiction of movement and change, multiple views of the same data, user interaction with maps, realism (through three-dimensional stereo views and other techniques), false realism (through fractal generation of landscapes), and the mixing of maps with other graphics, text, and sound. Geographic visualization using our growing array of computer technology allows visual thinking/map interaction to proceed in real time with cartographic displays presented as quickly as an analyst can think of the need for them. (MacEachren and Monmonier 1992)

The increased potential for human–map interaction that has become possible with current computer tools seems to be a critical component of GVIS as it contrasts with other kinds of map use. Friedhoff and Benzon (1989) make a similar point for scientific visualization in general.

As part of my efforts to organize a visualization working group under the auspices of the Map and Spatial Data Use Commission of the ICA, I found that the variety of ways in which visualization was defined by cartographers made discussion of the goals for a working group difficult, if not impossible. In response to the divergence of views, I developed a graphic characterization of visualization to be offered as an initial organizing concept for the visualization working group. The generally positive reaction to this characterization by a number of colleagues at the ICA meeting led to its use as a framework for linking contributions in this book. The characterization is based on treating cartography (or at least map use) as a cube — thus the (Cartography)[3] heading for this chapter section (Fig. 1.3).[3]

To make sense of how "scientific" visualization links with cartography, I start with the view that "visualization", like "communication" is not just about making maps, but about using them as well. As a communication approach has been dominant in cartography (particularly English language cartography) for at least two decades, it seemed that any attempt to delineate the territory of visualization (facilitated by maps) would have to consider how it relates to communication

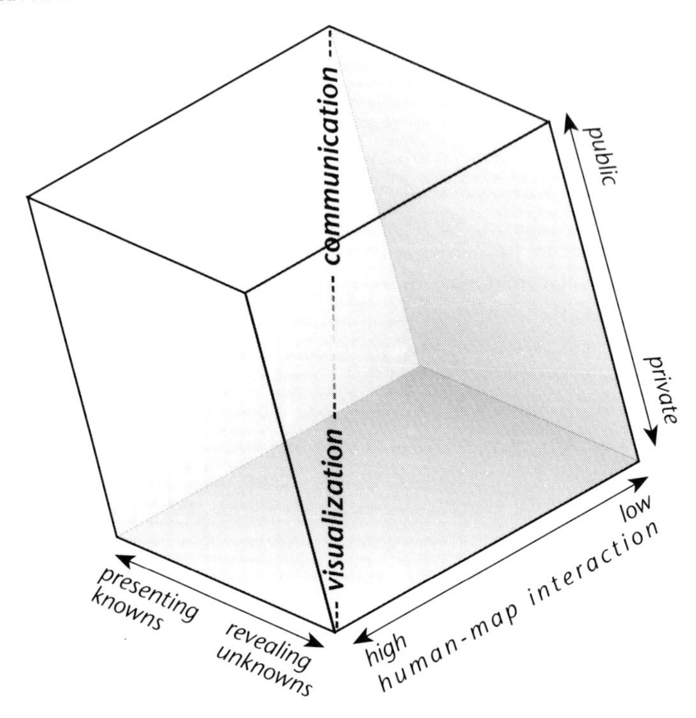

FIG. 1.3. (Cartography)[3] — a representation of the "space" of map use and the relative emphasis on visualization and communication at various locations within this space. This representation deals, not with kinds of maps, but with kinds of map use. Thus a particular category of map (e.g. a topographic map) might occupy any position within the space, depending upon what a user does with the map for what purpose.

(via maps). DiBiase's view places communication within the realm of visualization (as one of four components of visualization tool use). Similarly, Taylor's perspective presents visualization as the integration of three cartographic research streams. Both perspectives, in essence, imply that "visualization" equals "cartography" (with communication a subcomponent of the whole), a view leading to the conclusion that visualization offers nothing new. This conclusion, in turn, creates the potential for the *visualization revolution* in science to pass us by, while we sit on the sidelines thinking that cartography has done it all before.

The approach presented here *defines visualization in terms of map use* (rather than in terms of map-making or research approaches to cartography). The fundamental idea is that map use can be conceptualized as a three-dimensional space. This space is defined by three continua: (1) from map use that is private (where an individual generates a map for his or her own needs) to public (where previously prepared maps are made available to a wide audience); (2) map use that is directed toward revealing unknowns (where the user may begin with only the general goal of looking for something "interesting") versus presenting knowns (where the user is attempting to access particular spatial information); and (3) map use that has high human–map interaction (where the user can manipulate the map(s) in substantive ways — such as effecting a change in a particular map being viewed, quickly switching among many available maps, superimposing maps, merging maps) versus low interaction (where the user has limited ability to change the presentation) (Fig. 1.4).

There are no clear boundaries in this use space. There are, however, identifiable extremes. The space has the private, revealing unknowns, and high interaction ends of the continua meeting in one corner and the public, representing knowns, and low interaction ends meeting at the other corner. GVIS (as defined here) is exemplified by map use in the former corner and cartographic communication is exemplified by the latter. In my opinion, it is not interaction, private map use, or a search for unknowns that (individually) distinguish visualization from other areas of cartography, it is their combination.

I want to be clear here on what I am not saying. I do *not* suggest that research on cartographic communication is irrelevant – many maps are designed to communicate particular messages. Similarly, I do *not* suggest that the dividing line between visualization and communication is sharp (in fact, I think it is becoming fuzzier all the time). Communication is a component of all map use, even when visualization is the main object. Correspondingly, even the most mundane communication-oriented map can serve as a prompt to mental visualization. I view my definitions, then, as a convenience that allows us to emphasize the difference

	high interaction		low interaction	
	revealing unknowns	presenting knowns	revealing unknowns	presenting knowns
private	use of Ferreia and Wiggin's (1990) 'density dial' to manipulate class break points on a choropleth map in an effort to identify and enhance the spatial patterns.	use of a hypermedia interface to access a map collection – see Andrews and Tilton (1993) for discussion of a system designed to access the American Geographical Society Collection.	use of a 'closed' graphic narrative generated in response to a 'user profile' to take a 'guided tour' through a set of data (see Monmonier chapter).	use of a plat map to retrieve information concerning size of lot, rights-of-way, etc. for a piece of property
public	use of SimCity to allow students to assess the implications of public policies -or- use of MOSAIC via the Internet to give groups of scientists shared access to interactive simluations	use, by a TV meteorologist, of sketched annotations to a weather map (e.g., flow lines, position and direction of the jet stream, etc.) while explaining a storm situation.	use of Pike and Thelin's (1991) digital terrain map of the U.S. to explore geomorphic features and anomalies at varied scale.	use of you-are-here maps by the general public to figure out where they are in a shopping mall and how to get to particular stores.

FIG. 1.4. Exemplars of the eight extremes of map use space defined by the cube of Fig. 1.3. Each exemplar corresponds to a corner of the cube.

in goals (and design principles) for maps whose *primary* function is to facilitate transfer of knowledge from a few people to many people, versus maps whose *primary* use is to help individuals (or small groups of individuals) to think spatially. With respect to human–map interaction, no map use can take place without some level of interaction (although at times this interaction might be confined to visually scanning the map). In addition, higher levels of interaction do not require computers (e.g. you can draw lines of maximum gradient on a topographic map as an aid to mentally visualizing the runoff pattern of a drainage basin). "Interactive" computer tools, however, expand the possibilities for "interaction" with maps and thus the possibilities to facilitate visual thinking, perhaps in qualitative as well as quantitative ways.

Whether the terms visualization and communication (and the continua used to define them) are ultimately adopted by the discipline as I present them here is not important.[4] What counts is that cartographers become aware of the profound differences in goals (with the corresponding differences in approach to map design and evaluation of designs) that the distinction implies.

I share DiBiase's view that a key part of a visualization perspective on cartography is an increased attention to the role of maps in private data exploration – thus to the role of maps in research. Krygier (1994) contends that this maps-in-research emphasis is creating a fertile environment for a renewal of a "geographic" cartography in which cartographers are as concerned with the multiple roles of maps in geographic research as they are with how well people understand individual maps or map symbols. This shift in emphasis should draw cartographers closer to their geographic colleagues, after several decades of drifting apart. Krygier's view is echoed by Taylor (1993).

I have gradually come to the conclusion that visualization has implications for cartography that go beyond renewed attention to visual thinking and collaboration with our geographic colleagues for whom visualization tools can be developed. These developments in themselves are significant, and I do not want to diminish their importance. Beyond these, however, I think that GVIS can be to cartography what GIS has been to geography – a reinvigoration of an old, often taken for granted discipline whose relevance is increasingly recognized outside the discipline because it can help tackle important interdisciplinary issues.

Structure of the Book

Depending on your perspective, visualization is either a subset of communication, the opposite pole of a continuum (the other end of which is communication), or a superset that (if the qualifier "cartographic" is added) just becomes a more cumbersome equivalent to "cartography". As indicated in the above, I support the middle position. My emphasis is on visualization as a kind of map use. This is, however, only one of several contextual issues that should be considered as we explore the topic of visualization in modern cartography.

To give a more complete picture of the implications of visualization for modern cartography, my introductory comments are followed by a section (*The Context for GVIS Development*) that relates GVIS to our cartographic roots and to developments in the broader arena of spatial knowledge representation. Wood puts visualization in an historical cartographic context by presenting an argument that "visualization", if defined as the interactive use of maps to facilitate visual thinking, has gone on for as long as we have made maps. In this essay, he introduces the notion that an understanding of cognition (particularly of imagery) is essential as a basis for exploring questions of visualization in cartography. Peterson follows with specific cognitive issues relevant to visualization. He considers development of the cognitive perspective within psychology and application of that perspective by cartography. From this base he explores the links between "mental visualization" and dynamic visualization tools designed to facilitate visual thinking. Artimo concludes the section by placing visualization development in a technological context that includes GIS and what she defines as cartographic information systems (CIS).

If we are to use and design map-based visualization systems, there are a variety of factors that we must consider from perspectives not previously required by a more static cartography. The next section of the book: *Issues for Tool Design: Technology, Symbolization, and Human–Tool Interaction*, emphasizes three such factors. For each, two chapters address complementary topics.

Cartwright provides an overview of multimedia hardware and software as they relate to dynamic interaction with geographical information. He gives particular attention to the role of multimedia tools in the creation of spatial decision-support systems. Slocum and coworkers follow with an assessment of eight software environments that can be used to develop exploratory scientific visualization applications. Their chapter combines insight on the capabilities of several visualization development tools with an approach to selecting or designing visualization software (suited to geographic/cartographic applications) that is general enough to outlast the "half-life" of the specific software considered.

In relation to symbolization, two key issues were identified for consideration: color and sound. Computer-based GVIS environments today all use color. If analysts (or students) are to generate maps on the fly in response to database queries, there will be no time (even if the user had the expertise) to carefully consider the design of each map displayed. Interactive visualization requires much more precise specification of cartographic rules than has been required by traditional designers. Brewer provides a careful analysis of the issues involved when linking color schemes to data categories and proposes a set of guidelines that can allow visualization system designers (as well as designers of individual communication-oriented maps) to put the principles developed into practice. A second symbolization issue that few cartographers have considered is the sonic representation of data. Paper maps do not come with sound effects, but the hardware and software of visualization and multimedia environments makes sound a display tool that cannot be ignored. Krygier provides a brief review of the relevant "sonification" literature, describes a set of sonic variables analogous to Bertin's graphic variables, then offers several examples of how they might be used on maps.

The section concludes with a pair of chapters devoted to issues of user interaction with visualization environments. Lindholm and Sarjakoski provide an approach to the design of user interfaces that borrows from the computer science literature, but adapts principles to the unique demands imposed by geographically referenced information. McGuinness provides an introduction to expert–novice issues that cartographers must begin to face as we design visualization tools for narrowly targeted user groups. She also offers a synopsis of an experimental study that demonstrates some of the likely expert–novice differences in information access demands, and how to empirically assess them.

After setting the context and presenting some critical issues to consider in all visualization system design, we move on, in Section 3 (*Linking the Tool to the Use: Prototypes and Applications*), to emphasize specific GVIS tools and their applications. Prototype systems are described and guidance is provided on designing visualization tools targeted to applications in environmental assessment, education, and planning.

The section begins with Monmonier's discussion of graphic narratives as a base from which to build visualization tools that facilitate environmental risk assessment. Asche and Herrmann follow with an account of issues in developing prototype GVIS applications on a Macintosh platform using commercial multimedia authoring tools. They discuss applications in both planning and education. In contrast with Asche and Herrmann's use of commercial authoring tools, Koussoulakou describes a prototype system developed using lower level programming/graphics languages. The system is designed to facilitate transportation planning associated with air pollution amelioration in Athens, Greece. Where Koussoulakou emphasizes two-dimensional display and animation, Kraak's chapter focuses on three-dimensional depictions and direct interaction, again with an environmental application emphasis. The final two chapters of Section 3 share an emphasis on symbolization issues. In both, the focus is the development of symbolization to meet particular visualization application needs. DiBiase and his colleagues provide an overview of literature from several disciplines dealing with the problem of multivariate data representation. They go on to describe a prototype visualization system that allows analysts to look for relationships among climate variables (both actual and model derived). Van der Wel *et al.* then provide a review of recent efforts to visualize data quality and present some solutions to data quality visualization in the context of a transportation development plan.

Finally, my collaborator on this project, Fraser Taylor, offers some perspectives on the future of visualization in modern cartography. These perspectives include attention to the place of visualization in cartographic theory and practice.

My characterization of cartographic visualization as a kind of map use that emphasizes the private, high interaction, exploratory corner of map use space influenced the choice of chapters outlined above. Each chapter, however, stands on its own merits (and some actively counter my definition of visualization). However visualization is ultimately defined within cartography, the contributions included here chart a new territory that cartographers should find exciting to explore.

References

Andrews, S. K. and D. W. Tilton (1993) "How multimedia and hypermedia are changing the look of maps", *Proceedings, Auto-Carto 11*, Minneapolis, pp. 348–366.

DiBiase, D. (1990) "Visualization in the earth sciences", *Earth and Mineral Sciences, Bulletin of the College of Earth and Mineral Sciences, PSU*, Vol. 59, No. 2, pp. 13–18.

Ferreia, J. and L. Wiggins (1990) "The density dial: a visualization tool for thematic mapping", *GeoInfo Systems*, Vol. 1, pp. 69–71.

Friedhoff, R. M. and W. Benzon (1989) *Visualization: the Second Computer Revolution*, Harry Abrams, New York.

Hearnshaw, H. M. and D. J. Unwin (1994) *Visualization in Geographical Information Systems*, John Wiley & Sons, Chichester.

Krygier, J. (1994) "Visualization, geography, and landscape: the role of visual methods in a study of landscape change, derelication, and reuse", *PhD Dissertation*, The Pennsylvania State University.

MacEachren, A. M. (in collaboration with Buttenfield, B., J. Campbell, D. DiBiase and M. Monmonier) (1992) "Visualization" in Abler, R., M. Marcus and J. Olson. (eds.), *Geography's Inner Worlds: Pervasive Themes in Contemporary American Geography*, Rutgers University Press, New Brunswick, pp. 99–137.

MacEachren, A. M. and J. H. Ganter (1990) "A pattern identification approach to cartographic visualization", *Cartographica*, Vol. 27, No. 2, pp. 64–81.

MacEachren, A. M. and M. Monmonier (1992) "Geographic visualization: introduction", *Cartography and Geographic Information Systems*, Vol. 19, No. 4, pp. 197–200.

Philbrick, A. K. (1953) "Toward a unity of cartographical forms and geographical content", *Professional Geographer*, Vol. 5, No. 5, pp. 11–15.

Pike, R. J. and G. P. Thelin (1991) "Mapping the Nation"s physiography by computer", *Cartographic Perspectives* , No. 8 (winter, 1990–91), pp. 15-14 and map insert.

Rhind, D. (1993) "Mapping for the new millenium", *Proceedings, 16th International Cartographic Conference*, Cologne, Germany, pp. 3–14

Taylor, D. R. F. (1991) "Geographic information systems: the microcomputer and modern cartography" in Taylor, D. R. F. (ed.), *Geographic Information Systems: the MicroComputer and Modern Cartography*, Pergamon Press, Oxford, pp. 1–20.

Taylor, D. R. F. (1993) "Geography, GIS and the modern mapping sciences: convergence or divergence?" presentation to the *Canadian Association of Geographers Annual Meeting*, Ottawa, May 1993.

Endnotes

[1] The terms "cartographic" visualization and "geographic" visualization are both used to refer to spatial visualization in which maps are a primary tool. Although my initial publication on visualization as it relates to cartography (MacEachren and Ganter 1990) used the term "cartographic" visualization, I now favor the use of GVIS. The latter term implies a broader range of possible activities than the former. Cartographic visualization seems to exclude visualization in which remotely sensed images, photographs, diagrams, graphs, etc. are used together with maps to illuminate geographic questions. Visualization in modern cartography implies an integration of spatial display tools that the term GVIS seems to encompass.

[2] In a recent statement on visualization, Taylor (1993) has moved beyond the characterization of visualization as a field of computer graphics. In this paper, he emphasizes a revitalized approach to the cognitive level of spatial thinking that is fostered by attention to visualization. In particular, he stresses the potential for visualization to draw cartography and geography back together.

[3] The graphic model presented was formulated during the 16th ICA Meeting in Cologne, May 1993. It was influenced by participants in the initial meeting of the visualization working group (at which panelists Janos Szegö, Menno-Jan Kraak, Mike Wood, Mike Peterson, and Daniel Dorling discussed various aspects of visualization and its role in cartography). The seeds of the idea were sown in my initial collaboration with John Ganter and our discussions

with David DiBiase (in 1989–90). The idea began to germinate at a Workshop on Visualization in GIS organized by David Unwin and Hilary Hearnshaw at Loughborough University in July 1992. At that workshop, Ian Bishop, Anthony Gatrell, Jason Dykes, Mitchell Langford, Daniel Dorling, and I sketched out a draft section on "Advances in visualizing spatial data" for a book titled *Visualization in GIS* (Hearnshaw and Unwin 1994). Components of the scheme presented here can also be found in the chapter on visualization that I produced (with the collaboration of Barbara Buttenfield, James Campbell, David DiBiase, and Mark Monmonier) for Abler, Marcus, and Olson (eds.) *Geography's Inner Worlds* and the introductory essay (written with Mark Monmonier) for the special geographic visualization issue of *Cartography and Geographic Information Systems*. The insights offered by all of these colleagues had a substantial impact on my approach to visualization – but, of course, only I can be held accountable for the arguments offered here.

[4] Among the responses to my (cartography)^3 ideas were several critiques of various aspects of the model. Terry Slocum, for example, pointed out that the three-dimensional depiction of the model implies that there are only three axes to map use space (but that there are probably more) and that the cube also implies orthogonality among the continua (when this may not be the case). Mike Wood and Mike Peterson both objected to excluding the communication corner of the model from the purview of visualization; Wood because such a restriction does not match common usage in computer science (where all computer-generated displays have come to be called "visualizations") and Peterson because he contends that all maps can prompt mental visualization (which I agree with) and that all acquisition of knowledge from a map is fundamentally an act of communication (which I do not agree with). Beyond the critiques from other authors, six students in my graduate seminar on GVIS also provided a range of constructive criticism. In particular, they helped clarify the distinction between interaction with maps and the interactivity of computer software, and prompted development of the matrix of map use examplars presented in Fig. 1.4. Together, the various critiques tempted me to back off from my contentions and adopt the terms "visual thinking" and "presentation" as the labels for ends of the key diagonal. On reflection, however, I decided to retain my original terms, comment on the objections, and await responses. There were, of course, positive reactions to the (cartography)^3 idea, or I would have discarded it. Among them, Cindy Brewer provided a useful sketch of the three-dimensional map use space that helped to refine my own rather crude preliminary drawings. She also offered suggestions on the kinds of map use that occupy the corners of the model. Menno-Jan Kraak went so far as to suggest that I locate each chapter as a dot in its appropriate three-dimensional model position. I seriously considered this, but decided that I would never achieve consensus on the relative position of various chapters – I leave it to the reader to determine which chapter is the prototypic example of geographic visualization.

The Context for the Development of Geographic and Cartographic Visualization

CHAPTER 2

Visualization in Historical Context

MICHAEL WOOD*

Centre for Remote Sensing and Mapping Science
Department of Geography, University of Aberdeen
Elphinstone Road, Aberdeen AB9 2UF, UK

Introduction

In the new era of visualization in scientific computing (ViSc) and geographic information systems (GIS), when increasingly powerful graphic tools are being devised to support scientific investigation, the significance of traditional (pre-computer) cartography and map use might too easily be forgotten. Not only have general computer visualization systems inherited a spectrum of well-established map-related techniques for description and analysis, but most of their adopted symbols have rich cartographic provenance. Although originally associated with the wider field of geography, mapping is essential to many sciences. As demonstrated by the journalist Stephen Hall (1992) in *Mapping the Next Millenium*, exploratory analysis and explanation in astronomy, remote sensing, geology, climatology, medicine and fundamental physics have benefited greatly from maps. The power of maps in a broader sense has been superbly analysed by Wood (1992), and their potential significance at a more local level exemplified by Aberley (1993).

This review is in two parts. The first examines the process of cartographic visualization: its nature, elements and the evidence of its use in pre-computer spatial studies. The second part focuses on pre-computer maps themselves. Not only do they function, as sources, within spatial analysis, but they contain graphic (historical) evidence of both the core techniques and symbolism being developed within computer systems today. Visual mapping, recognized as one of the most

*e-mail: M.Wood@aberdeen.ac.uk

powerful analytical tools, is moving to new levels of importance, released from its "manual" shackles by GIS and ViSC.

Cartographic Visualization: A Fundamental Investigative Tool

Basis of Cartographic Visualization: Imagery and Spatial Knowledge

Graphic mapping, like written language and mathematical symbolism, is an outward expression of thinking. Even an unclassified satellite image "map" is such an expression, having its origins in human concepts and creativity. To be complete, therefore, a contextual review of cartographic visualization must examine some of the cognitive elements (especially imagery and knowledge) which lead to externalized mappings, and which have been fundamental to mapping and map-related thinking throughout history.

Although not all creativity and innovation need to involve imagery, there is significant evidence of its existence in such activities. When something is visualized (made visible), even if only by illumination in a dark room, the viewer is brought closer to understanding it. Shepard's (1978) description of the experience and use of mental imagery by creative thinkers is especially germane to this review.

Considerable attention has been focused by psychologists on the part played by internal (mental) imagery in thinking and problem-solving (Kosslyn 1983) (see the Peterson chapter in this book for a more extensive review of the relevant psychological literature). Not only is such imagery common in human experience, but scientists in particular have reported its contributions to their own research (Shepard 1978). James Clerk Maxwell, the theoretical physicist, made a mental picture of every problem he tackled, while Nikola Tesla (the inventor of fluorescent lighting) was prone to imagery of such an "extraordinarily concrete, three-dimensional and vivid character" that he claimed to be able to "construct" and test machines mentally before eventually creating them in reality. One of the few known examples in cartography is reported by Learmonth (1988), "....Professor P. H. Mahalanobis FRS claimed that he could visualize a map of India, say of the 400 districts, while reading a (statistical) table....". In most of these cases, imagery is either reported by the individual, or assumed to have preceded and contributed to the production of externalized concrete inventions or abstract theories. In this sense all maps, in common with other tangible creations, could be described as "external embodiments of what were originally purely internal images" (Shepard 1978), e.g. (1) memories of a journey, where the landscape, recorded informally as procedural experiences, is plotted out; (2) imagined coverage of a region or country by standardized survey maps, before its execution (e.g. General Roy's 18th century map of Scotland); (3) imagined mapping solutions to help solve problems in activities such as navigation and wayfinding (charts and orienteering maps); and (4) cartographic expressions of hypotheses, beliefs (e.g. scientific, religious) and plans.

On the other hand, there are many examples of images which have been externalized, normally through drawing, not as final products but at a distinctly intermediate phase in the creative or investigative process. This might either be to capture fleeting ideas or to allow further enrichment and more detailed development of previously imagined structures or concepts (McKim 1970). Cartographic visualization is assumed to operate in this manner.

Mental imagery relies heavily on existing knowledge. The development of geographical knowledge has been examined by various workers, including Piaget (Piaget and Inhelder 1956), Downs and Stea (1977) and Kuipers (1978). The relationships between declarative knowledge (of facts), procedural knowledge (from travel) and configural knowledge (geometric) have been examined more recently by Mark (1993), who further classifies spaces as haptic (defined by bodily interaction), pictorial (mainly from visual perception) and transperceptual (a cognitive construct of geographical space, developed over time). These ideas are then related to those of the linguist, Zurbin (Mark *et al.* 1989), who has proposed a model of spatial objects and concepts encompassing: (1) small manipulable objects (e.g. a book); (2) objects larger than humans (e.g. a large tree); (3) spaces scanned from a single viewpoint (e.g. a garden); and (4) spaces which cannot be perceived as units (such as extended geographic space).

Mark (1993) describes a map as a category (1) object representing category (4) geographic space. Zurbin's model not only focuses on absolute and relative size, but also on what are described by Mark as "models of human interaction and sensing". He also observes that "the power of the map lies in the way it allows spatial reasoning methods from the familiar haptic, table-top world to be applied to transperceptual (geographic) spaces", but warns that, due to its abstracted nature, the map "does not represent the world as it is experienced". All map users, therefore, should be critically aware of the implications of this characteristic to their studies.

An appreciation of the relationship between internal (cognitive) experiences and their external expression can help our understanding of maps and mapping in exploration and research. Mental imagery is vital, but the importance of real (externalized) maps should not be underestimated. Indeed, developing Mark's observation on "the power of the map" it could be argued that all real maps function as stepping stones towards greater understanding of transperceptual space. Mark and McGranaghan (1986) observed that "for most people, configural knowledge of transperceptual spaces is derived mostly from maps", and only supported by other procedural knowledge. Lloyd (1992) has since demonstrated that for distance and direction information (configural knowledge), for example, 10 minutes of map study was more effective than 10 years of living in the environment. The potential of cartographic visualization derives from both internal imagery and real maps.

Visualization in Cartography: A New Definition?

The term "visualization" in cartography has been adopted by different workers to describe processes from cognitive imaginings to concrete cartographic products (Visvalingam 1991; Buttenfield and Mackaness 1991), but the new, more focused and use-related definition suggested in the introductory chapter (exploration; the search for "unknowns", through cartographic or map-related techniques) seems to provide a link between these two extremes. Indeed, it could almost be described as an instinctively natural process. As suggested in the preceding section, this form of "visualization", although primarily internal (mental) is frequently externalized. The images produced are normally intended for specialized "viewers" (sometimes the investigator alone) and thus need not conform to fixed standards of symbolism, geometry or even scale. In either mode (mental or physical) visualization in cartography contributes to exploration and analysis rather than presentation and explanation.

The separation of visual thinking from the presentation of fully formed ideas to others is convincingly argued by McKim (1970) using examples primarily from art, creative product design and scientific invention. He emphasizes the important relationships between seeing, thinking and drawing. As most people lack adequate "acuity" in their mind's eye they must clarify and develop their thought images with sketches. Many map users might support the claim of James D. Watson (who discovered the structure of DNA) that drawing (for them, mapping) and thinking are frequently so simultaneous that the graphic image appears almost an organic extension of mental processes (Watson 1968). The term "graphic ideation" is coined by McKim to describe the visual thinker's exploratory and developmental processes in which ideas are conceived and restructured. This starts with "idea sketching" where inner thoughts are captured, rapidly, as a series of drawings, before the ideas are lost to awareness. These sketches, created to support the thought processes of the sketcher, focus initially on main themes rather than details, which are attended to at a later stage. The procedure is summarized as what is called a feedback loop, "ETC" (express, test, cycle) (Fig. 2.1): ideas are identified, evaluated and then recycled to continue the investigative process. In this context McKim also predicted the growing importance of what he called the "interactive graphic computer" for the enrichment and acceleration of these processes.

Although not directly applicable to all the spatio-geographic problems tackled

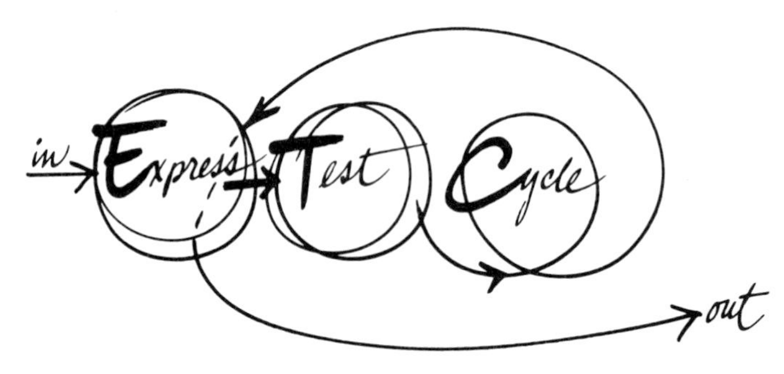

FIG. 2.1. Feedback loop (ETC; express, test, cycle) of McKim (1970).

through maps, McKim's conceptual ideas exemplify the very essence of cartographic visualization as defined in this text. In much scientific research, however, compact, data-limited topic areas (such as the product designs of McKim's text) are rare. Modern earth and atmospheric sciences, for example, depend on the cartographic representation of huge data sets and not just on sketches for significant exploratory investigations.

The high-speed production of different maps of one region, for example, only became possible with the arrival of computer technologies (e.g. Warren and Bunge 1970), but the value of examining a range of alternative cartographic images as part of the decision-reaching process has been appreciated and practiced in the past. Wood (1992) cites Bertin (1981), a strong advocate of graphic information processing, as declaring that the graphic image (presumably including the map) "...is not 'drawn' once and for all; it is 'constructed' and reconstructed until it reveals all the relationships constituted in the interplay of the data....A graphic is never an end in itself, it is a moment in the process of decision-making".

Like McKim, he too separates the analysis of data from the presentation stage.

Evidence (or Lack of Evidence) for Pre-computer Cartographic Visualization

If a significant aspect of cartographic visualization, in the pre-computer environment, comprised informal investigative routines, it is likely that only final summary or presentational maps will be preserved. Even then, only publication will normally guarantee their continued existence. Retention of all the working material of a project may be a matter of luck, unless it is filed within a structured archive for future reference.

Most direct evidence of the use of pre-computer cartographic visualization may thus have disappeared or, at best, may survive only as "lost" or deeply stored archives (and hence largely inaccessible). Where might the indirect evidence lie? In which fields of study might the techniques have been most visibly applied? Since the early 19th century, archaeologists, searching for explanations of both the absolute locations and the spatial relationships of their "finds", have made use of simple visualization techniques. The cartography of disease, however, stretching back over three centuries, may offer a richer source of evidence.

As the death and illness of individuals are essentially associated with point locations, the dot map has been the preferred technique in the geographical study of disease. Some of the earliest maps (later referred to as "sanitary maps"), relating sickness with the environment, date back to 17th century Italy. Wallis and Robinson (1987) refer to them as "crude" but as visualizations (cartographic ideations) they may have been "perfect". More "worthy" maps, presumably because they had been well executed and published, began to appear at the end of the 18th century. These, and their successors, are important for this review because they may represent evidence of investigations which had used cartographic visualization procedures. Some, such as the 1790 yellow fever maps of the east coast of the USA

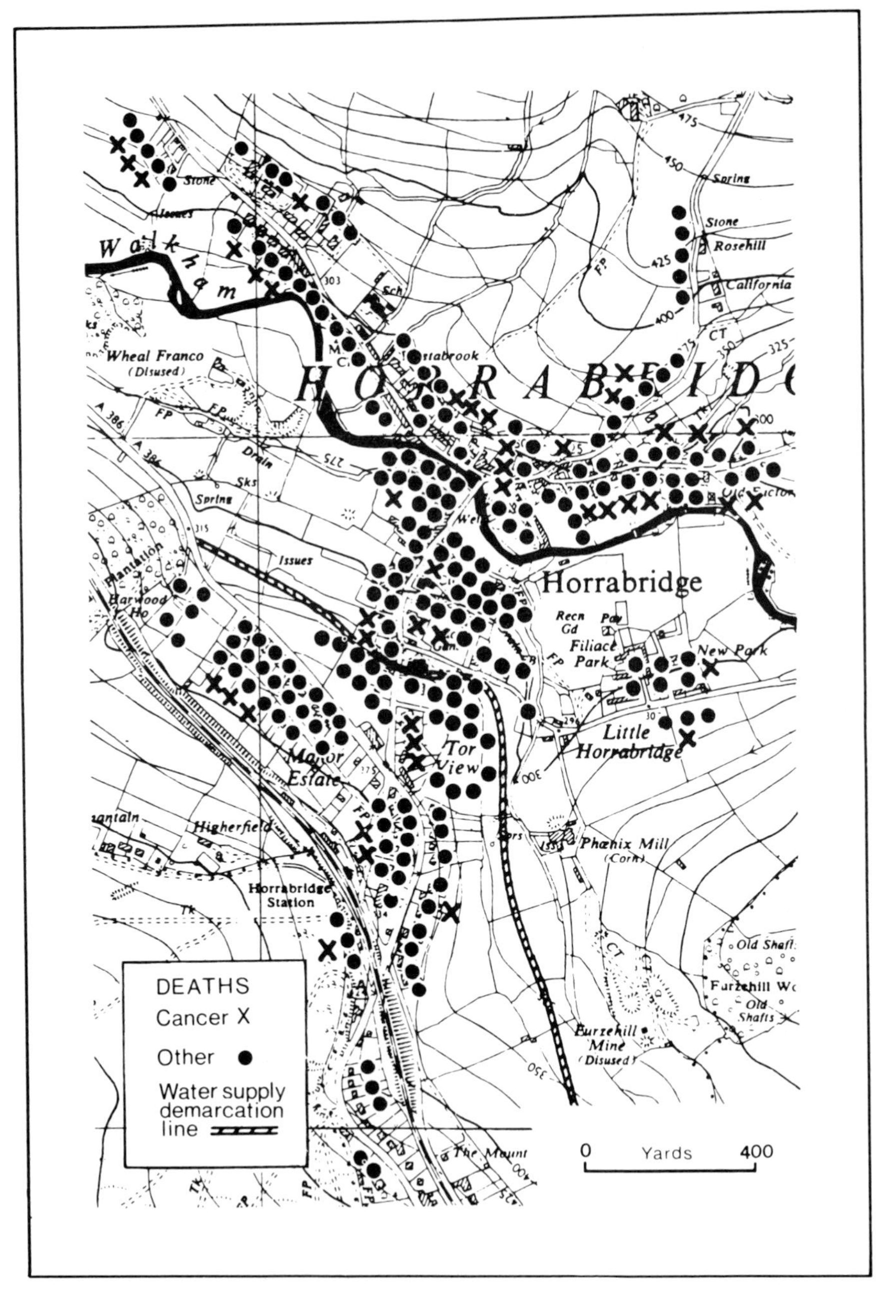

FIG. 2.2. Cancer deaths, Horrabridge, Devon. The areas of three different sources of water are separated by the river and the hatched line.

(relating the disease with sewerage, drainage and street paving) were also used presentationally, as propaganda, by groups who believed, separately, in both the contagious and endemic nature of the problem. The basic techniques of comparative mapping to allow visualization of spatial relationships have continued in disease geography through to the present day. As Learmonth (1988) observes "....even quite simple mapping....does make a case for the importance of within-country variations".

The oft-cited example of how mapping may have contributed to the explanation of a specific cholera outbreak is John Snow's contemporary research of 1853–4 (Gilbert 1958). Its use in the context of this present review could, of course, be criticized. The published map, showing the relationship between incidents of the disease and a contaminated, centrally sited well in the Soho area of London, may have been created for presentational rather than analytical purposes. However, the almost obligatory dependence on published maps to suggest the investigative techniques which led to them illustrates perfectly the problem of lack of evidence of visualization processes. As Snow did finally present a spatial explanation, the hidden (or lost) evidence of his cartographic research might well have included both mental spatial visualizations (similar to the examples recorded by Shepard 1978) and associated cartographic ideations for more detailed analysis. The crudeness of his working maps may have led to their disappearance. Even if most of the visualization was carried out internally, with spatial images of the relevant relationships, the process of cartographic visualization would still, in essence, have taken place. The published map would merely represent the results of his mental deliberations.

A more contemporary example of similar work is that of Allen-Price (1960). Like Snow, he suspected causal relationships between disease (cancer deaths) and water supply within the village of Horrabridge in Devon. By using initially mental and then physical mapping of the incidences on a large-scale map (Fig. 2.2) he was able to confirm the details of his suspicions. He, too, later used publication quality versions of his maps in a campaign for changes in water supply. The importance of cartographic visualization in the study of the ecology of disease is further revealed in the statement by Learmonth (1988) that "....mapping....seems to be the beginning of wisdom rather than an end in itself....". He also disputes claims by some statisticians that if correlations between variables seem obvious there should be no need to map them. If (for instance) "sociological variables are indeed crucial, there may, nevertheless, be significant regional or spatial factors at work" (Learmonth 1988). These will only emerge from mapping.

Traditional Cartography and Cartographic Visualization

Most extant pre-computer maps may not be direct examples of cartographic visualization. However, just as they can provide indirect evidence of the process, maps can also act as visual "table-top" laboratories (Mark 1993). In addition, especially in thematic cartography, they provide a historical record of symbolic innovation and evidence of spatial information processing.

The Traditional Map, Past and Present

Mapping has early origins, dating, on the basis of physical evidence, at least from 5000 years BC. Many more recent maps also exist in manuscript form, but, since the invention of printing in the 15th century, maps have increased rapidly in numbers. What functions have these maps fulfilled? Keates (1982), discussing modern products, identifies three categories which could help answer this question: (1) general maps — which appear to record selected landscape features, physical and cultural; (2) special purpose maps, e.g. designed for route finding (charts); and (3) special subject maps, e.g. geology.

Although it is tempting to classify many early maps as "general", the term "special purpose" may be more suited to most, e.g. for navigation, military or, indeed, to express territorial superiority. But, as observed by Robinson and Petchenik (1976) "there is no clear line between various types".

Special subject maps, sometimes subsumed within the term "thematic", emerged more recently (mainly post-17th century) and were revolutionary in character. The mapping of new data sets, and for new tasks, presented design challenges to the early cartographers and many of the map (and graphic) symbols, commonplace today, originated during the growth period of thematic mapping (Robinson 1987).

A review of the whole range of special purpose and special subject maps might suggest a further subdivision into major works (e.g. geology, marine charts) and "minor" examples. The former, representing extensive (and frequently state-supported) surveys and scientific field projects, have played a major role as data sources in contemporary scientific research. The latter group (e.g. accompanying papers and reports) may, as noted previously, contain most of the available indirect evidence of cartographic visualization.

The process of map production (of an artistic manuscript or a fully printed map sheet) is never trivial, costing dearly in both time and money. Although some processes, especially for the duplication of black and white images, are relatively cheaper today, adequate justification is still required for the higher costs of colour printing. Ultimately, the "maps that actually get made are maps that individuals or groups are willing to pay for" (Petchenik 1992, personal communication with D. Wood, quoted in Wood 1992). The majority of printed maps (i.e. maps in the modern archive), therefore fall into a special category in which their original purposes were regarded to be of sufficient importance for the finance involved. Although suffering the inevitable disadvantages of their genre (static, out of date, restricted by scale), these maps also provide rich and accessible sources of information that have been used by spatial scientists to prompt mental visualization.

Are Traditional Maps Suitable Vehicles for Cartographic Visualization?

Should scientists seeking solutions to spatially-related problems put their faith in printed cartographic works? Inevitably, a well-produced map (like an elegantly typeset text) can create an impression of accuracy and neutrality far superior to its true quality. Maps are artefacts and, despite official protests to the contrary, they do display the effects of human shortcomings, as well as other cultural or political characteristics of their times. This is especially true of small-scale maps in which space is limited and the non-exact science of cartographic generalization has been applied. Even a superficial comparison of the recent topographic maps of Europe (e.g. at 1:50,000 scale) will reveal significant disparities in data quality and abstraction, in the nature and density of information content and in symbol codes. Serious users of such documents soon become aware of these frustrating peculiarities and either learn to adjust to them or generate their own more appropriate visualization tools.

The Warnings of the "Deconstructionists"

The process of identifying inconsistencies and unravelling cartographic fact from fiction accelerated in the late 1980s. A new academic movement within the history of cartography claimed, initially, to have made radical "discoveries" of distortion and misrepresentation in maps of all types, previously assumed by the investigators to be value-free and exemplars of the "cumulative progress of objective science" (Harley 1989). Many of these so-called discoveries, however, were common knowledge to practising cartographers. Borrowing, sometimes over-selectively (Belyea 1992), from the works of post-modern thinkers, they introduced a new veneer of terminology to historical cartography. Against what could be described as a caricature, that cartographers are "obsessed with accuracy....uninterested in (and)....barely conscious of the distinction between map and reality" (Godlewska 1989), they developed a new genre of map criticism and interpretation. Grandiose terms were used, such as cartographic "silences" (for the sometimes inevitable and unsurprising omissions from maps) and "hidden agendas" (e.g. where state maps, created at great cost and effort to governments, are discovered to be as important as expressions of power and claims to territory as they are records of topographical accuracy).

Some of their findings, however, are recognized as having shed light on aspects of maps as sources. Blakemore, for instance, identifies deconstructionists as offering a revival of "interest in the human and social processes that underpin cartography" (Blakemore 1989); that map products can have social implications.

But who are really to blame for the so-called biases uncovered? Have "cartographers" been unjustly targeted? It must be remembered that, certainly over the last century or so, cartographers were not the decision-makers when it came to content or emphasis. As Keates (1982) has observed, ..."In many instances the initiator of the map is also the map user, or some body or institution representing his interests and needs....virtually all national topographic map series have been and are produced to satisfy the requirements of government departments, military agencies, local authorities and a variety of "pressure groups" ranging from engineers to tax collectors, "most of whom" may understand little or nothing of how the map should be made....". It might thus be observed that in the majority of cases, especially of "official" maps, cartographers are just facilitators and that responsibility for the more serious distortions/misrepresentations identified by the deconstructionists must, in the end, be blamed on human frailty and misjudgement largely beyond the cartographers' domain.

The Map Viewer and Cartographic Information

"Map reading is a creative process" (Eastman 1985). Even the simplest map is a magical thing when interacted with by a reader (Morris 1982). Sign, symbol code, culture and myth mingle, separate and regroup according to each reader's peculiar structures of knowledge, education, experience, beliefs and motivation. Discourses emerge which vary with the percipient and even, perhaps, with the occasion! Some are mundane and restricted (perhaps derived solely from the legend), others imaginative (e.g. patterns of ley-lines "emerging" from the symbolized landscape). The latter results from what cognitive psychologists refer to as the viewer's "top-down" or concept-driven processing — his or her knowledge and beliefs superimposed on a particular set of spatial patterns. While, by its content, it may display or conceal bias, a map without reader interaction, a reader who can make interpretations and judgements, is totally passive, cannot communicate, does not offer a discourse. The magic of discovery or invention is hidden until the map is touched by the sight and mind of the viewer. Everyone, scientist and lay-person alike, should become aware of these processes before turning their criticism on the map.

There are many ways in which spatial researchers have used existing maps as a data source or visualization tool. A form of cartographic ideation has been used, through sketching of either hypothesized or field-observed features on the surface of a printed map, followed by analysis of the new relationships created. But, accepting the limitations of any particular maps, the possibilities of applying cartographic visualization are as wide as the creative powers of the investigator and the restrictions in data quality will allow.

Although having been created for other purposes, therefore, many existing maps can serve (or have served) as visualization tools. To be of maximum value to the investigator, however, they should not only be data-rich and accurate, but also eminently legible. This is achieved most effectively through the introduction of "visual hierarchy", by which related features and distributions are designed to appear as graphic "layers". A well-designed map should also provide what Tufte (1990) describes as "micro-macro readings": the visual analysis of detail being accompanied by the possibility of obtaining clear overviews of the region depicted. Sadly, however, not all maps display this design quality and the investigator may not only have to check the provenance of the source data but also struggle, visually, to extract the information contained.

Cartographic Visualization and the Emergence of New "Thematic" Symbols

Although identical cartographic symbols can be used for both analysis and presentation and can seldom be categorized as unique to either "side", some specific developments in design can be observed. The needs of investigators to create external visualizations of new topics and relationships prompted the emergence of thematic cartography. This included the invention of new symbolization techniques to facilitate these visualizations. Some examples are given below (Wallis and Robinson 1987):

- Isarithms, initially used to depict both compass declination and depths in navigable waters, first appeared in the 16th century, although the sub-category,"isopleth", for representing statistical surfaces, is much more recent
- The first dasymetric map, of possibly the first world map of population density, is attributed to Scrope (1833)
- Flowlines first appeared as quantitative symbols in 1837 (Robinson 1955)

As noted in the previous section, the authors of *Cartographical Innovations* (Wallis and Robinson 1987) were mainly concerned with good quality examples of maps and "visualization" need not always generate such products. Some of the earliest applications of new symbols may well have grown from the graphic "doodles" of great minds. It would seem reasonable to assume that the ocean current or wind direction flowlines of Hondius or even the early isarithms of the Portuguese chartmaker Luis Teixeira (showing equal compass declination), were cartographic solutions, either invented, or at least refined, through the application of what we are now defining as cartographic visualization processes. Other than by backward extrapolation from published works, we can now only speculate about how and when these symbolic techniques might have been used in research.

The Cartographic Provenance of the Symbolism of Modern Scientific Visualization Systems

The general term, scientific visualization, has been described as "complex generalized mappings of data from raw data to rendered image" (Haber and McNabb 1990), and "....concerned with exploring data and information graphically — as a means of gaining understanding and insight into the data" (Earnshaw and Wiseman 1992). If "cartographic visualization" represents the geographical expression of this wider field, a commonality of symbol systems and methods might be expected to exist. It does, and most, if not all, graphic visualization techniques used within ViSC software are firmly rooted in cartography. Although overlooked by many leading authors, one paper, "Data visualization — has it all been seen before?" (Collins 1992) offers firm evidence of this fact. Some of the jargon of scientific visualization used by Collins may confuse the cartographer but similarities in both terminology and application are obvious. Here are some examples (ViSC terms in italics):

For *scalar* (ordered) data

Contour maps, Height maps (terrain-like images possibly representing non-terrain data)

Colour maps (including choropleth, dasymetric and chorochromatic techniques)

For *vector* (aligned) data

Streamlines (e.g. for ocean currents), *Arrow plots* (e.g. for wind directions)

Therefore, symbols and symbol techniques with a long cartographic history now form the basis of some of the most powerful investigative graphic systems available.

Conclusion

Cartographic visualization, although a new term, may well describe some of the oldest investigative geographical processes. Many visualizations were devised in the pre-computer past, but dependence on the manual generation of graphic elements, especially maps, has often restricted their successful application. A researcher, for instance, might see a need either to plot many hundreds of distribution maps or to create multiple cross-sections of a complex geological region. But, with tight deadlines to be met, this preferred approach to the task would be impossible. Scientists knew that the limitations of traditional cartography could restrict "the ways that (they thought) about their models and thereby limit potential insights" (Palmer 1992).

Poor interaction between investigator, graphic tool and data has been one of the main handicaps to the more effective use of pre-computer maps as visualization rather than presentation tools. Interaction is possible, but with manual techniques takes place so slowly that the resulting knowledge revealed is often dated by the time a suitable graphic is produced. One of the most important advantages of the new visualization and GIS technologies is that "changing the pace of interaction can result in qualitative as well as quantitative increases in productivity" (Palmer 1992). The nature of the investigative working environment is distinctively different. The facilitating role of GIS and ViSC not only enhances rates of activity, but opens new possibilities for innovation in the development and use of cartography as an exploratory tool.

References

Aberley, D. (1993) *Boundaries of Home: Mapping for Local Empowerment*, New Society Publishers, Philadelphia.

Allen-Price, E. D. (1960) "Uneven distribution of cancer in West Devon, with particular reference to the diverse water supplies", *The Lancet*, Vol.1, pp. 1235–1238.

Belyea, B. (1992) "Images of power: Derrida/Foucoult/Harley", *Cartographica*, Vol. 29, No. 2, pp. 1–9.

Bertin, J. (1981) *Graphics and Graphic Information Processing*, de Gruyter, Berlin.

Blakemore, M. (1989) "Deconstructing cartographers via deconstructing the map" in Dahl, E. H., Commentary, Responses to J. B. Harley's article, "Deconstructing the map" published in *Cartographica*, 26 (2), *Cartographica*, Vol. 26, No. 3, 4, pp. 90–93.

Buttenfield, B. P. and W. A. Mackaness (1991) "Visualization" in Maguire, D., M. Goodchild and D. W. Rhind (eds.), *Geographical Information Systems: Principles and Practice*, Longman, London. pp. 427–443.

Collins, B. M. (1992) "Data visualization — has it all been seen before?" in Earnshaw, R. A.

and D. Watson (eds.), *Animation and Scientific Visualization: Tools and Appplications*, Academic Press, London, pp. 3–28.

Downs, R. M. and D. Stea (1977) *Maps in Minds: Reflections on Cognitive Mapping*, Harper and Row, New York.

Earnshaw, R. A. and N. Wiseman (eds.) (1992) *An Introductory Guide to Scientific Visualization*, Springer, Berlin.

Eastman, J. R. (1985) "Cognitive models and cartographic design research", *The Cartographic Journal*, Vol. 22, No. 4, pp. 95–101.

Gilbert, E. W. (1958) "Pioneer maps of health and disease in England", *Geographical Journal*, Vol. 124, pp. 172–183.

Godlewska, A. (1989) "To surf or swim" in Dahl, E. H., Commentary, Responses to J. B. Harley's article, "Deconstructing the map" published in *Cartographica*, 26(2), *Cartographica*, Vol. 26, No. 3,4, pp. 96–98.

Haber, R. B. and D. A. McNabb (1990) "Visualization idioms: a conceptual model for scientific visualization systems" in Nielson, G. M., B. Shriver and L. J. Rosenblum (eds.), *Visualization in Scientific Computing*, IEEE Press, London, pp. 74–93.

Hall, S. (1992) *Mapping the Next Millenium: the Discovery of New Geographies*, Random House, New York.

Harley, J. B. (1989) "Deconstructing the map", *Cartographica*, Vol. 26, No. 2, pp. 1–20.

Keates, J. S. (1982) *Understanding Maps*, Longman, London.

Kosslyn, S. M. (1983) *Ghosts in the Mind's Machine: Creating and Using Images in the Brain*, Norton, New York.

Kuipers, B. (1978) "Modelling spatial knowledge", *Cognitive Science*, Vol. 2, pp. 129–153.

Learnmonth, A. (1988) *Disease Ecology*, Basil Blackwell, Oxford.

Lloyd, R. (1989) "Cognitive maps: encoding and decoding information", *Annals of the Association of American Geographers*, Vol. 79, pp. 101–124.

McKim, R. H. (1970) *Experiences in Visual Thinking*, Brooks/Cole, Monterey.

Mark, D. M. (1993) "Human spatial cognition" in Medyckyj-Scott, D. and H. M. Hearnshaw (eds.), *Human Factors in Geographical Information Systems*, Belhaven Press, London, pp. 51–60.

Mark, D. M. and M. McGranaghan (1986) Effective provision of navigation assistance for drivers: A cognitive science approach in *Proceedings, Auto-Carto London*, Vol. 2, 14–19 September, London, Royal Institute of Chartered Surveyors, London, pp. 399–408.

Mark, D. M., A. U. Frank, M. H. Egenhofer, S. M. Freundschuh, M. McGranaghan and R. M. White (1989) "Languages of spatial relations: initiative two specialist meeting report, *Technical Report 89-2*, National Centre for Geographic Information and Analysis, Santa Barbara, CA.

Morris, J. (1982) "The magic of maps: the art of cartography", *Unpublished M. A. Dissertation*, University of Hawaii at Manoa.

Palmer, T. C. (1992) "A language for molecular visualization", *IEEE Computer Graphics & Applications*, Vol. 12, pp. 23–32.

Piaget, J. and Inhelder, B. (1956) *The Child's Conception of Space*, Routledge and Kegan Paul, London.

Robinson, A. H. (1955) "The 1837 maps of Henry Drury Harness", *Geographical Journal*, Vol. 121, pp. 440–450.

Robinson, A. H. (1987) *Early Thematic Mapping in the History of Cartography*, The University of Chicago Press, Chicago.

Robinson, A. H. and B. B. Petchenik (1976) *The Nature of Maps: Essays Toward an Understanding of Maps and Mapping*, University of Chicago Press, Chicago.

Scrope, G. P. (1833) *Principles of Political Economy, Deduced from the Natural Laws of Social Welfare, and Applied to the Present State of Britain*, Longmans, London.

Shepard, N. S. (1978) "Externalization of mental images and the act of creation" in Randhawa, B. S. and W. E. Coffman (eds.), *Visual Learning, Thinking and Communication*, Academic Press, London, pp. 133–189.

Tufte, E. R. (1990) *Envisioning Information*, Graphics Press, Cheshire, CT.

Visvalingam, M. (199) "Towards a wider view of visualisation", *Cartographic Information Systems Research Group (CISRG) Discussion Paper 9*, University of Hull, Hull.

Wallis, H. M. and A. H. Robinson (eds.) (1987) *Cartographical Innovations: an International Handbook of Mapping Terms*, Map Collector Publications (1982) in association with the International Cartographic Association.
Warren, G. and W. Bunge (1970) *Field Notes: Discussion Paper No. 2: School Decentralization*, Detroit Geographical Expedition and Institute, Detroit.
Watson, J. D. (1968) *The Double Helix*, New American Library, New York.
Wood, D. (1992) *The Power of Maps*, Routledge, London.

CHAPTER 3

Cognitive Issues in Cartographic Visualization

MICHAEL P. PETERSON

Department of Geography and Geology
University of Nebraska at Omaha, USA

Introduction

Cartographic visualization may be viewed as a logical extension of cartographic communication. The explicit interest in the map as a form of communication has fundamentally changed cartography over the past 40 years. Instead of being simply concerned with the techniques of map construction and an endless adaptation to technological innovations, cartographers could legitimately examine the function and purpose of maps. As a result, cartography was transformed into the science of communicating information through the use of a map (Morrison 1978).

The specific research in cartographic communication had two major phases. Beginning in the 1950s, cartography incorporated the research methods of psychophysics and examined the stimulus–response relationship of individual symbols. In the late 1970s, cartographic research developed in the direction of cognitive psychology and became concerned with how maps were mentally processed and remembered. This emphasis on cognitive issues in map use was stimulated in large part by Robinson and Petchenik's, *The Nature of Maps*, published in 1976.

A number of cartographic studies examined maps from the perspective of cognition (Dobson 1979; Olson 1979; Gilmartin 1981; Steinke and Lloyd 1981; Eastman 1985; Lloyd and Steinke 1985a,b; Peterson 1985). However, the emphasis on cognition in cartography never became as dominant as the previous concern with psychophysics (for a review of both, see Kimerling 1989). In part, this was because cognition did not present the same set of well-defined research projects to cartographers. Cognitive experimentation with maps was more complicated and the results were more difficult to interpret. Perhaps the major reason, however, was the computer. By the early 1980s, the interests of cartographers were occupied by microcomputers and the development of geographic information systems (GIS). In a sense, these developments represented the same kind of "adaptation to technological innovation" that had previously hindered the development of cartographic theory and the concern with communication. But, in this case, the adaptation was necessary because the integration of these tools introduced new forms of cartographic communication, including those associated with visualization.*

By the end of the 1980s, an interest in cognition had resurfaced. Researchers in GIS turned to cognitive science to answer a variety of questions that involve our use of spatial data (see Mark and Frank 1991). Cartographers, now viewing the computer not only as a tool to make maps on paper but as a medium of communication, have a renewed interest in mental processes. As a result, we once again turn to cognition to gain a greater insight into visual mental processing, particularly of the dynamic displays associated with visualization.

Cognition, defined as the "intelligent processes and products of the human mind" (Flavell 1977), includes such mental activities as perception, thought, reasoning, problem solving and mental imagery. An interest in such phenomena is relatively new, even in psychology. Dominated by behaviorism from the 1920s to the early 1970s, psychology concerned itself only with physically observable phenomena. According to the behaviorist, mental processes that occur between stimulus and response, including all aspects of thought, are not directly observable and therefore irrelevant. An important element in the development of cognition was a variety of technological advances that enabled researchers to externalize mental events in more objective ways.

A particularly important cognitive construct to visualization is the mental image. Indeed, the verb "to visualize" implies the construction of a mental image. Computer visualization could be defined as the use of the computer to create images of complex things that are beyond the capabilities of the human mind. There is, of course, a strong relationship between maps and both the human and computer forms of visualization. In discussing maps, for example, Muehrcke and Muehrcke comment

> The pictorial devices serve the need for visualization which is the very foundation of intuition and creativity. Environmental visualization may be strictly mental and intangible, or it may take the more concrete form of a physical map or chart. (Muehrcke and Muehrcke 1992, p. 5)

Visualization can be interpreted as both the internal and external creation of an

*The view of cartographic visualization presented here differs somewhat from that expressed in Chapter 1. Here cartographic visualization is seen as a more dynamic form of map use, consisting primarily of increased levels of interaction and the use of animation. Cartographic communication is also interpreted differently. It is not merely as the communication of a particular message. Rather, cartographic communication concerns itself with improving the effectiveness of maps as a form of communication.

image. A conception of the associated cognitive processes is important in understanding both forms of visualization and the relationship between them.

This chapter is divided into three parts. Firstly, different theories of human information processing are reviewed. In the second part, we take a closer look at the concept of the mental image that seems so central to the process of visualization. Results of experiments will be presented that demonstrate both the existence and functional properties of mental images. There is evidence to indicate that mental images are not copies of reality, like "pictures in the head" but are generalized depictions analogous to maps. Finally, a critical element of visualization is the use of animation, particularly in depicting processes of space and time. The last section examines theories concerning the processing of moving images.

Cognition and Human Information Processing

One theory of human information processing envisions a series of memory stores, each characterized by a limited amount of information processing (Klatzky 1975). Like all models, this "stage model" is overly simplistic. However, the model serves a purpose in identifying distinct and largely verifiable memory stores in information processing. Three memory stores are generally referred to: (1) the sensory register; (2) short-term memory (STM); and (3) long-term memory (LTM). For visual information processing these three memory stores are referred to more specifically as iconic memory, the short-term visual store (STVS) and long-term visual memory (LTVM).

Figure 3.1 depicts these information processing stores in the recognition of a state in the USA. According to this model, human information processing of visual information begins with iconic memory that is thought to hold information in

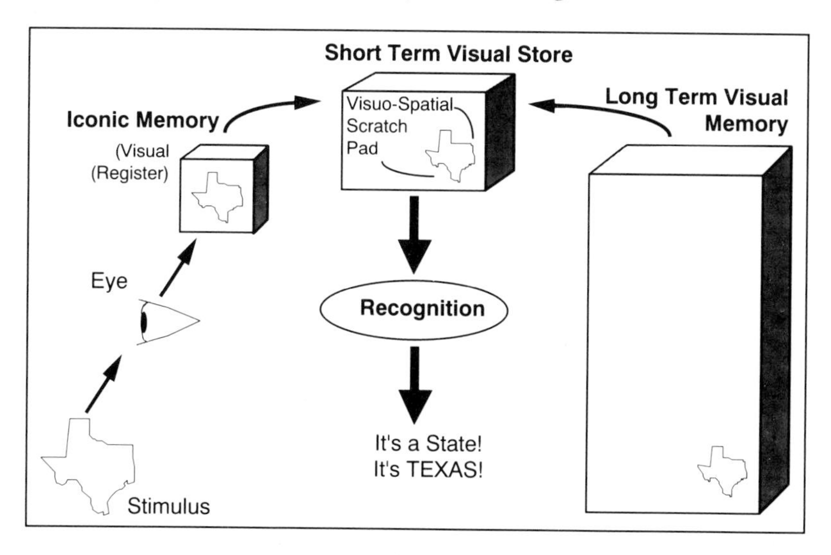

FIG. 3.1. Recognition of a state (after Peterson 1987).

sensory form for about 500 ms (Humphrey and Bruce 1989: 196), long enough for it to be initially recognized. Iconic memory is a type of physical image within the retina of relatively unlimited capacity and unaffected by pattern complexity. The STVS is of much longer durations but is of limited capacity and therefore affected by complexity.

Moving information from iconic memory to the STVS requires attention. This process simply refers to the human capability to "tune-in" certain sources of information and reject all others. An example of the process would be the visual identification of an animal that is partially obscured by vegetation. Attention makes it possible to "mentally focus" on the animal and reject the surrounding stimulus. This focusing is based on information in LTM. Kosslyn and Koenig (1992: 57) point out that stored information is used to make a guess about what we are seeing and this guess then controls the attention process.

Visual perception is based on the recognition of patterns. Visual pattern recognition converts the contents of iconic memory into something more meaningful through a matching process with previously acquired knowledge stored in LTM. There are at least three models of visual pattern recognition:

1. Template matching. With this model, the image in iconic memory is matched to LTM representations and the one with the closest match indicates the pattern that is present (Humphreys and Bruce 1989: 61).
2. Feature detection. This model specifies mini-templates or "feature detectors" for simple geometric features. An example of this type of model is the Pandemonium model proposed by Selfridge (1959; see also Lindsay and Norman 1976) that incorporates a series of "demons" or intelligent agents that work on a pattern by breaking it down into subcomponent features (see Fig. 3.2).
3. Symbolic description. A third model proposes that objects are represented symbolically as structural descriptions. A structural description can be thought of as a list of propositions or facts that describe properties of the individual parts of an object and their spatial relationships. For example, a propositional model for the letter "T" is "a horizontal line that is supported by a vertical line, and that this support occurs about half way along the horizontal line" (Bruce and Green 1990: 186). However, this model does not clearly specify how a pattern is recognized based solely on its structural description (see Pinker 1985). Also, it is difficult to see how a propositional model could work for complex shapes such as the outline of a state.

Visual information that has been recognized may be sent into the STVS. Information is held here by a process called rehearsal that serves to both recycle material in the STVS so that information does not decay and to transfer information about rehearsed items into LTVM. Just as an "articulatory loop" is thought to provide additional storage for verbal material (the continuous internal repetition of a phone number, for example), a visual-spatial scratch pad (VSSP) is proposed as the equivalent "shelf" on which visual information is temporarily stored within the STVS (Baddeley and Hitch 1974). Kosslyn and Koenig use the term "visual buffer"

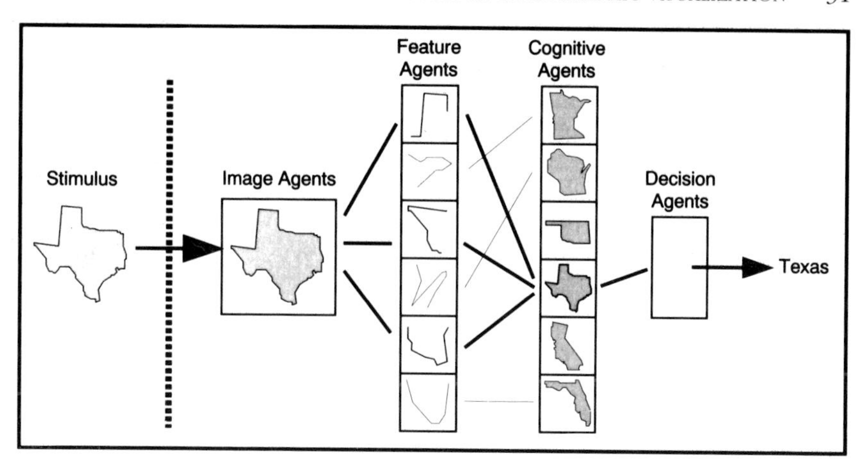

FIG. 3.2. The Pandemonium model in recognizing the outline of Texas (after Peterson 1987).

to describe short-term visual memory and point out that it can be activated from iconic memory or LTM. Whatever the source of information in the STVS, it requires an active rehearsal for its maintenance and plays an important role in active visualization and imagery.

A number of theories exist concerning LTVM. One theory suggests that LTVM is essentially a permanent storehouse — nothing is ever lost. Forgetting, it is argued, is a retrieval problem in which the appropriate link to a piece of knowledge is disrupted. Studies have shown an apparently limitless capacity to remember pictures. For example, Shepard (1967) showed subjects several hundred pictures from various sources. Later, subjects were asked whether they had seen the picture before. A 98% accuracy rate was discovered. Another theory about LTM is that information is stored in a form compatible with sensory perception to facilitate the complicated process of pattern recognition.

Finally, there is evidence suggesting that locational information is coded into LTM without attention or conscious awareness (Mandler *et al.* 1979). This phenomenon has been often noted. For example, we might remember the location of an article in a newspaper without intending to do so, or the location in a notebook of an answer to a question without remembering the answer. Although it is possible that we may be consciously aware of this sort of spatial information, it is unlikely that we intend to commit it to memory. This indicates that the human system encodes, and perhaps even uses, a great deal of locational information without conscious awareness.

Computational Models of Information Processing

The development of computer models of information processing has been a major emphasis in cognitive psychology during the past decade (Landy *et al.* 1991). A great deal of interest has been shown in the building of neuron-like models of visual perception. These so-called "connectionist" models have the advantage of being biologically plausible and provide a mechanism to incorporate parallel computations in visual information processing. They are based on assumptions about patterns of neural connections and the way synapses in the brain are modified. The models go beyond the simple linear approach to information exemplified by the stage model.

In general, computational theorists in cognitive psychology view the process of perceiving as one of inferring from ambiguous "clues" in the retinal image. Such inferences are based on additional knowledge from memory that is brought to bear in the process of recognition. A slightly different way of thinking about image interpretation is given by Marr (1982). Marr's theory incorporates the computational approach and is an attempt to explain visual information processing in a framework that cuts across the boundaries of physiology, psychology and artificial intelligence.

According to Marr, it is very general knowledge about the physical world that is used to interpret the retinal image. Perception, he argues, is based on assumptions that are generally true of the world. These assumptions include the fact that surfaces are generally smooth, or that light comes from above — assumptions that a visual system would not have to discover from experience but is "hard-wired" (i.e. innately specified).

An important part of Marr's theory is that a number of different representations must be constructed from the information in the retinal image. The first representational stage, called the *primal sketch*, captures the two-dimensional structure of the retinal image. The next representation, called the *2 1/2 D sketch*, describes how surfaces are oriented with respect to a viewer. Finally, Marr argues that a representation of the three-dimensional shape of the object must be constructed. This last representation is called the *3D model.*

Marr and Nishihara (1978) present a theory of object recognition based upon this general theory. They argue that objects or their components can be constructed from generalized cones or cylinders and that the lengths and arrangements of the axes of these components, relative to the major axis of the object as a whole, could be used in object recognition. They present the example of discriminating between a human and a gorilla. If each is viewed as consisting of generalized cones that are used to represent the head, trunk, arms and legs, the two can be discriminated based on the relative lengths of the axis of these components.

Marr and Nishihara (1978) go on to suggest that a configuration of cones and cylinders derived from an image could be matched against a catalog of different object models, distinguished on the basis of the number and disposition of their component parts. A major advantage of this approach is that descriptions at different spatial scales can be constructed. Humphreys and Bruce (1989: 72) point out that each entry in the catalog could point to a hierarchically organized set of three-dimensional models at different scales and the catalog could be accessed at any one of these levels (see Fig. 3.3).

Marr's theory of visual information processing and object recognition is based on

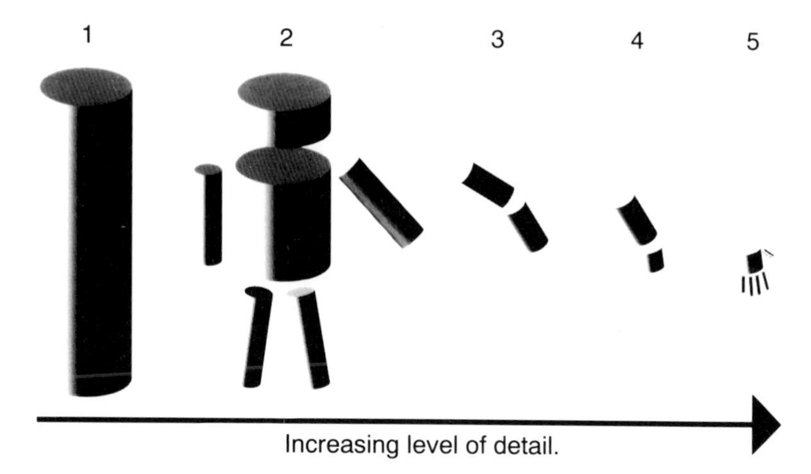

FIG. 3.3. A hierarchy of three-dimensional models with increasing levels of detail (after Humphreys and Bruce 1989: 71).

breaking down the retinal image into its component parts. These parts are not propositional descriptions but simpler shapes. The process is analogous to generalization in cartography. Maps are based on a simplification of reality. It seems the mind is engaged in much the same process. This simplification process is also confirmed in studies on the mental image.

Mental Imagery

Visual imagery has always been a central topic in the study of cognition. The mental image is defined as an internal representation similar to sensory experience but arising from memory. It may be viewed as a form of internal "visualization". The purpose of research in visual mental imagery has proceeded in three basic directions: (1) to determine the existence of mental images; (2) to define their properties; and (3) to establish how they are used in thinking.

Existence of Mental Images

Several studies have established the existence of visual codes in memory. In an often cited study, Shepard and Metzler (1971) measured the time it took for subjects to decide whether two complex block structures depicted the same three-dimensional shape (see Fig. 4). It was found that the amount of time taken to make the judgments increased linearly with the difference in rotation between the two objects. The rate of rotation was found to be 60° per second. The researchers concluded that the subjects were rotating the visual image of one figure into congruence with the other and the further the objects had to be mentally rotated,

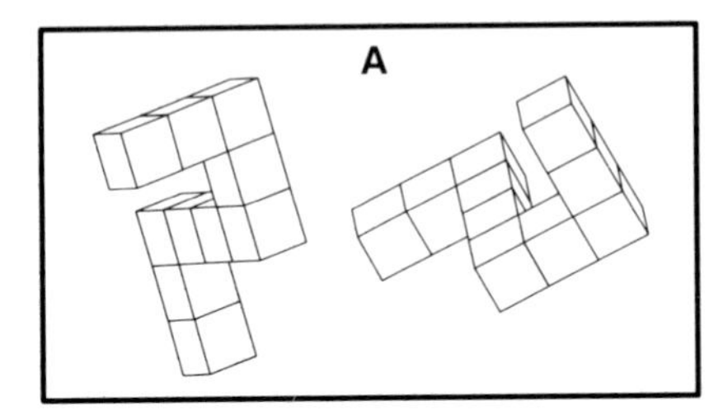

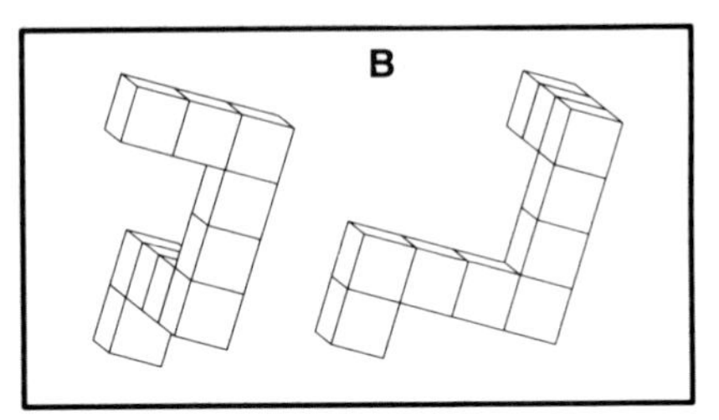

FIG. 3.4. The figures in A are identical but are rotated by 80 degrees. Figures in B are not the same (after Shepard and Metzler 1971).

the more time was required for the decision. Like physical rotation, mental rotation takes longer the further the distance involved, suggesting an internal imagery space with properties analogous to physical space.

Another line of research concerning the existence of visual codes in memory suggests a difference between visual and linguistic memory. Paivio (1971) first postulated that visual imagery and verbal symbolic processes represent alternative coding systems in memory. Moreover, visual representations are not restricted to material that is intuitively spatial, such as a map; rather there is a distinction between material that is concrete and readily imageable and material that is abstract and cannot readily be associated with an image (McTear 1988: 106). Paivio (1986) showed that the image evoking value of nouns, referred to as "imagery concreteness" (e.g. "apple" leads to a more concrete mental image than the more abstract "distance") correlates highly with the ability to remember. It was shown that words coded under both image and verbal mediators will be more easily remembered than words that are only verbally coded as more "links" exist to those words in memory. It seems that our ability to "visualize" an object improves our ability to remember it.

The existence of mental imagery is not universally accepted in psychology (Tye 1991). Interpreting the results of mental rotation experiments conducted by Shepard and Metzler (1971), Pylyshyn (1981) argues that imagery consists of the use of general thought processes to simulate physical or perceptual events. He argues that we know implicitly that physical objects cannot change orientation instantaneously. We also know that the time required for a rotation is equal to the angle of rotation divided by its rotation rate. Pylyshyn (1981) proposes that when we mentally rotate objects, we perform the relevant computation then wait the corresponding amount of time before indicating our response.

Properties of Mental Images

A variety of studies in psychology have attempted to define the function and properties of mental images in human thought. Researchers have theorized that imagery uses representations and processes that are ordinarily dedicated to visual perception. Kosslyn (1980) theorizes, in particular, that mental images are analogous to displays on a graphic computer terminal or cathode ray tube (CRT) on which graphic displays can be generated. In this view, the mental image is a data structure consisting of two parts: (1) the "surface representation" or quasi-pictorial entity that we experience; and (2) the "deep representation", the information in memory from which the surface image is derived. It follows from this CRT metaphor that mental images have two important structural components: (a) they have analog properties, i.e. they preserve relative interval distances between their component parts; and (b) they have limited spatial extent.

In a series of experiments, Kosslyn (1980) demonstrated that mental images are characterized by these two structural components. In experiments involving visual scanning, the time required to scan between objects in mental images was used to determine whether mental images have truly analog properties. In one of these experiments, a fictional map with seven locations (house, well, tree, lake, etc.) was presented to subjects. Subjects were repeatedly asked to draw the map on a blank piece of paper until objects were depicted to within 0.25 inch (0.635 cm) of their actual location. Next, subjects were asked to image the mental map and mentally "stare" at a named location. A second location was named and the reaction time measured for the subject to scan to the object. It was found that the time to scan increased linearly with the distance to be scanned (Kosslyn 1980: 44; Tye 1991: 44). Kosslyn concluded that mental images, like maps, depict information about interval spatial extents.

If mental images have limited spatial extent, as theorized, then the mental image must contain a limited number of features (in the same sense that a map is limited in the number of features it can reasonably show). Kosslyn (1983) devised a procedure in which size was manipulated indirectly by asking subjects to image a "target" animal, such as a rabbit, as if it were standing next to either an elephant or a fly (see Fig. 3.5). The assumption was that if a mental image takes up a fixed amount of space and most of this space is taken up by the elephant, then the adjacent scaled image of the rabbit should be subjectively small and not as detailed. Alternatively, if a rabbit is imaged next to the image of a fly, then the rabbit should

FIG. 3.5. Animals used in Kosslyn's selection of features experiment. On the left a rabbit and an elephant are imaged simultaneously. On the right a rabbit and a fly are imaged together.

seem much larger with more defined features. The results of the experiment showed that people required more time to mentally "see" parts of the subjectively smaller animals, i.e. the animals imaged next to the larger animals. Subjects explained that they had to "zoom in" to see the properties of the subjectively smaller images. Kosslyn concluded that a selection of features had occurred in the original smaller (before "zoomed") image because the mental image has a fixed size and resolution (Kosslyn 1983: 56).

Kosslyn has shown that images, like maps, represent relative interval distances. More significantly, he showed that mental images are not copies of sensory impressions, like "pictures in the head", but rather they seem to be intellectually processed and generalized representations, analogous to maps. Mental images can be defined as the internal counterpart of maps in the sense that both are characterized by a selection of features.

Visual Cognition

Visual cognition is defined as the use of mental imagery in thinking. Mental images can be derived from material stored in memory or they can be mental "drawings" of things we have never seen. Kosslyn and Koenig (1992: 129) state that imagery can be used in the following ways:

1. Reasoning. Reasoning with imagery is often reported by scientists or inventors. It is based on the creative generation of images. This may be accomplished by combining familiar elements in new ways or from scratch using only elementary components. The use of imagery in reasoning involves not only generating and inspecting imaged objects but transforming and retaining images long enough to work with them (Kosslyn and Koenig 1992: 146).
2. Learning a skill. Learning a new skill can be aided through the use of imagery. In this case, an image is used to define physical movements (Kosslyn and Koenig 1992: 155). The technique is often used in sports training. The shooting of baskets, for example, can be improved by using imagery as a part of practice. This aspect of visual cognition may be used to understand the effect of a map in controlling our movements in space.
3. Comprehending verbal descriptions. A mental image can be used to interpret verbal descriptions. The description may consist of a floor plan to a house or instructions on how to travel to a certain place. A mental image seems to be vital in the interpretation of the description.
4. Creativity. Finke (1990) shows how images can stimulate the discovery of unexpected patterns, new inventions, and creative concepts.

The way imagery is used in thinking is only now beginning to be understood. In one way or another, all of these aspects of visual cognition can be related to cartographic visualization.

Mental Imagery in Cartography

The concept of the mental image has influenced recent research in cartographic communication. The ability to rotate an image is clearly important in map orientation. The mental rotation of maps was examined by Steinke and Lloyd (1983). Goldberg *et al.* (1992) looked at image transformations of three-dimensional terrain maps. The general effect of orientation on map learning and the resultant rotation requirements has received considerable attention in both cartography and psychology (Shepard and Hurwitz 1984; MacEachren 1992).

Other research has concentrated on the relationship between maps and images and how images are used in map use. Peterson (1985) proposed the concept of "imagability" to describe the quality of the mental map derived from a map and examined the quality of a mental image derived from a variety of point symbol maps. Later, Peterson (1987) examined the overall role of mental imagery in cartographic communication. Lloyd (1989) examined how mental images in the form of cognitive maps are used to estimate distance and direction. Finally, MacEachren (1991) looked at the role of maps and images in spatial knowledge acquisition. Through an experiment, he found evidence for independent but interrelated verbal and spatial components of regional images. The findings support the claim of Paivio (1971) that the two components represent alternative coding systems in memory.

The mental image is not universally accepted in cartography. Mark and Frank (1991: 3) question whether the the concept tells us anything about the "architecture" of spatial reasoning. Head (1991) proposes a more linguistic approach to our understanding of how maps communicate, but concedes that there are significant differences between cartographic communication and natural language (Head 1991: 260).

Visual Processing of Dynamic Images

A great majority of the work in cognitive psychology has involved the mental processing of static images. Of course, the world is dynamic and the mind must be able to process moving images. In addition, it has been pointed out that even if an observer views a scene containing only motionless objects, the observer's eyes will be in continuous motion. Heckenmuller (1965) found that if the effects of such eye movements are neutralized, by artificially stabilizing the image on the retina, we become blinded. Motion, either of objects in the environment or created artificially by the movement of the eyes, is essential to perception. Psychologists have taken at least three separate approaches to the understanding of motion processing.

Apparent Motion

It has long been known that the perceptual system can be made to perceive motion when none actually exists. This phenonmenon has been used as one approach to study the perception of motion. Apparent motion can be explained as follows:

> If an observer is presented with a display in which two lines in different locations are alternatively exposed, at certain interstimulus intervals (which depend on the spatial separation, luminance and duration of lines) the observer reports seeing the first line move across the position of the second, i.e., the observer sees not two lines alternately appearing in different locations, but one line which moves slowly from place to place. (Bruce and Green 1990: 160)

The phenomenon provided a justification for analyzing the perception of motion by determining how correspondences are established between the "snapshots" of the changing retinal patterns that are captured at different instances.

The major conclusion of these studies is that correspondences between individual elements in successive displays are established on the basis of the primitive elements of figures (primal sketch) rather than between whole figures (Julezs 1971; Marr and Poggio 1976). Ullman (1979) provides a computational account for how correspondences are made using the principle of "minimal mapping". He defines an affinity measure between pairs of elements in a display. To solve the correspondence process in a display with several elements, a solution is found that minimizes matches with poor affinities and maximizes those with strong affinities. He goes on to demonstrate how three-dimensional structure can be recovered from motion.

The phenomenon of apparent motion seems to indicate that the mind constructs the perception of motion from a series of static images. This principle is the basis of both film and displays on a CRT. Film consists of 24 still pictures per second (Hochberg 1986: 22–23). Because the interruption between each frame would be visible as an annoying flicker, projectors are equipped with rotary shutter to interrupt the projection more frequently and bring the interruption rate above the average viewer's so-called critical flicker frequency. The rotary shutter brings the flicker rate up to about 72 Hz. Similar rates are used to refresh the screen of a CRT.

The phenomenon of apparent motion is also used to implement a type of computer animation called color cycling (Gersmehl 1990). With this approach, a path is defined and broken up into small segments. Then, the colors are shifted down each segment of the path to give the impression of movement. The technique is commonly used in television weather forecasts to depict the movement of the jet stream.

Ecological Approach to Visual Perception

The importance of motion perception is one of the major influences in the development of Gibson's (1979) theory of "direct" perception. He argues against the predominant view that visual perception involves the recognition of static retinal snapshots. Instead, Gibson claims that information to specify the structures in the world and the nature of an observer's movement is detected "directly" from dynamic patterns of optical flow. In other words, perception is dependent on a change in the display.

Gibson argues that it is the flow in the structure of the total optic array that provides information for perception, not the individual forms and shapes of the "retinal image". He emphasizes the perception of surfaces in the environment. Surfaces consist of texture elements and different objects in the environment have different surfaces. The structure that exists in the surfaces structures the light that reaches the observer. Gibson argues that it is this structure in the light, rather than stimulation by light, that furnishes information for visual perception.

According to Gibson, analysis of the movement of objects must begin with analysis of the information available in the continuously changing optic array. To explain the phenomenon of apparent motion, Gibson argues that the stimuli preserves the transformational information that is present in the real moving display. He maintains that the stimulus information contains transformations over time that specify the path of the motion.

Gibson's theory stands apart from the mainstream of thought in cognitive psychology, especially his contention that perception is based on the direct processing of dynamic information in the optic array. However, his ideas have a great deal of applicability to the processing of dynamic displays and have influenced work on the mental processing of computer-generated displays (see, for example, Wanger *et al.* 1992).

Evidence from Physiology

It is has long been theorized that the visual system has a special sensitivity to moving images. Recent physiological evidence points to special neural subsystems in the brain that handle the visual processing of movement. The first evidence of these subsystems was based on people who had suffered bilateral damage to a certain part of the cortex. One patient, for example, had an impaired ability to see continuous movement (see Zihl *et al.* 1983). This person had difficulty crossing a road because she could not judge traffic speed. She had difficulty with conversations because she could not follow facial expressions. She also reported that when pouring tea into a cup, the tea appeared frozen "like a glacier". It was also difficult for her to know when to stop pouring because she could not detect the rising liquid in the cup. Despite these problems, she remained able to recognize stationary objects.

These results suggest a more physiological model of visual mental processing as opposed to the strictly cognitive models. This interpretation is analogous to the stage model of information processing but relies on a more anatomical explanation. For example, the visual register corresponds to the eye where receptors, stimulated by light, connect to ganglion cells that connect to individual fibers of the optic nerve, called axons. The axons transmit their messages to the brain as electrical impulses. The impulses proceed to the primary visual cortex in the back of the brain after passing through a synapse, called the lateral geniculate nucleus, about half way along the journey to the cortex. Axons from the right half of the visual field of both eyes go to the left side of the brain, while those from the left fields go to the right half of the brain. These two sides of the visual field are reunited at the same place in the cerebral cortex.

The cerebral cortex plays an important role in the visual system. It has been found in monkeys that over half of this part of the brain is involved in the processing of visual information. This "visual" part of the cortex seems to correspond to the short-term visual store. It is here that the images presented to the eye are actually "visualized". The cortex is made up of an astounding number of cells. There are about a million ganglion cells in the human retina that connect to a million axons that reach the brain from each eye. There are about 100 times this many cells in the visual cortex (Barlow 1990: 12).

At one time it was thought that the visual cortex was used to store knowledge about the environment (Herrick 1928). A more recent theory suggests that the visual cortex performs inverse optics — it reconstructs the objects that cause visual images from the images themselves (Poggio *et al.* 1985). Barlow points out that these proposed functions are very similar (Barlow 1990: 8). Stored knowledge is needed to make reconstruction of the image possible. He contends that this represents *understanding*, the most fundamental function of the visual cortex.

An early discovery about visual motion processing was made by Barlow and Levick (1965). They documented the existence of neurons in rabbits that are selectively sensitive to speed and direction of retinal image motion. It has been generally verified that motion is detected in the lower animals, particularly insects, by neurons in the retina. In primates, however, the perception of motion can be traced directly to higher levels of processing in the cortex (Movshon 1990: 126).

It seems that there are two main groups of retinal neurons that feed into separate pathways that can be traced directly to the cerebral cortex. One of these "visual pathways" appears to specialize in analyzing motion. The other seems to be more concerned with the form and color of visual images (Movshon 1990: 127). Evidence in support of a motion pathway was determined through the functional analyses of neural activity. There is strong evidence that the visual processing of motion is accomplished separately from the processing of the attributes of objects, such as form and color.

Conclusion

Visualization is a mental process. It is dependent on the recognition of patterns in both static and dynamic displays (MacEachren and Ganter 1990: 66). Cognitive psychologists believe that perception and pattern recognition are closely linked to memory. A great deal of evidence has accumulated that the representations in memory have analog qualities — that they can be described as a form of mental image. An important observation, however, is that these representations are not like "pictures in the head". Rather they are intellectually processed and generalized representations, more analogous to a map than to a photograph. A possible implication of these findings for cartographic visualization is that displays should consist of different levels of generalization to better conform to how maps are remembered and mentally processed.

The function of mental images in the process of cartographic communication can be modeled in the form of a human GIS, as proposed by Peterson (1987). It seems that much of the information derived from a map is in image form — sometimes encoded without conscious awareness — and may be used long after the map has ceased to be a visual stimulus. It seems probable that the mind is a storehouse of images from which previously unconsidered spatial information may be derived. The process of cartographic communication, therefore, continues in the absence of the map.

Visual motion processing is not completely understood. It seems that parts of the visual cortex specialize in motion processing. That this is accomplished in the cortex and not in the eye indicates that training and experience are factors in our ability to detect motion. This suggests that greater experience with dynamic displays would result in a greater ability to interpret these displays. In referring to the effect of video games, Peterson (1991) suggests that they require a form of spatial thinking and memorization on the part of children to which their parents have never been exposed. Indeed, it seems that children require less time to become proficient at these games than adults. Although considered a waste of time by most, these games may serve the function of training the visual cortex to process dynamic displays. This ability may later be used in more constructive ways in scientific visualization or in the processing of dynamic map displays. It can also be postulated that greater experience with dynamic displays will create individuals who are more capable of interpreting these displays.

The visual processing of motion is central to perception. Every movement that we make, including the quick saccades of the eye, is dependent on the processing of motion. Motion processing is a natural activity of the eye–brain system. We should not be concerned about whether people can process dynamic displays. Rather, the viewing of static displays may be more difficult for the brain, requiring a greater degree of training.

References

Baddeley, A. D. and G. Hitch. (1974) "Working memory" in G. Bower (ed.), *The Psychology of Learning and Motivation, VIII*, Academic Press, New York, pp. 47–89.

Barlow, H. (1990) "What does the brain see? How does it understand?" in Barlow, H., C. Blakemore and M. Weston-Smith (eds.), *Images and Understanding: Thoughts about images, Ideas About Understanding*, Cambridge University Press, Cambridge, pp. 5–25.

Barlow, H. B. and W. R. Levick (1965) "The mechanism of directionally selective units in the rabbit's retina", *Journal of Physiology*, Vol. 178, pp. 477–504.

Bruce, V. and P. R. Green (1990) *Visual Perception: Physiology, Psychology and Ecology*, Lawrence Erlbaum, Hillsdale.

Dobson, M. W. (1979) "Visual information processing during cartographic communication", *Cartographic Journal*, Vol. 16, pp. 14–20.

Eastman, J. R. (1985) "Cognitive models and cartographic design research", *Cartographic Journal*, Vol. 16, pp. 95–101.

Finke, R. (1990) *Creative Imagery: Discoveries and Inventions in Visualization*, Lawrence Erlbaum, Hillsdale.

Flavell, J. H. (1977) *Cognitive Development*, Prentice Hall, Englewood Cliffs.

Gersmehl, P. J. (1990) "Choosing tools: nine metaphors of four-dimensional cartography", *Cartographic Perspectives*, Spring (5), pp. 3–17.
Gibson, J. J. (1979) *The Ecological Approach to Visual Perception*, Houghton-Mifflin, Boston.
Gilmartin, P. P. (1981) "The interface of cognitive and psychophysical research in cartography", *Cartographica*, Vol. 18, pp. 9–20.
Goldberg, J. H., A. M. MacEachren and X. P. Korval (1992) "Mental image transformations in map comparison", *Cartographica*, Vol. 29, No. 2, pp. 46–59.
Head, C. G. (1991) "Mapping as language or semiotic system: review and content" in Mark, D. M. and A. U. Frank (eds.), *Cognitive and Linguistic Aspects of Geographic Space*, Kluwer Academic, Dordrecht, pp. 237–262.
Heckenmuller, E. G. (1965) "Stabilization of the retinal image: a review of method, effects and theory", *Psychological Bulletin*, Vol. 63, pp. 157–169.
Herrick, C. J. (1928) *An Introduction to Neurology*, Saunders, Philadelphia.
Hochberg, J. (1986) "Representation of motion and space in video and cinematic displays", in Boff, K. R., L. Kaufman and J. P. Thomas (eds.), *Handbook of Perception and Human Performance*, Vol. 2, Wiley, New York, pp. 22-1–22-64.
Humphreys, G. W. and V. Bruce, (1989) *Visual Cognition: Computational, Experimental and Neuropsychological Perspectives*, Erlbaum, Hove.
Julezs, B. (1971) *Foundations of Cyclopean Perception*, University of Chicago Press, Chicago.
Kimerling, A. J. (1989) "Cartography" in Gaile, G. L. and C. J. Willmott, (eds.), *Geography in America*, Merrill, Columbus, pp. 686–718.
Klatzky, R. L. (1975) *Human Memory: Structures and Processes*, Freeman, San Francisco.
Kosslyn, S. M. (1980) *Image and Mind*, Harvard University Press, Cambridge.
Kosslyn, S. M. (1983) *Ghosts in the Mind's Machine*, Norton Press, New York.
Kosslyn, S. M. and O. Koenig (1992) *Wet Mind: the New Cognitive Neuroscience*, Free Press, New York.
Landy, M. S. and J. A. Movshon (1991) *Computational Models of Visual Processing*, MIT Press, Cambridge.
Lindsay, P. H. and D. A. Norman (1976) *Human Information Processing*, Academic Press, New York.
Lloyd, R. (1989) "The estimation of distance and direction from cognitive maps", *The American Cartographer*, Vol. 16, No. 2, pp. 109–122.
Lloyd, R. and T. R. Steinke (1985a) "Comparison of quantitative point symbols/the cognitive process", *Cartographica*, Vol. 22, No. 1, pp. 59–77.
Lloyd, R. and T. R. Steinke (1985b) "Comparison of qualitative point symbols: the cognitive process", *American Cartographer*, Vol. 12, 156–168.
MacEachren, A. M. (1991) "The role of maps in spatial knowledge acquisition", *The Cartographic Journal*, Vol. 28 (December), pp. 152–162.
MacEachren, A. M. (1992) "Learning from maps: can orientation specifity be overcome?" *The Professional Geographer*, Vol. 44, No. 4, pp. 431–443.
MacEachren, A. M. and J. H. Ganter (1990) "A pattern identification approach to cartographic visualization", *Cartographica*, Vol. 27, No. 2, pp. 64–81.
McTear, M. (ed.) (1988) *Understanding Cognitive Psychology*, Ellis Horwood, Chichester.
Mandler, J. M., D. Seegmuller and J. Day (1979) "On the coding of spatial information", *Memory and Cognition*, Vol. 5, pp. 10–16.
Mark, D. M. and A. U. Frank (1991) *Cognitive and Linguistic Aspects of Geographic Space*, Kluwer Academic, Dordrecht.
Marr, D. (1982) *Vision*, Freeman, San Francisco.
Marr, D. and H. K. Nishihara (1978) "Representation and recognition of the spatial organization of three-dimensional shapes", *Proceedings of the Royal Society of London, Series B*, Vol. 200, pp. 269–294.
Marr, D. and T. Poggio (1976) "Cooperative computation of stereo disparity", *Science*, Vol. 194, 283–287.
Morrison, J. L. (1978) "Towards a functional definition of the science of cartography with emphasis on map reading", *The American Cartographer*, Vol. 5, pp. 97–110.
Movshon, A. (1990) "Visual processing of moving images" in Barlow, H., C. Blakemore and M. Weston-Smith (eds.), *Images and Understanding: Thoughts About Images, Ideas About Understanding*, Cambridge University Press, Cambridge, pp. 122–138.

Muehrcke, P. C. and J. O. Muehrcke (1992) *Map Use: Reading, Analysis and Interpretation*, 3rd edn, JP Publications, Madison.

Neisser, U. (1967) *Cognitive Psychology*, Appleton-Century-Crofts, New York.

Olson, J. M. (1979) "Cognitive cartographic experimentation", *Canadian Cartographer*, Vol. 16, No. 1, pp. 23–44.

Paivio, A. (1986) *Mental Representations: A Dual Coding Approach*, Oxford University Press, Oxford.

Paivio, A. (1971) *Imagery and Verbal Processes*, Holt, Rinehart & Winston, New York.

Peterson, M. P. (1985) "Evaluating a map's image", *The American Cartographer*, Vol. 12, No. 12, pp. 41–55.

Peterson, M. P. (1987) "The mental image in cartographic communication", *The Cartographic Journal*, Vol. 24, pp. 35–41.

Peterson, M. P. (1991) "The ideas of Nu Cartoman", *Cartographic Perspectives*, Vol. 9, pp. 16–17.

Pinker, S. (1985) "Visual cognition: an introduction" in Pinker, S. (ed.), *Visual Cognition*, MIT Press, Cambridge.

Poggio, T., V. Torre and C. Koch (1985) "Computational vision and regularization theory", *Nature*, Vol. 317, pp. 314–319.

Pylyshyn, Z. W. (1981) "The imagery debate: analogue imagery versus tacit knowledge", *Psychological Review*, Vol. 88, pp. 16–45.

Robinson, A. H. and B. P. Petchenik (1976) *The Nature of Maps: Essays Toward Understanding Maps and Mapping*, University of Chicago Press, Chicago.

Selfridge, O. G. (1959) "Pandemonium: a paradigm for learning", in *The Mechanization of Thought Processes*, HMSO, London.

Shepard, R. N. (1967) "Recognition memory for words, sentences and pictures", *Journal of Verbal Learning and Behavior*, Vol. 6, pp. 156–163.

Shepard, R. N. and S. Hurwitz (1984) "Upward direction, mental rotation, and discrimination of left and right turns in maps", *Cognition*, Vol. 18, pp. 161–193.

Shepard, R. N. and J. Metzler (1971) "Mental rotation of three-dimensional objects", *Science*, Vol. 171, pp. 701–703.

Steinke, T. R. and R. E. Lloyd (1981) "Cognitive integration of objective choropleth map attribute information", *Cartographica*, Vol. 18, No. 1, pp. 13–23.

Steinke, T. R. and R. E. Lloyd (1983) "Images of maps: a rotation experiment", *Professional Geographer*, Vol. 35, No. 4, pp. 455–461.

Tye, M. (1991) *The Imagery Debate*, MIT Press, Cambridge.

Ullman, S. (1979) *The Interpretation of Visual Motion*, MIT PRess, Cambridge.

Wanger, L. R., J. A. Ferwada and D. P. Greenburg (1992) "Perceiving spatial relationships in computer-generated images", *IEEE Computer Graphics & Applications*, May, pp. 44–55.

Zihl, J., D. Von Cramon and N. Mai (1983) "Selective disturbance of movement vision after bilateral brain damage", *Brain*, Vol. 106, pp. 313–340.

CHAPTER 4

The Bridge Between Cartographic and Geographic Information Systems

KIRSI ARTIMO

Department of Surveying

Helsinki University of Technology

Helsinki, Finland

Introduction

The history of geographic information systems (GIS) starts from "computer-mapping" programs in the early 1960s. SYMAP, the famous ancestor of GIS software, was developed in close contact with geographical applications, especially research on spatial analysis (Steinitz 1993). It is evident that the needs of spatial analysis and the new potential of computers created these applications of computer-based geographic information processing. During the following decades some other applications also became dominant in the field of spatially referenced data processing, and also some aspects other than analysis became relevant. One of these was the idea of GIS as a centralized "databank". Databanks, including both spatial and attribute data, were established by starting huge data collection projects. Urban planning was one of the pioneers in the use of these databanks and integrated registers. The term "land information system" (LIS) was used by surveyors who, in addition to digital mapping, designed and maintained cadastres. The term LIS was officially accepted by the International Federation of Surveyors to describe a concept comprising "all data related to land" (Andersson 1981). Data organization was, however, the main problem until the database management boom of the 1970s which initially seemed to solve all problems. During the next 10 years it was found, however, that spatial data management is not the same as conventional data management. Relational database management systems have now been applied to GIS software for over 10 years; the disadvantages are generally agreed upon but the optimum solution to GIS data management is still missing. Much work is now done in the field of object oriented database management applied to geographical data. An introduction to the O–O approach to GIS is given, for example, in Laurini and Thompson (1992).

During the last few years increasing attention has been paid to the advanced analysis and visualization of geographical data. GIS has also become a well known concept in various diciplines. If geographers were those who started GIS and surveyors were those who brought digital techniques into large-scale mapping and engineering applications, today GIS applications can be found almost in every sector of human life. GIS is generally understood in a very broad sense as it is a generic concept for all computer-based systems which process "map data". However, in the literature the definition of GIS still includes the emphasis on spatial analysis, as in the beginning. In this chapter the use of a broadened "umbrella concept" definition of GIS is argued.

During the whole period described above cartographers have been worried about the development and future of the profession of map-making. GIS seems, according to any definition, to be in close relation to maps and it has clearly been seen that the new techniques do not automatically improve the quality of maps. In the beginnning of digital mapping the relatively simple output facilities such as pen plotters set the quality limits. Now the problem is that by using the modern systems anyone can produce impressive maps. The results, however, are not guided by the system; the user has to be able to design the map. Not everyone is a cartographer and the maps which are produced show it!

There are no simple solutions to the problems of geographical data visualization. A great variety of possibilities exist: two- and three-dimensional visualization, plotted and printed maps and maps on the screen, millions of colours to be selected, tens and hundreds of symbols to be used, varieties in layout possibilities, static and animated visualizations. The selection of computer-based visualization software starts from simple thematic map production programs extends to visions about virtual realities. A need for intelligent cartographic systems is evident, especially when maps and other visualizations of geographical data are becoming common outputs in various types of information systems in society. The substance of this chapter, in brief, is that it is cartographers who must take the lead in developing these intelligent cartographic information systems.

Cartographers deal traditionally with the problems of visualization of geographic data. Visualization can be said to be a modern synonym to cartography. Cartographers are able to use computer-based systems for their own map production purposes. Now a more generic definition of GIS is required and, in addition to this, a clear idea about the role of visualization in a GIS. In this chapter the concept of a cartographic information system is introduced to stand for a computer-based (sub)system for geographical data visualization functions among and in GIS.

This chapter begins with a discussion about the term GIS and an introduction of the term CIS (cartographic information system). Some other concepts such as digital map are also treated, as well as the variety of possibilities of visualization of geographical data. The changes in cartographic communication are dealt with as the background for the discussion about the changing role of cartographers. A claim of this chapter, that cartographers are the potential links between the traditions of cartography as the science of visualizing of geographic data and the modern techniques, is argued by the above topics.

Defining GIS

GIS has traditionally been defined in relation to the functions that a system might perform. An alternative, and broader, definition will be proposed here that defines GIS in relation to the data handled rather than the functions that might be performed on that data.

The term GIS was first used in 1962 (Price 1992). As mentioned in the introduction, GIS was invented to be a tool for spatial analysis in geography. The term "information system" (IS) on which the term GIS is based was a popular term during the 1960s and 1970s. We could compare GIS, for example, with MIS (management IS), which also was a product of the decades mentioned and which was defined as "an IS whose prime purpose is to supply information to management" (Illingworth *et al.* 1986). Definition of an IS can be found, for example in *The Encyclopedia of Computer Science* (Ralston and Meek 1976) and it is as follows: "An information system can be defined as a collection of people, procedures and equipment designed, built, operated and maintained to collect, record, process, store, retrieve, and display information." According to several definitions an IS can also be manual (Lundeberg *et al.* 1979). In Ralston and Meek (1976) it is added that "sometimes these (information systems) are called computer-based information systems (CBIS) to distinguish them from manual systems."

Definitions of "information system" appear most frequently in the literature of information system design written during the 1970s. One of the developers of the so-called ISAC method for IS design (Lundeberg *et al.* 1979) defined the term "information system" as follows: "An information system is a system which has been developed to create, choose, collect, store, process, distribute and interpret information." Langefors (1971) says that "An information system is a system of information sets needed for decision and signalling in a larger system containing subsystems for collecting, storing, processing and distributing of information sets." All of the previous definitions of IS are based on the structural subparts which an IS should have, "people, procedures, equipment", as well as the functions which it might provide, "to collect, store, process, distribute, intepret etc."

In software engineering literature today the term "computer-based system" is more frequently used instead of information system. "Information system" as it is defined is no longer a characteristic term of all computer-assisted systems. In many modern applications so-called embedded software (Pressman 1992) is used and IS as a term does not fit very well. "Embedded software resides in a read-only memory and is used to control products and systems for the consumer and industrial markets (e.g. keypad control in a microwave oven, fuel control in an automobile)" (Pressmann 1992). An IS is more a characteristic of an information system of an organization, and also there is a slight tendency for it to be a decision support tool (like MIS). In the GIS field an example could be a vehicle navigation system with GPS positioning. A navigation system is not only an information system processing information for decision-support or distributing information for users as defined previously (it can, of course, do that too), but more widely it is a computer-based system that accomplishes a certain operation by processing geographical information. This type of system can also include some embedded software functions.

Pressman (1992) gives a definition of a computer-based system (CBS) as: "A set of arrangement of elements that are organized to accomplish some method, procedure or control by processing information." The elements of a computer-based system are: software (computer programs), hardware (electronic and electromechanical devices), people (users and operators), database, documentation (manuals, forms), and procedures. This definition is very similar to the definition of an IS being based on the structural parts and the functions of the system. However, the definition of a CBS is wider in defining the types of the functions and the way of performing them.

In this chapter we are going to formulate a definition of a geographical information system which is based on the term computer-based system. Aronoff (1991) gives a definition of GIS which is based on this term: "A GIS is a computer-based system that provides the following four sets of capabilities to handle georeferenced data: 1 input; 2 data management (data storage and retrieval); 3 manipulation and analysis; and 4 output." Aronoff's definition, as well as most definitions of GIS which can be found in the contemporary literature, is based on listing the functions which are necessary in a GIS. In addition to the main functions of any information system (Ralston and Meek 1976) in most cases they emphasize certain funtions such as data management and especially analysis functions as the distinguishing characteristics of a GIS.

According to Marble (1990) a GIS must include the following components: a data input subsystem, a data storage and retrieval subsystem, a data manipulation and analysis subsystem, a data reporting subsystem. Marble says "... to be considered a GIS, the software system must include all four of the stated functions." Most digitizing systems, remote sensing systems or thematic mapping systems are not GIS according to this. Analysis and database management are important functions of GIS today. However tomorrow may bring some new functions such as the use of artificial intelligence or visualization which will bring essential new features to future GIS. There is no reason to restrict the definition to contemporary technologies only. On the other hand, applications such as thematic map production and remote sensing, which are not GIS according to these definitions, will perhaps be developed in the near future into very intelligent applications, for example, utilizing knowledge-based generalization or pattern recognition in identifying objects.

Instead of defining a GIS by trying to refine the definition of a CBS by using certain functions the proposed definition in this chapter is based on the elementary and unquestionable characteristic of a GIS, the type of data. The concept of "geographical data" is widely understood in the way the following definition describes.

> Geographical data describes objects from the real world in terms of
> — the object's position with respect to a known coordinate system
> — the object's physical attributes associated with the geographic position,
> — the spatial relationship of the object with surrounding geographic features (topology). (Price 1992)

In general, we can say that geographical data consists of spatial and attribute data. Spatial data are presented by coordinates, geometry and topology. Attributes can be identifying, linking, temporal or describing (VHS 1990). Time as the fourth dimension, not only as an attribute, is getting increasing attention in the definitions of geographical data (Langran 1992). A short and broad definition of a GIS is as follows:

> A geographical information system is a computer-based system that processes geographical information.

According to this the definition of a GIS is an "umbrella concept" which can be used in connection with all systems which process geographically located data. Often innovations are made in a specific environment, but later spread to a much wider use. GIS is an example of this kind of innovation. Born in the environment of spatial analysis, it is now broadened to be a concept for various types of systems. The following systems are thus examples of GIS: land use planning, road design, environmental monitoring, electronic charts and navigation, traffic management, tourist guides, real estate management, marketing, weather forecasting and geographical research. The common feature of these applications is the need for geographical information and, in addition to this basic feature, in most cases the need for visualization of that data.

A final remark must be made about the competing definitions. One umbrella concept is given by Laurini and Thompson (1992). They propose the term "spatial information system" as a general concept to stand for an information system which processes spatial, not only geographical, data. However, GIS is in much wider use today than any other term, and thus it is proposed to be the generic concept.

Cartographic Information System

The structure of a computer-based system is often hierarchical (Pressman 1992). As a CBS, a GIS can also be seen as a hierarchical structure constructed by several subsystems on various levels. Each subsystem is specialized for certain functions. The set of required functions varies according to the type of GIS in question. Examples of a subsystem are data input functions and devices, data analysis functions and data management functions. Geographical data visualization can also be seen as a subsystem of a GIS. Of course, in some cases visualization can be the main function of the entire system, as in map production. We can also call the whole system a "stand alone" cartographic information system (CIS) and then it can include, for example, database management functions, data input functions and data editing and processing functions individually. In the remainder of this chapter the concept of a CIS is discussed and its role described.

Without going into a deep discussion on cartography, which is perhaps not essential in this context, we accept that cartography is the art, science and technology of making maps (ICA, 1973). The more up to date definition, which was given as a recommendation by the ICA working group (Anson and Gutsell 1992) "Cartography is the organization, presentation, communication and utilization of geo-information in visual, digital or tactile form" does not cause any contradiction. Based on the definition of cartography and the definition of a computer-based system the following definition of CIS is given:

> A cartographic information system is a computer-based system the goal of which is to produce maps, either printed maps, plotted maps or maps viewed on the screen. A CIS can be an individual system and thus have all characteristics of a CBS and GIS, but it can also be a subsystem of a GIS and utilize the shared services of data management, data input etc. with other subsystems of the GIS in question.

A CIS as an individual system can be a linear process-oriented system such as a map production system. A CIS can also include intelligent functions such as automated generalization, name placement, map symbol choice or colour design. A CIS as a subsystem can be a limited set of visualization functions such as map viewing on the screen in a navigation system. A CIS can provide interactive functions via the screen map or it can be an animated weather map on TV. Taylor (1991) says that "all GISs have a cartographic component". In this chapter this cartographic component is called a CIS.

A CIS needs digital map data. As mentioned, the data files can be "owned" by the CIS or they can be shared with the main GIS in question. Digital map data are stored in files or in a map database. A database is generally defined as "A large, organized collection of information that is accessed via software and is an integral part of system function" (Pressman 1992). The software which accesses and manages the data in the database is called a database management system. Map databases of cartographic information systems are often organized by simple file systems such as spaghetti files of CAD software (Aronoff, 1991). Many GIS software products are based on relational data management. A structured database approach is necessary when several applications use the data and when there are large amounts of attribute data linked to spatial objects. Representation of topological relations also requires a structured database approach.

An advanced CIS also utilizes knowledge in addition to data. A knowledge base contains a representation of the knowledge that is required to assess relations between entities in a database (Shea 1991). In this context a cartographic knowledge base contains knowledge about the data contents of maps, scales, projections, symbology, layout and colours; in general, cartographic representation of objects as well as generalization. Generalization and name placement are examples of the first applications of using knowledge in cartographic programs

(Lichtner 1979). The knowledge includes cartographic theory and the professional know-how of the cartographer. A knowledge base can be implemented by a shell where knowledge is formulated as rules and facts; knowledge can also be formulated by using an advanced programming language. Artificial intelligence is necessary in one form or another in a CIS when non-professional users are going to make their own maps using the system without the help of a cartographer (Müller and Wang 1990). The implementation of the knowledge in the system depends on the entire architecture of the GIS in question.

The concept of a CIS as discussed in this section underlies the visualization of geographical data. The technical implementation of it depends on the environment. As a CIS can be a subsystem of a GIS, some other types of GIS can also be a subsystem of another computer-based system.

Categorizing Maps as the Products of CIS

A CIS produces maps. Outputs can be plotted or printed paper maps or maps on the screen. Maps vary in their general appearance as well as in their dimensionality. Sometimes it is difficult to say whether an ouput is a map or a picture. In this section the variations of maps are discussed as well as the primary qualifications of a map.

Appearance of a Map

The map on the screen and the map on paper have many published name-pairs. The terms "temporary map" and "permanent map" were used fairly early in the history of computer-aided cartography. "Permanent map" and "temporary map" according to Riffe (1970) do not exactly match with map on the screen and paper map. Riffe also uses permanent map for screen maps if they are in "conventional" use. Temporary maps can include intelligence and allow for abilities which are not possible even today. Riffe also describes "non-map", which depicts data from sensors in applications such as, navigation and traffic. He says that the maps of tomorrow are temporal maps and non-maps. Morrison (1980) uses the term "temporary map" as a synonym for screen map and then uses "printed map" as the opposite. "Quick map" is also used for screen map. The terms "softcopy" and "hardcopy" are generally used in CAD applications and in graphic technology. In this chapter an analogical pair of terms for maps is used: softmap for a map on the screen and hardmap for a map on paper or other "hard" material (Makkonen and Sannio 1991). Tactual maps, reliefs and globes can also be made of hard materials and they are thus hardmaps.

The softmap has several variations. It can be static as a conventional hardmap but viewed on the screen. An interactive softmap enables map design and data queries to the database by using the map as a user interface. A softmap can also be animated: it can be "a flyby" in a digital three-dimensional model. This is also called a "viewpoint transformation" in (DiBiase *et al.* 1991), or a visualization of a quantitative or qualitative attribute of objects, for example, by using statistical data. The terms "animated" and "static" maps were used by Riffe (1970). Animated map can also be called dynamic, but dynamic can stand for interactive and not animated. Dynamic is an opposite to static. Animation brings the dimension of time to the map.

Hardmaps can be conventional maps, which means traditional/standard thematic or topographic maps which are produced by using map production software packages. Conventional maps are designed to be plotted or printed. They are typically hardmaps, but in the design and editing stage they are used as interactive screen maps. Three-dimensional relief models, globes and tactual maps (made of hard materials) are also conventional maps and they can also be produced by a CBS by utilizing NC (numerical control) methods. Hardmap can also be an output of a "non-conventional" softmap.

In comparison with the "real map" and "virtual map" of Moellering (1991) we can see many similarities. Real map and hard map are equal concepts. Softmap includes several characteristrics of the three virtual map types. The main difference is that the view expressed here is that a map must be viewable. This is also the reason for defining the new concepts of softmap and hardmap. Thus here all of Moellering's viewable virtual maps are called softmaps.

Dimensionality

A discussion about dimensions of map products is given in Kraak (1988). Kraak discusses use of the term "three-dimensional map." In most cases maps are only 2 or 2·5-dimensional. However, the term three-dimensional map is used for a map "when it contains stimuli which make the map user perceive its contents as three-dimensional" (Kraak 1988). Also, according to ICA (1973), "maps may be regarded as including all types of maps, charts and sections, three-dimensional models and globes..."; three-dimensional models can be considered as maps. When dealing with three-dimensional models there are several marginal questions to discuss. An urban three-dimensional model is a map (Kraak 1988), but is a three-dimensional model of one building a map? From a practical point of view it is very strange to call a perspective drawing of a building a map. However, if the building is zoomed in, for example into "the scale of ants", the house is the "world" and the drawing of the house is "the map of the world". This discussion could tend toward philosophy and away from useful concepts. Common sense is required at this point, but we have the view that somehow the scale has a certain role in the definition of a map. In addition to three dimensions, we have already discussed time, the fourth dimension.

Definition of a Digital Map

Digital maps include examples as different as static topographic depictions and three-dimensional animations. There is an interesting discussion in Vasiliev *et al.* (1990) about this question. The final result in that paper is that perhaps it is not so necessary to give any strict definition of the concept. However, the term "map" is intuitively understood correctly. The description given of a map includes air photos, satellite images, CRT screens with map-like images on them and digital files of coordinate data. It is hard to accept the last statement that digital files could be called maps. As mentioned earlier, we claim that a map must be viewable by human senses. In the following we try to argue another point of view to the question. If the concept "digital map" is studied thoroughly on a functional basis, it is found that a digital map is a combination of digital map data and its visualization. Neither a softmap nor a hardmap is a complete digital map as such. A softmap cannot be viewed without the data retrieval from the files. A hardmap is a product of a digital process. It is the final form of a complicated data flow which starts from the digital map files. There cannot be an output from a computer-based system without the source data. On the other hand, it is impossible for a human to "see" the digital files, at least in the form of a map. Thus it seems clear that the concept of a digital map must be defined by using two components, the map data files and the visualization method. In addition to this, "visualization" must be broadened a little. We must also include tactual (and sonic) maps. For the definition of a digital map we have now come up with: digital map data and perceptualization.

The difference between landscape drawings and maps is in the scale of the maps. Landscape drawings are not measurable, they do not have a scale and thus they are not maps. A map always needs coordinates, a projection and a defined accuracy. On the other hand, the map data content must be described, either intuitively or by using a legend. For example, video and digital air photos can be called softmaps if the data content has been interpreted and classified. If not, they are pictures.

The following informal definition of a digital map is given:

> A digital map can be defined as a combination of the digital map data and the perceptualization method. Perceptualization must be measurable by some scale and it must be interpretable either intuitively or by a legend.

Visualization of Geographical Information

Visualization means in general "to make visual", "to bring something as a picture before the mind" (Hornby 1985). In the field of scientific visualization the term received additional features and computer assistance is one of them. In the following we deal with the visualization possibilities of geographical data. Because this chapter is about GIS and CIS, the techniques which are treated are all computer-based.

Software for the visualization of geographical data varies from mapping software packages to virtual realities. In the following the main types of software are briefly described.

Conventional mapping software is the traditional tool for producing maps. Software packages for both thematic and topographic mapping exist: small surveying packages used in the field, statistical thematic software and topographic map production environments. Much tailored software for conventional mapping purposes exists in every country. This software aims for conventional two-dimensional hardmap production. However, softmaps are used in the editing and design stages.

Map publishing software enables map design on the screen. Softmaps can be plotted using colour plotters and digital colour separations can be produced to plot print films. Intergraph's Map Publisher is an example of map publishing software which is able to handle map coordinate data and provides cartographic quality in outputs. Aldus Photostyler and Designer are examples of software in a microcomputer environment. Microcomputer software can be used for small map products when the quality requirements are not so demanding. By using map publishing software both two- and three-dimensional presentations are produced.

GUI maps (Skog 1993) are softmaps which are used as graphical user interfaces (GUI) to geographical databases. Statistical databases can be provided with a map interface. Tourist guide information systems use softmaps in visualizing where hotels, monuments or other addresses are situated. A GUI map can also be animated and statistical information can be visualized by showing changes in quantity or quality over time. Time series and weather forecasts are practical examples of the use of animation. A typical representative of this group is a Windows application in the PC environment. Some so-called GIS software packages (more about GIS software later in this chapter) also offer end-user map interfaces to GIS databases. ArcView of ArcInfo is one example of this type.

CAD software can be used for the visualization of three-dimensional data (elevation and terrain models). Basic visualization functions can be performed in a PC environment with basic software, but more advanced animation requires a high quality software environment. Many GIS software packages include CAD software. In GIS applications AutoCAD (Autodesk) and MicroStation (Intergraph) are popular CAD software packages. CAD visualizations can be viewed on the screen as softmaps or plotted by colour plotter to hardmaps.

GIS software is an established name for software packages which in addition to data input and output functions support advanced analysis and data management functions. This name is based on the previously discussed term "GIS" as it is defined in the contemporary GIS literature. Unquestionable examples of GIS software are, for example, Intergraph MGE, ArcInfo, System9 and Gradis. GIS software utilizes softmaps as a visualization tool. Hardmaps are possible if desired.

A multimedia environment (see Chapter 5) provides new possibilities of linking diverse visual material and sound (see Chapter 8) with maps. Multimedia stands for "linking of video, with sound, still images and computer generated screens" (Deeson 1991). The softmaps in multimedia applications can be static, interactive or even animated. Animation can be generated by using a three-dimensional model or by utilizing video. Multimedia applications can be implemented by using hypermedia (see Chapter 12). Multimedia uses softmaps. When plotted it is no longer a multimedia environment.

To define "hypermedia" the concept of "hypertext" is required. Hypertext is "the technology of non-sequential reading and writing. Hypertext is technique, data structure and user interface" (Berk 1991). Hypertext is organized by linking pieces of text in an associative way as a semantic network. The reader reads the text by reading from one node to another, not by reading the entire text from the beginning to end. Hypermedia means broadening the concept of hypertext by multimedia; "hypermedia implies linking and navigation through material stored in many media: text, graphics, sound, music, video etc." (Berk 1991). Hypermedia applications are implemented by using software environments such as HyperCard, SuperCard and ToolBook. Huffmann (1993) has described the use of hypermedia in cartography. Atlases are a popular application implemented by hypermedia. Several educational applications in GIS and cartography have been developed in these environments (Mesenburg 1993). Hypermedia is a multimedia screen-oriented tool and utilizes softmaps.

Virtual reality (VR) is "the province of computer systems which are able to combine, with great effect to the user's senses, a mix of real world experience and computer generated material. Current systems have a 'data glove' for the user to wear; fitted with sensors and optic fibres, this links with the computer and acts as a highly complex input unit" (Deeson 1991). A practical approach to VR is given in "Adventures in virtual reality" (Hayward 1993). Descriptions of VR applications, also close to GIS, can be read in *Virtual Reality* (Rheingold 1991). Visualization in GIS is developing towards VR and this can be seen in the amount of three-dimensional and multimedia applications and in their visual quality (see Bishop in Chapter 6). Morrison (1989) describes the possibility of an "armchair geographer" walking down "any street in the world". He does not use the term VR, but obviously he means it when he says that it is only matter of time until this is possible. VR can also be called as a softmap. There is no use of hardmaps in this tool.

The software relevant to geographic visualization has in many cases been developed by professionals other than cartographers. Morrison (1989) says about his "armchair geographer" applications that "Other industries will be able to supply the auditory and olfactory senses with appropriate stimuli, but cartographers must be prepared to stimulate your vision." This is also true in the other software types. There is a great need for cartographers' knowledge about cartographic conventions and "rules of thumb" in software development projects. Cartographers are professionals of visualization and now there is a wide field of visualization software development where this know-how is required.

From Passive to Interactive Visualization Tools

Maps have traditionally been made by cartographers. Kolacny's cartographic communication model (Kolacny 1970) describes the process very well. In a traditional cartographic communication process the main roles are those of the map-makers and the map-users. There is a time lag between map production and map use, thus the realities of the map-maker and the map-user are never equal. In the traditional model map-making is a professional task and the map-user does not understand completely the language of the map-maker. The message of the map is not always understood in the way which was meant by the map-maker.

When the map develops it also has effects on the communication process. The modern map production and visualization environments have caused several changes to the traditional Kolacny model. MacEachren and Ganter (1990) see cartographic visualization as the core of computer-aided map use. They say that in the traditional communication process the message is known and only the optimum map is sought, but in the visualization-based model the message is unknown, the user is an analyst and the system assists him or her by visualizing retrieved data and supporting insight. This kind of map use is based on interaction in the CIS. It seems that we can outline a new interaction-based communication model in which the CIS seems to be the dominating environment. The main factors of the change in cartographic communication are features of the developed CIS: digital production process, database technology and knowledge base technology.

The digital production environment rationalizes and speeds up the map production process, thus there is less time lag between map production and map use. Electrostatic or other colour plotters make it possible to avoid off site printing. Digital processing makes it possible to make individual maps, "maps on demand", or "custom-made maps" (Riffe 1970) for separate users. This is also enabled by the use of high-quality colour plotters. The outputs can in many cases (in planning and research) replace the more expensive printed version. Map data collection is much quicker than before because of scanning techniques, the use of GPS and remote sensing. Updating and processing of data into the final map outputs is quicker because of computer-assisted map design and electronic publishing systems. Map data content, symbology, layout and colours can be designed on screen and the final colour separations can be made digitally and plotted to film. Every stage takes less time and effort than before.

Database technology makes it possible to have updated geographical data available for maps. The efforts to establish joint use of geographical data between organizations by developing geographical data dictionaries and standardizing data transfer formats have given good results (Moellering 1991). In the design and use of geographical databases both cartographers and users must learn a new language and a new approach to data, conceptual analysis and data modelling. Map-users and map-makers thus learn a common language which they can use when the question is about geographical data and its visualization. The terms "object" and "entity", "attributes" and "relationships" make it easier to discuss data contents of an application or a map. It is not required that every user is able to build database queries, but, for example, in the application design stage one might have to cooperate in conceptual modelling and thus be able to realize the world of database terminology.

The use of knowledge in a CIS enables the automatic production of maps. Non-professionals are also able to make acceptable map products when the system knows which map type suits the data in question, chooses symbols and colours and also designs the layout. These kinds of systems are not a reality today, but it can be seen that they are required in the future. If the role of the cartographer has earlier been one of a map-maker it will now change towards manager of the CIS. This means that the cartographer must be a specialist on the problems of spatial data management, including data conversions from a coordinate system and one format to another; he or she also is a specialist on cartographic knowledge and updates the knowledge base according to the requirements of the users. The user uses the CIS for visualizing some geographical data which he or she retrieved from some other GIS database.

Cartographic communication is developing towards the use of interactive visualization. This will lead to the development of computer-based systems which enable this kind of view and access to geographical data. A CIS is the tool for visualizing geographical data. Both the cartographer and the user need more education than before. The cartographer must be able to manage and maintain the CIS and the user must be able to use the system individually.

The Cartographer as the Bridge Between Cartographic and Geographic Information Systems

The education of a modern cartographer is much more demanding and broader than before. A cartographer must be able not only to design good maps, collect data and produce maps by digital methods, but also design, establish and maintain CIS. A cartographer must be familiar with GIS engineering, including hardware, software, databases, applications, data input and output techniques as well as their use.

Clarke (1990) considers the new fields that cartographers must become proficient in as disadvantages. He says that "the cartographer of the nineteen nineties must be a data base expert, a user-interface designer, a software-engineer, retain a sense a map aesthetics, and still produce maps." This should not be seen as a disadvantage, but a very good advantage for cartography. Without these new tasks the profession of cartographers would have already died.

Before the discussion about different roles in cartography we must be reminded that "cartographers" are not a homogenous group with one scientific background. There are at least two groups of cartographers with very different basic educations. There are geographer-cartographers and surveyor-cartographers. Surveyors are specialized in collecting map data and processing them into accurate and standard maps. Geographers are more map-users in the sense of spatial analysis. They are involved in different research and planning problems. Depending on the university in question, both surveyors and geographers today get more or less education on computer-based systems.

In the following there is a brief description of the different roles of people in a CIS environment. They differ in the amount of CIS specialization. Roles as user, developer and engineer can be distinguished. People in these roles can be called cartographers, but they can also be called planners or system analysts or software engineers. The field of each of the three roles is not completely that of a cartographer, but under good circumstances a cartographer can achieve one of these roles.

A CIS user is a person who utilizes a CIS in some application. A CIS user is a cartographer who wants to use geographic data for some planning, research or map production purpose. He or she is able to utilize a CIS, knows the data input and output procedures and is able to design a new digitizing procedure for a new map type. A CIS user is able to select the right output medium for the purpose in question. He or she knows about map editing and has an overall idea of the software structure of a CIS. A CIS user is able to learn about new software. A CIS user can be a professional (a cartographer, a geologist, a planner) or a non-professional user. Depending on his or her background (if he or she is a cartographer) he or she may be the one who is updating the cartographic knowledge of the CIS.

A CIS developer is a person whose role is the most similar to the traditional role of a cartographer. He or she is a user and application oriented person who is able to act as a project leader in a CBS engineering project. He or she is not a programmer but must be able to manage the system development project from the beginning to the end. He or she must have a good overview of the system market, must manage the cost–benefit point of view and must be able to communicate with potential users and various experts. Systems analysis, project planning and requirements analysis are the main topics in his or her information systems design education. A system developer must have enough information about spatial data processing such as vector and raster based solutions, terrain models, spatial data structures and algorithms as well as hardware architecture. Visualization methods are, of course, the most important field of contribution.

A CIS engineer is a person who works as a software consultant. A software consultant either develops software products for clients or builds applications by using some ready-made modules and by combining them with his or her own programming. A CIS engineer's speciality is software engineering applied to CIS. Compared with the profile of the CIS developer, the CIS engineer's role involves the software design, coding and testing. Familiarity with cartography is required and includes knowledge about spatial data modelling, data strucutres, algorithms and special requirements of visualization. A CIS engineer can be a computer scientist who has studied spatial data processing and cartography.

These descriptions of CIS roles can be compared with Ormeling's view about the role of the cartographer (Ormeling 1989). He also emphasizes the need for understanding computer based systems design. According to Ormeling "cartographers have a mediating role between computer scientists developing packages and users". Cartographers should receive education in the following topics: knowledge of spatial concepts (methods in mapping and analysis), data aquisition (remote sensing, photogrammetry, field work), data processing (integrating, for example, remote sensing data and digital cartographic data), design (communication between computer scientists and users), production and reproduction (digital methods), information policy and distribution, map use and

GIS. In addition to guiding most young cartographers to become users and developers, we should also encourage those who are interested in computers to specialize in more computerized roles. Individuals with these skills are increasingly required in the demanding tasks of research and development.

Conclusion

In this chapter the concept of GIS has been discussed and a broad definition proposed. The broad definition includes all computer-based systems which process geographical data. According to this definition CAD-based land use planning systems as well as file-based map production systems and vehicle navigation systems are GIS. It has been stated that every GIS has a cartographic information subsystem. A CIS is more a semantic concept than technical. A CIS can be a set of visualization functions in a GIS for a special application. On the other hand, a CIS can be an individual map production system with databases of its own.

CIS are computer-based systems and they produce digital maps on "hard" materials and on the screen. Digital maps can be two-, three- and four-dimensional, they can be static or interactive, even animations. Several software types can include CIS modules for the visualization of geographical data. At this time the user must, in addition to technical abilities, also have a lot of knowledge about cartography, pictorial communication and graphical user interfaces. Knowledge is available in the literature, but much new knowledge is needed. Both softmaps and hardmaps require research and development as visualization media. Multimedia, hypermedia and CAD software packages have already been used in the visualization of geographical information. Virtual reality technology is waiting for GIS users. Both technologically and visually well educated persons are now required for development of intelligent CIS.

The entire field of GIS and the development of CIS is not a monopoly of cartographers, nor geographers and surveyors. However, the specialists in geographical data processing and visualization, cartographers, have an important role in the supply of cartographic tradition to the development of GIS. CIS comprise cartographic knowledge. They can be used for the various applications of GIS and they are an essential part of them, but they are still a product of cartography and cartographers have a duty to take care of them.

References

Andersson, S. (1981) "LIS, what is that? An introduction", *Proceedings of FIG XVI International Congress*, Montreaux, Switzerland.

Anson, R. W. and B. V. Gutsell (eds.) (1992) *ICA Newsletter*, No. 19, May.

Aronoff, S. (1991) *Geographic Information Systems: a Management Perspective*, WDL Publications, Ottawa.

Berk, E. (1991) "A hypermedia glossary, Appendix" in Berk, E. and J. Devlin (eds.), *Hypertext / Hypermedia Handbook*, McGraw-Hill, New York.

Clarke, K. C. (1990) *Analytical and Computer Cartography*, Prentice Hall, Englewood Cliffs.

Deeson, E. (1991) *Collins Dictionary of Information Technology*, Harper Collins, London.

DiBiase, D., A. M. MacEachren, J. Krygier, C. Reeves and A. Brenner, (1991) "Animated cartographic visualization in earth system science" in Rybaczuk, K. and M. Blakemore (eds.), *Proceedings of the 15th International Cartographic Conference*, Bournemouth, pp. 223–232.

Hayward, T. (1993) *Adventures in Virtual Reality*, Que Corporation, USA.

Hornby, A. S. (ed.) (1985) *Oxford Advanced Learner's Dictionary of Current English*, Oxford University Press, Oxford.

Huffmann, N. H. (1993) "Hyperchina: adventures in hypermapping" in Meesenburg, P. (ed.), *Proceedings of the 16th International Cartographic Conference*, Cologne, pp. 25–42.

Illingworth, V., E. L. Glaser and I. C. Pyle (eds.) (1986) *Dictionary of Computing*, Oxford University Press, Oxford.

International Cartographic Association (1973) *Multilingual Dictionary of Technical Terms in Cartography*, ICA.

Kolacny, A. (1970) "Kartographische Informationen — Ein Grundbegriff und Grundterminus der modernen Kartographie", *The International Yearbook of Cartography*, Vol. 10, pp. 186–193.

Kraak, M.-J. (1988) "Computer-assisted cartographical three-dimensional imaging techniques", *PhD Thesis, Delft University of Technology.*

Langefors, B. (1971) "Theoretical analysis of information systems", *Studentlitteratur*, Lund.

Langran, G. (1992) *Time in Geographic Information Systems*, Taylor & Francis, London.

Laurini, R. and D. Thompson (1992) *Fundamentals of Spatial Information Systems*, Academic Press, London.

Lichtner, W. (1979) "Computer assisted processes of cartographic generalization in topographic maps", *Geo-Processing*, Vol. 1, pp. 183–199.

Lundeberg, M., G. Goldkuhl, and A. Nilsson (1979) "A systematic approach to information systems development", *Information Systems,* Vol. 4, Pergamon Press, Oxford, pp. 1–12, 98–118.

MacEachren, A. and J. H. Ganter (1990) "A pattern identification approach to cartographic visualization", *Cartographica*, Vol. 27, No. 2, pp. 64–81.

Makkonen, K. and R. Sainio (1991) "Computer assisted cartographic communication", in Rybaczuk, K. and M. Blakemore (eds.), *International Cartographic Conference* Bournemouth, pp. 211–222.

Marble, D. F. (1990) "Geographic information systems: an overview" in Peuquet, D. and D. F. Marble (eds.), *Introductory Readings in Geographic Information Systems*, Taylor & Francis, London.

Mesenburg, P. (ed.) (1993) *Proceedings of the 16th International Cartographic Conference*, Cologne.

Moellering, H. (1984) "Real maps, virtual maps and interactive cartography" in Gaile, G. L. and C. J. Willmott (eds.), *Spatial Statistics and Models*, D. Reidel, Boston, pp. 109–132.

Moellering, H. (ed.) (1991) *Spatial Database Transfer Standards: Current International Status*, ICA Elsevier Applied Science, Barking.

Morrison, J. (1980) "Computer technology and cartographic change", in Taylor, F. (ed.), *The Computer in Contemporary Cartography*, Wiley, Chichester, pp. 5–24.

Morrison, J. (1989) "The revolution in cartography in the 1980s" in Rhind, D. W. and D. R. F. Taylor (eds.), *Cartography, Past, Present and Future,* ICA Elsevier Applied Science, Barking, pp. 169–185.

Müller, J. C. and Z.-S. Wang (1990) "A knowledge based system for cartographic symbol design", *The Cartographic Journal*, Vol. 27, pp. 24–30.

Ormeling, F. (1989) "Education and training in cartography" in Rhind, D. W. and D. R. F. Taylor (eds.), *Cartography, Past, Present and Future,* ICA Elsevier Applied Science, Barking, pp. 127–138.

Pressman, R. S. (1992) *Software Engineering, a Practitioners Approach*, MacGraw-Hill, Maidenhead.

Price, K. (1992) "Introduction to GIS", *ABC's of Geographic Information Systems*, ASPRS Sponsored Workshop, organized by Price, K. and B. O'Neal; Dallas, Texas.

Ralston, A. and E. L. Meek (1976) *Encyclopedia of Computer Science*, Petrocelli/Charter, New York.

Rheingold, H. (1991) *Virtual Reality*, Summit Books, New York.

Riffe, P. D. (1970) "Conventional map, temporary map or nonmap?" *The International Yearbook of Cartography*, Vol. 10, pp. 95–103.

Shea, K. S. (1991) "Design considerations for an artificially intelligent system", in Butternfield, B. and McMaster, R. (eds.), *Map Generalization*, Longman Scientific and Technical, Harlow.

Skog, K. (1993) "Interaktive kart (GUI-kart) for geografiske databaser" in *Proceedings of AM/FM GIS Conference*, Lillehammer. (AM/FM Nordic Region, Norges Karttenkniske Forbund.)

Steiniz, C. (1993) "GIS: a personal historical perspective", *GIS Europe*, Vol. 2, No. 5, pp. 19–22.

Taylor, D. R. F. (1991) *Geographic Information Systems: the Microcomputer and Modern Cartography, Modern Cartography*, Vol. 1, Pergamon Press, Oxford.

Vasiliev, I., S. Freundschuh, D. M. Mark, G. Theisen and G. J. McAvoy, (1990) "What is a map?", *The Cartographic Journal*, Vol. 27, pp. 119–123.

VHS (1990) "Representation of geographical information", *State Administrative Standard VHS 1041, Finland* [in Finnish].

Issues for Tool Design: Technology, Symbolization and Human–Tool Interaction

CHAPTER 5

Interactive Multimedia for Mapping

WILLIAM CARTWRIGHT*

Department of Land Information
RMIT University, 124 LaTrobe Street
Melbourne, Victoria 3000, Australia

Introduction

Multimedia is not really a new concept for cartographers. They have always designed and produced maps and associated products using a plethora of devices and media elements. Looking at any atlas, we can see the innovative use of maps, graphics, text and photographs. The "toolbox" which is already used contains computer graphics, photogrammetry, statistical analysis, computer aided drawing (CAD), geographical information systems (GIS) and printing technology.

Multimedia offers cartographers the opportunity to develop mapping products using a variety of delivery platforms. The ability to integrate picture, sound and movement gives the ability to assemble almost any combination of presentation package required for a particular spatial data presentation, including graphic solutions using space–time data.

This chapter looks at multimedia for developing mapping packages and gives an overview of the types of storage medium available. The architecture of multimedia is described and design issues particular to multimedia are discussed. A number of mapping applications which have been developed for multimedia are given as examples of how powerful multimedia mapping packages can be when used for applications such as decision support.

---

*e-mail: rlswec@minyos.xx.rlm.oz.au

Multimedia

Multimedia has been described as an " 'interactively woven' simultaneous addition to 'mixed media' of sound, music video and text; and that multifunctionality (*sic*) is the force behind Digital Signal Processing (DSP) 'multimedia and mixed media" (Davis 1993).

VERBUM Interactive, a CD-ROM based product, was the world's first fully integrated multimedia magazine. Produced in August 1981 as a co-production of *VERBUM* magazine, *MOOV* design and *GTE ImagiTrek*, this project uses a point and click interface to lead the user through an array of text, sound, graphics, animations, product demonstrations, talking agents and music (Uhler *et al.* 1993).

Since then the world of publishing has embraced multimedia using, mainly, CD-ROM as an alternative to paper. Book abstracts and bibliographic references have been published on optical devices. In the USA CNN News and ABC News are publishing interactive newspapers using Apple's Quicktime movies. CNN's product is Newsroom Global View from Compact Publishing, Inc. With Newsroom, users select video essays from the simulated video monitors and see CNN's news coverage. It includes more than one hour of narrated video. Users can explore different points of view from articles, charts and maps drawn from worldwide publications. Users can select the GLOBE and zoom in on detailed maps of every country and region, obtain complete descriptions and data on all of them, examine information and analyse trends from a database of world statistics and build their own colour charts and graphs to display their results using graphics tools. Many other commercial multimedia products include maps (both static and animated) as important components. Table 5.1 gives some representative examples and their main features.

According to the US market analyst, Market Intelligence, multimedia hardware and software sales will multiply by five times over the next five years, from around US$5 billion in 1992 to over US$24 billion by 1998 (AudioVisual International 1993a).

At present the greatest application of multimedia is in corporate communications, particularly in the training areas, and in education and entertainment. Market Intelligence predicts that the mass market will emerge later in the decade, as multimedia gains acceptance from low end users and interactive software applications begin to appear (AudioVisual International 1993a).

In the USA a 12 corporation coalition, First Cities, was formed to deliver real multimedia services. The companies are drawn from the computing, communications and media fields. It is their intention to develop a US infrastructure which will develop applications in multimedia conferencing, shopping services on demand, distance learning and healthcare. This is the first concentrated effort to properly establish and implement a true multimedia industry (Advanced Imaging 1993).

When we look at the companies involved — Microelectronics & Computer Technology Corp., Apple Computer, Kaleida Labs (Apple/IBM) and Tandem Computer from the computing industry; Bellcore, Southwestern Bell, US West and

Corning from the communications area; and Eastman Kodak and North American Philips from the media supply arena — we can expect fairly dramatic results in a fairly short space of time.

Another consortium has been formed to develop the links between multimedia and teleconferencing: the Multimedia Communications Community of Interest (MCCI) in France. Its members are France Telecom, Deutsche Bundepost Telekom, Northern Telecom, Telstra Corporation, IBM and INTEL (Audio Visual International 1993b).

Architecture of Multimedia

When looking at multimedia, developers need to consider not just the "look" of, say, a map or a video screen display, but also things such as storage devices, computer platforms, communications and authoring languages and enabling software. These items can referred to as the "architecture of multimedia". Knowledge about the individual components of multimedia is essential when developing a mapping package using multimedia.

Multimedia requirements are (Hedberg 1993):

- large storage spaces
- inexpensive CD units
- CD-ROM burners (CD-R)
- interactivity, rather than user "page turning" (i.e. make the user think)
- provision of a sufficient structure through which the user passes, and yet at the same time allowing the user to make sensible and creative choices

Storage Devices

Optical discs allow information to be recorded in such a way that it can be read by a beam of light. The compact disc, more commonly referred to as the CD, was jointly developed by Sony of Japan and Philips of The Netherlands in 1982. Optical disc storage can either be analogue or digital. Analogue mediums are either videodisc or analogue WORM (write once read many). Digital formats are CD-ROM, digital WORM, rewritable CDs, digital video-interactive (DV-I) and Sony's Minidisc. Some videotape formats do offer an inexpensive way of using the medium of video as a storage and interactive viewing device when linked to image retrieval software. Removable cartridges also have applications due to their large storage capacity. CD-ROM and videodisc are read-only mediums, whereas digital and analogue WORM and rewritable CD can be written to once or many times.

Shandle (1990) saw three multimedia formats taking shape: CD-ROM/videodiscs with Apple; CD-I with Philips, Matsushita and Sony; and DV-I with IBM and Intel. Philips, Matsushita and Sony are backing compact discs with associated CD-I players and software, aimed at home users and using domestic television. Apple and third-party multimedia producers are targeting both home and business users

TABLE 5.1

Some Commercial Multimedia Publications Incorporating Map Products

Product	Producer	Medium/platform	Contents
Compton's Family Encyclopedia with World Atlas	Compton's New Media Inc.	CD-ROM	10,000 images, maps and graphs 30 minutes of sound Interactive, multiple-window world atlas
Compton's Interactive Encyclopedia for Windows	Compton's New Media Inc.	CD-ROM	Thousands of pictures, drawings and photographs 10 research paths Interactive world atlas with links to 121,000 related pictures and articles Enhanced sound and full-motion Video for Windows (VfW) Merriam-Webster on-line dictionary and thesaurus
Fodor's '94 Travel Manager	GeoSystem	Apple Newton Message Pad	View locations in a number of US cities and retrieve restaurant or business information and navigate around the city using a map or narrative directions Address, telephone number and a brief description by clicking individual site icons
The New Grolier Multimedia Encyclopedia	The Software Toolworks	CD-ROM	33,000 articles with more than 250 full colour maps Boolean search logic and hypertext-like linking of articles with on-line help and bookmark features Electronic bookmark, Note pad and "Knowledge Tree" assist users of the package

The Pacific Rim Discovery CD-ROM	John Wiley & Sons Canada	CD-ROM	Multimedia presentation containing thousands of articles, photographs, maps, graphs, literary references of nine regions and more than 30 countries of the Pacific Rim Full hypertext links between related subjects and back-up material in the form of text related to maps, graphs, charts, photos and video
Mammals: A Multimedia Encyclopedia	The National Geographic Society	CD-ROM	700 full-screen captioned colour photographs, 45 full-motion video clips, 155 animal vocalizations, 150 range maps and statistics screens, essays on each mammal
Picture Atlas of the World	The National Geographic Society	CD-ROM	900 captioned photographs, audio and video clips 800 interactive maps Essays on 200 countries Animations on latitude and longitude and map projections

TABLE 5.2
Storage Devices Available for Multimedia Packages

Medium	Storage	Usage	Comments
CD-ROM Digital Worm	600 Mb	Low cost distribution Document storage Continuous data logging	Digital
Kodak Worm	10.2 Gb	Large volume storage	
Analog Worm	24 mins PAL video	Low cost distribution	Analogue
CD-ROM XA (eXtended Architecture)	600 Mb+	Interactive packages using different types of audio Multi-session recordings (e.g. Kodak PHOTO-CD)	Stores hi-fi stereo, mono, mono/speech audio
CD-R (Compact Disc — Recordable)	600 Mb+	Rewritable, using magneto-optical technology which detects different polarity angles on the optical disc which correspond to 1 or 0 bits	Fully rewritable and removable Ideal format for prototyping, the final production for limited distribution and archiving i.e. low volume production (Udell 1993)
CD-I (Compact Disc Interactive)	Up to 74 minutes of video, animation (at 16 frames/second audio (in hi-fi, stereo or mono — 10.5 hours of mono or 72 minutes of high fidelity stereo digital audio), text If text only is required, a disc can hold 120 million words	Developed by Philips, Matsushita and Sony as an extension to their CD-Audio System	Becoming the *de-facto* standard as a training and information delivery vehicle (Dickens 1992) Based on published standards for software and hardware Display is via a television set and a CD-ROM XA "bridge" An ISO/MPEG international standard is in place (Shaw and Standfield, 1992)
PHOTO-CD	100 photographs per disc (domestic)	Store and display scanned photographs for display through a domestic television and controlled via a hand-held remote device	Aimed at the domestic market (Richardson and Pemberton 1992) Can be accessed using Kodak PHOTO-CD, CD-I or CD-XA readers
CDTV (Commodore Dynamic World Vision)	Up to 100,000 A4 pages of text or up to 65 minutes of video or up to 17 hours of AM quality audio or up to 7000 colour still pictures or up to 660 Mb of software or combinations of these	Consumer multimedia system	Designed around proprietary hardware of the Commodore Amiga computer Many see it as having the potential of becoming the next VCR (Shapiro 1991), but competing formats of Commodore and Philips may see problems similar to VCR VHS/Beta formats

DV-I (Digital Video Interactive)	Up to 72 mins of full-screen motion video	Compression of text, audio, image and moving image	Developed by Intel specifically for PC control (Cartwright and Miller 1992) Initial development made by RCA in 1982, sold to GE in 1986 (who gave it it's first public display in 1987) and acquired by Intel in 1990 Marriage of two distinct fields — digital computing and analogue video (IBM 1992a)
CD-V/Video CD (Karaoke CD) (Compact Disc — Video)	74 minutes of all-digital full-motion video and audio	Same capacity as a laser disc	Launched in the UK by Philips in 1988 Video CD format is a standard agreed upon by Sony and Philips to replace CD-V Standard CD-ROM players can be used with an adaptor kit
Sony Minidisc (MD DATA)	140 Mb	Interchangeable between PCs, Macs and other supported platforms	Adaption of Sony's 2 1/2" Audio Minidisc for computer data storage Uses magneto-optical format and can convert the magnetic layer of ferrite cobalt using only 1/3 of the magnetic coercivity as conventional MO drives data transfer rates same as CD-ROMS (150 KPS)
80 mm format CDs	1/3 storage capacity of "standard" CDs 36 photographs of Kodak disc	Digital data, scanned photographs	Developed by Panasonic and Kodak
Videodisc	36 minutes of PAL or NTSC read-only video for standard disc or 60 minutes for extended play disc	High volume production runs	12" diameter Laservision double-sided discs Two audio tracks
Videotape — Removable Cartridges	90-150 Mb	Prototypes	Used before committing a project to videodisc or CD-ROM

via Apple products. IBM and Intel are promoting applications using their jointly developed DV-I chip set and software using artificial intelligence tools, TV emulators, video cards and DV-I compressed video libraries operating under MS-DOS or OS/2 for business users.

Options which need to be canvassed from users when choosing a suitable storage medium/media are:

- is one single established format essential?
- is full motion video needed?
- is a combination of music, speech, sound/visual effects, animation, full motion video, graphics and text required in every multimedia mapping application?
- is a full motion video display needed on, say, high-definition television (which even the 6 Mb/s chips with MPEG II may not be able to accommodate) ?

Often, the answer to the first question is "no". Owing to the range of storage types, multimedia designs must frequently produce different versions of products for several kinds of storage device. Table 5.2 lists storage devices available for multimedia packages.

At present many mediums are available for multimedia. If an optical disc is to be the storage basis for still and dynamic video and digital data, which format should be adopted? CD-I, DV-I [the " 'soft paper', which incorporates full-motion video, photos and stereo audio; and overcomes the limitations of videodisc and can be assembled with cut-and-paste simplicity, duplicated with the ease of videotape dubbing, stored on a network server and incrementally (sic) updated with new segments over time" (IBM 1992b)] and Kodak's Photo CD.

Computer Platforms

Computer platforms used for multimedia can be PCs, the Amiga, Apple Macintosh varieties and hybrid systems. Initially, multimedia needed large workstations or Macs to operate, but now high-end desktop PCs are being used more widely, as are a variety of hybrid systems. Table 5.3 gives details of a variety of platform configurations.

Communications

Since the first "videophones" were displayed at the World's Fair in New York in 1964, the idea of seeing who was at the other end of the telephone line has been a goal for the developers of telecommunications systems. Now, "desktop video" allows the integration of full motion video on desktop computers, making it possible to have live video conferences, to attach video clips to data files and to view video files from the desk: all adding-up to "content-rich" video.

The current telecommunications revolution has been the result of the three streams of technology converging: optical fibres, the broadband integrated service digital network (B-ISDN) and computers (The Age 1992).

Related to today's range of devices and systems is the increased use of sophisticated terminals, microcomputers and a highly innovative industry producing software packages — all increasingly being thrust towards the domestic market, or at least that sector of the domestic market which marketing analysis has targeted as being "ripe" for fairly intensive selling campaigns.

In 1992 the Federal Communications Commission (FCC) in the US gave permission for telephone companies to offer video services, allowing for and opening up possibilities for cable home banking, shopping and movies. The cost to wire all homes in the USA was estimated to be between US$100 and 400 billion. It is expected that highly concentrated developments of offices or shopping centres will be connected first and later the system will move into homes. However, advances in cellular telephone technology and high frequency microwaves may provide a solution by offering a non-physical alternative. Future alliances are seen between telephone companies and newspapers (Kindleberger 1993). Newspapers and magazines, with huge printing and distribution costs, may well be one of the very areas of mass media which utilizes multimedia to disseminate popular information directly into the home. Those involved in graphics need to be attuned to the changes in communications technology, what it offers and what graphic communicators can do with such a powerful mass media system.

Davis (1993) is an advocate of using video teleconferencing to link both multimedia applications and mixed media applications, "built" around and onto a base of specialized hardware (audio, video), a host CPU with digital signal processor (DSP), an appropriate operating system and multimedia extensions in the form of Quicktime or Video for Windows. He defines multimedia applications as still image and animation, motion video, MIDI, high quality audio and hypertext; and mixed media applications as fax, modem, telephone answering machine, speech recognition and voice annotation.

To function in an interactive video environment, video computers must have fast 32 bit processors, video input from NTSC/PAL formats to VGA, video output from VGA to NTSC/PAL, a video camera (usually mounted on top of the VDU, as on the Indy workstation), DSP chips for processing digital video, compression/decompression in silicon, a fast path to memory and hard disk and ready access to digital commands (Reinhardt 1993).

Desktop video allows the user to watch CNN news while working on a spreadsheet, or to view a training videotape for a new software package while using the program in a separate window. Improvements are still needed in the areas of cost, quality, connectivity and applications; lower costs for computing services; increases in quality levels (about 10–15 frames per second using 1/4 screen Windows); digital telephony and video-enabled applications such as OLE 2.0 (Reinhardt 1993).

TABLE 5.3
Computer Platforms

Platform	Attributes	Functions
PC	Fast processing Large memory High resolution graphics CD-ROM reader Video disc player Video capture boards	Digitally recorded and playback high quality, full-motion video, photographic images, animation and hi-fi sound Compress and decompress video to MPEG standards
Amiga	Four audio channels with built in stereo speakers Software, like AmigaVision Version 1.3g	Multimedia authoring and programming (Nelson 1992)
Apple	Low end systems with Mac Classic II with a 68030 chip, System 7.1 and Quicktime Starter Kit (Pournelle 1993) Or the Mac Quadra 700, with CD-ROM, good caching software like Speedcache+, a SoundBlaster Pro card and Photo-CD software (Pournelle 1993) High-end system is the Quadra 840 AV — a Quadra 800 chassis and a 40 MHz 69040 processor	Low-end to high-end multimedia systems
Sony Multimedia CD-ROM (MMCD) Player	A self-contained portable CD-ROM XA play-back device which resembles a tiny laptop computer (7″x6″x2″) DOS-based Has miniature keyboard, LCD screen, audio-out jack, TV-out jack	Plays audio CD and CD-ROM XA discs. When linked to a television, 256 colours can be displayed. IBM's "LinkWay Live!" is used as an authoring tool. Many business titles exist already (training, negotiation etc.) (IBM 1992c; Udell 1993)

FM-TOWNS	Dedicated multimedia system from Fujitsu	Has a standard CD-ROM player, supporting full motion video as a 240x240 window (Mansfield 1992)
Tandy VIS	Uses a modified ROM version of Windows, Modular Windows	Simplified interface on a television screen with infra-red control Uses CD-ROMs and gives NTSC output (Udell 1993)
Philips CD-I 360 Portable	Similar to Tandy VIS CD-ROM/NTSC output device	Connects to television set or may be used as a stand-alone device (Udell 1993)
IBM Utimedia Touch Acitvity Center	Information "kiosk"	Displays full-motion videos and photographic imagery, voice overs, graphics and animation via touchwords, pictures and symbols Can be used for tourist information, public access information and "point of sale" information Units can include credit card transaction facilities (IBM 1993h)

TABLE 5.4
Some of the Authoring Packages Available

Product	Use	Comments
Authorware Professional	CD-ROM and videodisc authoring	Icon-based authoring package
Macromind Director	Animation program	Creates, records, stores and plays back animation Includes a 32 bit colour paint program and accepts a wide range of graphics, video and sound files. Director supports MIDI devices and accepts PICS files
Icon Author	Use of videodiscs, video tape players and sound blaster cards to be used for presentations	Text, graphics, video and audio packages can be assembled in a similar way to that used in Authorware Professional. (AimTech Corporation 1992)
PC-Motion	Incorporation of full-motion video (30 frames per second), images and graphics. Videodiscs, DV-I and video signals (broadcast, cable and camera) can be used	
PhotoMotion	Uses software, rather than a chip to compress and decompress images. It has been termed "software-only motion video"	From IBM. It offers a lower motion video frame rate, lower resolution and fewer colours than chip-based systems. The National Geographic Society's *Picture Atlas of the World* was produced using PhotoMotion (IBM 1992d)
LinkWay Live	Works through a video adaptor (like IBM's "M-Motion") and uses electronic pages and folders as building blocks to create personalized multimedia presentations	From IBM. Text can be written with a built-in word processor, artwork scanned-in, audio files captured, artwork created using it's paint tool, screen images captured or imported from other applications and buttons etc created to provide access to pages in electronic folders. LinkWay Live supports DV-I and PhotoMotion (IBM 1992e)
TouchVision	Video editing tool for DV-I	The video editing tools allow for video editing, audio mixing, special effects generation and text/graphics overlays to be produced (IBM 1992f)
New World	DV-I application kit designed for C and C++ programmers	It's applications are mainly point-of-sale products, displayed via mode-independent graphics layers (IBM 1992g)
ActionMedia II	"Real time" capture and playback of full-motion video	In 1990 IBM and Intel formed a partnership to develop ActionMedia
Indeo	Compress/decompress software	Intel has developed it's "Indeo" compress/decompress software and licensed the code to Microsoft and Apple (for QuickTime video). The software gives 24 frames/sec and $^{1}/_{4}$ screen video playback on a 486 PC and 15 frames/sec and 1/10 screen video playback on a 386 PC. Indeo uses C-Cube's Motion JPEG (Joint Photographic Engineering Group) or MPEG (Motion Picture Engineering Group) (Advanced Imaging 1993)

Video for Windows	Incorporation of video with screen displays	Package allows the incorporation of video with Windows 3.1 NT. The software includes Media Player for cutting and pasting and playback of video sequences, VidEdit which edits and compresses digital clips, VidCap for simple image capture and Converter which enables Apple Quicktime video to be converted for PC/Windows playback. JPEG compression software marketers are collaborating to standardise their products for VFW, which will be known as JPEG-DIB (Device Independent Bitmap) (Advanced Imaging 1993)
Quicktime	Multimedia extensions for Apple System 7	Movies can be made from digitised video from, say a VCR with Quicktime interface. It supports two file formats: MOVIE, which manages different forms of dynamic data (video and digitized sound plus developments to handle MIDI and SMPTE) and an extended version of PICT, which allows images to be compressed as well as cut and pasted. Quicktime is divided into three sections — Movie Manager, Image Compression Manager and Component Manager

Authoring Languages and Enabling Software

Each microcomputer platform (e.g. Mac, PC-DOS, PS-Windows, Amiga, etc.) has a variety of software choices for multimedia development (see Table 5.4 for descriptions of some of the products).

Davis (1993) sees future multimedia relying heavily on Quicktime and Microsoft's Video for Windows and says that "Quicktime and Microsoft's VfW could well be the 'killer multimedia enabling platforms' for merging audio and video — bringing digital signal processing (DSP) into the mainstream of personal computing". Quicktime can provide both a software layer of integration between multimedia applications and multimedia hardware resources such as (DSP), audio and videoboards, and synchronizes independent audio, video, still image, animation and voice over data tracks within its "movie" data files. However even though DSP PCs are on the way, higher levels of integration are still years in the future.

The future of software packages for multimedia development will be in their ability to run on any, or the most widely accepted, platforms. Script X from Kaleida Labs is a device-independent multimedia programming language, produced as a joint venture between Apple and IBM (Pournelle 1993). Such joint software developments illustrate that cross-platform development is now possible.

Design Issues for Multimedia and Mapping

The design approach to maps using multimedia, and hence areas where user testing needs to be considered are:

- production of storage media to accepted standards
- imagery and image compression
- authoring systems for constructing navigation strategies for "search", "locate" and "display" routines
- map displays — design/adaptation of individual maps

Standards

Standards exist for image input for desktop imaging (TWAIN; the standard without a name), for paper-based compression (for example CCITT T-6), file header standards (the *de facto* being TIFF — tagged image file format) and for video [NTSC (National Television Standards Committee), the US video standard; PAL (phase alternate line) the European standard; and SECAM (sequential colour and memory) the French video standard, as well as numerous *de facto* and committee standards].

Proposed standards, which also affect publications using multimedia are

- SFQL (Structured Full-Text Query Language)
- CD-RX (CD-ROM Read-Only Data Exchange)
- DXS (Data Exchange Standard) [Standard 168 of the European Computer Manufacturer's Association (EMCA) , the so-called "Frankfurt Specification" for CD-ROM file systems]

Philips and Sony have continually worked on their CD-ROM Colour Book Standards, which complement ISO 9660. These standards have been published as standard "books".

Book	Medium
Red Book	CD Audio
Yellow Book	CD-ROM, CD-ROM XA
Green Book	CD-I
Orange Book	MO, CD-R Part 1 CD-ROM MO Part 2 Single and multi session CD-R (with ECMA standard 168)

Image Compression

Compression is imperative when we consider that a true colour 800x600 pixel image needs 1.44 Mb of disk space and an uncompressed 10 second video clip played at 30 frames per second, with a resolution of 320x200 pixels, in true colour, needs 57.6 Mb of storage. Map producers will need to become more familiar with image compression, moving image compression and movement attributes.

Joint Photographic Experts Group (JPEG)

JPEG is a compressed bit map file format developed by the Joint Photographic Experts Group. It is supported by PICT, QuickTime, NEXTstep 2.0 and variations of Graphics Interchange Format (GIF) (Andrews and Tilton 1993). JPEG compressions should be accepted by the ISO as the standard for image compression. JPEG can compress images anywhere between 2:1 and 160:1 using a symmetrical compression algorithm. JPEG is considered to be a "lossy" compression scheme, as reconstructing an exact replica of the original is not possible. (A "lossless" compression scheme is LZW, an algorithm developed by Lempel, Ziv and Welch, which works on the principle of substituting more efficient codes for data; Andrews and Tilton 1993.)

A discrete cosine transformation (DCT) breaks an image into 8x8 (64) pixel blocks and determines which image data can be deleted without damaging the appearance too much. It mathematically changes the pixel colour attributes into a frequency attribute. Compression and decompression take approximately the same time, giving effective compression ratios of up to 25:1. JPEG/DCT algorithms suffer from serious problems at higher compression ratios. As uncompressed files increase in resolution the JPEG/DCT compressed files either increase in size or display images with decreased quality. A "blocky" image results at high compression ratios (~100:1) and "ripples", spreading out from the edges (Gibb's phenomenon), are noticeable at higher frequencies — an unavoidable aspect of DCT. The assumption that higher frequencies are unimportant does not hold if the image has sharp edges

TABLE 5.5

Codecs for Videoconferencing. Reproduced with Permission from Byte *(1993: 72)*

Codec	Source	Applications	Adopters
Captain Crunch	Media vision	Video playback, videoconferencing, CD-ROM	Cirrus Logic, Weitek
CINEPAK	SuperMac Technology	Video playback, videoconferencing, CD-ROM	Apple, Atari, Cirrus Logic, Creative labs, Microsoft, Sega, 3DO
Px64/H.261	CCIT	Universal videoconferencing over digital 'phone lines and LANs	AT&T, British Telecom, CLI, Motorola, NEC, Picturetel, Video Telecom
Indeo	Intel	Video playback, CD-ROM	Apple, Microsoft
JPEG	Joint Photographic Experts Group	Still-image compression and transmission	Widespread
MotiVe	Media Vision	Video playback, videoconferencing, CD-ROM	Microsoft
MPEG	Motion Pictures Experts Group	Video playback, videoconferencing, CD-ROM	Philips, many others

e.g. for maps. Finally, compressed images are resolution dependent — that is, any display of an image at a higher resolution will result in a "blockiness" due to the need to replicate pixels, therefore the images may need to be re-scanned if higher resolution graphics cards are used at a later stage (Anson 1993).

Fractal Image Compression

Fractal image compression is seen by some as an alternative to JPEG image compression. Microsoft uses fractal image compression in ENCARTA, their multimedia encyclopedia. Here images are modelled as a collection of fractal segments (Fractal is a term coined by Benoit Mandlebrot, meaning a fractured structure possessing similar looking forms at many different sizes) plus appropriate three-dimensional affine transformations for each, allowing for a fractal image format (FIF) file to be written for each image. Compression takes much longer than JPEG/DCT, approximately eight minutes on a 386/33 PC compared with 41 seconds using JPEG/DCT, but decompression takes about seven seconds compared with 41 seconds (Anson 1993).

Joint Bi-level Image Experts Group (JBIG)

JBIG is the standard being developed to complement JPEG for images smaller than six bits per pixel.

Motion Picture Experts Group (MPEG)

MPEG is the compression standard for digital motion video. It is based on the same DCT as used by JPEG, plus motion compensation. MPEG works by a process of intraframe coding which removes redundancies within individual frames. If a background of a video clip stays the same from frame to frame, for example, MPEG will save the background only once and store only the differences between frames. Compression rates of up to 50:1 are possible.

Px64

Px64 is the video conferencing standard for video compression algorithm in video telephony and videoconferencing, promoted by the Consultative Committee International Telegraph and Telephone (CCITT). Some video systems use their own standards, but offer Px64 as an option. Px64 is important for interactive networked multimedia systems, as nearly all developers are focusing on ISDN or local area networks (LANS) as a minimum requirement for acceptable video quality (Halfhill 1993a).

Table 5.5 summarizes the codecs for desktop videoconferencing.

Authoring System

The key features of an authoring system for multimedia are the development of enquiry and reporting techniques and browsing facilities. The "tools" of a multimedia designer are image queries and navigational strategies.

Image Queries

Access may be required by the content of the image. This may be achieved by making queries about attributes of images in terms of patterns, colours, textures, shapes and related layout and position information. Image content query problems which designers need to address are:

- how to specify a query — either by visual queries or text-based queries
- methods used to perform matching — using neural net technology (automatically learning and indexing patterns in an image) or feature-based methods (by computing features of images and image objects and performing retrievals based on similarity of feature values)
- how to retrieve similar patterns?
- user interfaces and the role of the user in the system (Niblack and Flickner 1993)

This is seen to be achieved by either full-screen or object queries.

Navigation

The real power of multimedia is the sheer flexibility that it offers to cartographers in the design of multimedia "information packages". The imagination of those involved in the design and packaging of the products is all that limits the effective use of multimedia. Component design should take place on two levels; one including content elements and a holistic plan for their relationships and another containing a series of tools which can be applied as required through the exploration of the content (a third phase is to examine the proposed navigation system instructions from the user's perspective) (Hedberg and Harper 1992).

For mapping, navigation design describes how the system will be used, including:

- ordering of user-selected maps (those chosen "at random" by the user during an unstructured search or "browse" through the map collection)
- cartographer-guided user-selected map "sets" ("sets" which have been assembled according to a pre-determined design strategy by the cartographer/multimedia package designer) (see Chapter 11)
- pre-sequenced map "sets"

An effective design strategy will allow users easy and structured access to comprehensive map and graphics sets and complementary textual, audio and video support materials. Using multimedia, users can aggregate or disaggregate the screen image and select displays containing both analogue and digital outputs — allowing a display to be "assembled" by the user which best suits their own particular need or the need of the map-use task at hand. Users can see exactly what they want to see; the system becomes transparent and can be used without the need to consider usage restrictions. (See Chapter 9 for related ideas on interface design.)

Proper navigation strategies can improve the efficiency and effectiveness of multimedia systems. Hedberg and Harper (1992), when looking at multimedia navigation systems applied to learning, indicate that there are a number of problems with existing research, including the cognitive demands of different navigation systems and the extent to which current design models address the issue and the importance of navigation in achieving improved learning outcomes. Therefore they recognized the need to establish a context and then use metaphors which are graphic and holistic "....the more quickly recognised visual links might save time and limit misinterpretation" (Hedberg and Harper 1992: 219).

With navigational systems, conceptual clarity is important. Hedberg and Harper (1992) suggest that the structure of a navigational system can be designed around the use of four tools:

1. Guide metaphor, creating a character and using it to link ideas and visual travel through the hypermedia materials.
2. Sequential navigation, using cues to show how far along a path a user is; the cues vary from a simple screen number of the total or some conceptual description of the sequence.
3. Visual navigation, using a plan of the possible paths.
4. A hybrid navigation system (mixtures).

Information presented to the user can give both internal and external cues. External cues include the use of colour to identify areas or paths, positioning elements on the screen to indicate location in the relation to the underlying metaphor *or* the separation of positional navigational choices from the functional options, the simple use of contextual cues, the regular use of a standard format to indicate links with other sources of information, written directions which reinforce underlying concepts and simplified icons to provide standard and immediately comprehensible support for navigation or learning. Internal cues require the development of search strategies and links to metaphorical path maps, the modification of path maps to highlight paths which have been travelled and the creation of new links (by the user) using package tools — adding new information or linkages (Hedberg and Harper 1992).

Multimedia and Map Displays

The design of maps as part of a multimedia presentation requires new strategies. Design concepts from video production will need to be embraced if true multimedia mapping innovations are to result, rather than traditional mapping designs in another guise.

The production of multimedia mapping packages needs to include the use of videotape and digital video facilities and techniques. Storyboards need to be created and the ideas and presentation modes tested on potential map users. Putting together a multimedia package requires assembling a sequence of video images (still or motion), graphics, or a combination of both which allows the user of the package to interactively choose and use appropriate images or image sets. System design should allow for animation, the construction of digital graphics and maps (with or without an analogue "backdrop"), the inclusion of pictures (still or motion) and database outputs as tabular screen images, pie charts or graphs. Contemporary data, archival information and future projections based on historic and current data can also form part of the information package.

Early in the design process individual maps need to be designed with regard to how they will work as a "stand-alone" piece of graphics, as part of a set of maps, or as one of a sequence of moving images. Individual graphics can be composed as analogue or digital video images, as computer graphics outputs or as analogue /digital combinations. Designers need to consider the length of image viewing time and whether a user needs to zoom, pan or generally move about the multimedia information page(s).

If the recording of existing printed maps, say topographical maps, is needed it is best achieved by recording a series of screen "scales". Examining a series of trials filling the screen with the complete map, quarters of the map, eighths and so on, is the only real way to see how individual maps will reproduce on a composite screen. This part of the process is time consuming, but much quicker than drawing new versions of particular maps from scratch. Some maps, however, have been found to be completely unusable and unsuited to viewing in a format other that the original paper version (Cartwright 1990).

Ways in which conventional paper maps were designed may be completely inappropriate for presentation using multimedia and will need to be reconsidered. Based on an empirical evaluation, Gooding and Forrest (1990) conclude that "the use of conventional maps presented as video images produces a lower level of performance in simple map reading tasks when compared to their use in paper form. This would suggest that if video maps are to be used effectively, they should be developed specifically for that medium." Once new design ideas have been formulated and graphics completed, maps then can be evaluated or "proofed" using a multimedia monitor. The actual video image can vary immensely from that of contemporary hard copy or CAD outputs. Colours incorporated within the original presentations can be altered at a late stage in production, as analogue video editing and manipulation of the video image offers many alternatives via editing suites.

Perhaps one of the most powerful "tools" of multimedia for the generation and display of space–time information is the use of animation. Computer-generated animation allows space–time attributes to be properly seen and comprehended. Animation can portray space–time data in different ways by accessing different data from the data model, by manipulating variables which describe the graphic model or the sequence (DiBiase *et al.* 1991). Software packages such as Animator, which runs on a mainframe computer, can generate animation "stills". A Cray computer was used to produce the animation for the movie Tron. Now "fly-throughs" generated on fairly sophisticated CAD packages and animation on PCs combined with MPEG-based video sequences allow the combination of "real" video sequences with computer graphics animations using multimedia systems which are in (economic) reach.

Variables of the data set can be used to select data to allow images to be generated and displayed to depict a change in an object's location, a change in value, a change in object type, a change in the number of objects or changes in combination of objects. Data processing can be utilized to generate new data which illustrates changes in classifications or changes in geometrical reference systems, such as the use of different projections.

By varying the appearance of graphic objects using changes in the position, direction, shape, size, hues, textures and intensity, different perspectives on the various types of data can be depicted. Specification of the virtual camera position allows for the depiction of elements with respect to vertical and horizontal position, distance (using zoom), angle (perspective depiction) and the position of light sources and shadows. The sequence of the objects can change the order, rhythm and speed of the presentation. Acoustic animation variables can manipulate both sound and the sequence of sounds (see Krygier's coverage of acoustic variables elsewhere in this book).

Mapping Applications using Multimedia

Many mapping products have already been produced using many of the "building blocks" of multimedia. Among the first was the United Kingdom's Domesday project, jointly produced by the BBC, Acorn Computers and Philips to commemorate the 900th anniversary of William the Conqueror's tally book. This double LaserVision videodisc system, driven by a BBC computer and incorporating the software on the disk itself is still looked upon as an innovative package which, in the view of the author, has yet to be matched in terms of coverage and innovation (Mounsey 1988).

Two general categories of multimedia systems have been developed for work with spatial information. Some, like the Domesday system, function primarily as spatial storehouses. Others are more narrowly targeted tools for spatial decision support.

Another example of the flexible spatial data storehouse is a system developed by Multimedia Australia for the City of Woodville, South Australia. This system provides multimedia information kiosks (which employ touch-screen, digital and video technology running on IBM PS/2 computers linked to the city council's main computer). Users can access community information files that tell a continual story about the city and the general workings of the council. Other systems with similar public information access goals have been produced in several countries [see, for example, the Shell CD-ROM package of Germany (Dornier/Deutche Aerospace and Mairs Geographischer 1993), the State of Kentucky Multimedia Public Information System (GIS World 1993b) and the 1992 Property Revaluation Project developed for Franklin Township, New Jersey (Barlaz 1993; Barlaz and Gottsegan 1992)].

In contrast with systems acting as a spatial data storehouse accessible to the general public, there has been considerable development of multimedia-based tools for spatial decision support systems (SDSS) for use by professionals in making decisions or in communicating the implications of those decisions at public meetings. Examples include a prototype SDSS used to support the decision-making process for siting the Wilson Bridge Crossing in the USA (Armour 1992) and a Georeferenced Pavement Information System produced by a Tokyo-based company that markets their system to decision-makers (Sekioka *et al.* 1993).

One of the most innovative applications of multimedia tools to spatial decision support is an effort currently underway by the Planning Support Systems Group at the Department of Urban Strategies and Planning at MIT. They have developed a multimedia environmental impact assessment system for use with major developments. The system is based on Shiffer's collaborative planning system and allows for navigation among chunks of related information gathered during collaborative urban planning meetings. It has been implemented to demonstrate the capabilities to present proposals and conclusions, to undertake geographically/visually/textually related searches and to visualize large amounts of data such as traffic projections or shifts in demographics by using multiple representation aids. It was tested on information from the National Capital Planning Commission, Washington DC.

Displays include maps (with overlays), descriptive video images, sounds and text. The hardware used is a Macintosh Quadra 900 linked to a network of PCs and Unix workstations, a video projector, an infrared pointing device and microphones for user interaction. The software used is Apple Quicktime for digital video and Aldus Supercard for media integration. Information stored is both digital and analogue (video tape for audio and visual "note-taking" on field trips). Communication is real-time and asynchronous (for generating shadow diagrams).

Planning-related support in the MIT system is used for land use analysis, traffic analysis, assessment of visual elements and the illustration of proposed changes to visual elements. The land use display consists of colour-transparent polygons overlaying an aerial image. As a user points to any number of buildings on the map an interactive height indicator displays a height profile of the buildings selected (including drawings/photographs of individual buildings), in relation to the capital. Interactive traffic map "displays" represent data in a number of ways: numeric information, as a dynamic bar, digital motion video and an audio-level of traffic noise played back at the level experienced in the field. Visual environments can be assessed using digital video which allow for a 360° axial view, whereby the viewer can pan around the scene, fly-through or drive-through using "navigation shots" and aerial or ground-level images superimposed over the map/airphoto screen. Access to images of the artist's renderings and models appearing at appropriate geographical locations allow the user to navigate around the virtual model or step through a series of images, as in a slide show.

Shiffer (1993) says that "representation aids can influence the display of various types of data so that they may be more readily understood by users, using multimedia tools so that human–machine interaction becomes so engaging that the computer essentially becomes transparent to the human." For example, highlighting the impact of a proposed transit station on a surrounding neighbourhood using one graph to display changes in land values, another to reflect shifts in demographics and a video to display physical changes allows for visualization from several perspectives. Such visualization tools can "empower groups and individuals who have traditionally been informationally (*sic*) disadvantaged due to a lack of sophistication" (Shiffer 1993).

Shiffer's MIT project is an example of how a multimedia package linked to what can be best described as a GIS can allow users/decision makers to properly visualize problems, data and changes through the use of both visual information in the form of static graphics, overlays and dynamic graphics, and aural information (adding information from another sensory domain for presentation).

The use of a GIS incorporating multimedia devices as a tool for the visualization of geographical relationships can be seen as perhaps one of the ways in which spatial decision support (where decisions are based on the evaluation and consideration of data which is spatially unique and geographically referenced) can be effectively made available. The hardware and software developments of both multimedia and GIS have now reached a stage where the technological issues have been resolved. The linking of such powerful systems allows for the presentation of spatial data, which is spatially accurate and timely, and presented in such a way that it supports the decision-making process.

The addition of multimedia elements to a GIS can improve the user's visualization of reality when it is displayed graphically as three-space data and time. As spatial data about natural and cultural objects change over time, in terms of position, weighting and dominance, it is important that the display of the quantitative information (the information, when given an accurate four-dimensional position, can be termed geographical information) correctly gives the user a narrative of space and time which captures the essence of what is being depicted and hence aids visualization.

The Future

Kindleberger (1993) sees future multimedia development linked to the development of new hardware, software and communications advancements:

- new chips operating at between 100 and 200 mps, compared with 10–20 mps
- faster networking protocols (e.g. fiber distributed data interface and fast ethernet distribution, with claims of speeds of 100 mps)
- high-definition television
- "media-smart" computers which select and filter programs
- "universal information appliances"
- "digital highways"
- the use of personal digital assistants such as Apple's Newton
- video games industry companies such as Sega and Nintendo will influence the manner in which user interface design develops; for example, the Sega CD, an interactive entertainment system which connects to their 16 bit Genasis system

Video approaches or computer approaches? John Hartigan from Philips Professional Interactive Media Systems sees a debate between video and computer developers of multimedia. Film/video developers would argue that a movie should not be stopped or interrupted, whereas those from the computer side see programs which include moving images and things such as finesse of colour, contrast and composition as a waste of time. He says "Are we really facing, in Multimedia, 'the image that ate the computer industry' — or 'the data that ate the video industry' " (Hartigan 1993).

The sheer growth in the use of optical storage devices is certain to promote the usage of multimedia because of the easy access and interactive manipulation of screen images, and sound. One illustration of the acceptance of CD-ROM for publishing are the figures for CD-ROM published titles and sales of CD-ROM related products. In 1992 commercial and in-house titles in the USA grew from 3500 to 5000. Installed drives in the USA doubled to over five million. In 1991 20% of the 17,000 computer-related retail outlets in the USA carried CD-ROM products; in 1992 the proportion was 80% (US Bureau of Electronic Publishing figures). In 1992 the US government produced 2000 titles (National Technical Information Service) (Byte 1993).

It has been predicted that by the turn of the century PCs will be operating to up to 1 GIPS, hold 4 Gb of data (with access to terabytes of networked data) and enough bandwidth and speed to take graphical interfaces beyond today's menu and mouse-driven applications; allowing for speech and three-dimensional gesture recognition, virtual reality, agents and eye-tracking access to a whole range of applications (Friedman 1993).

Jakob Neilson of Bellcore, speaking at the 5th International Conference on Human–Computer Interaction, said that the future of human–machine interaction could include operations where the PC automatically changed the size of the screen

"Inspire" "Persuade"

CULTURAL Facts: Disagree Values: Disagree	POLITICAL Facts: Agree Values: Disagree
Facts: Disagree Values: Agree LEGAL	Facts: Agree Values: Agree COMPUTATIONAL

"Verify" "Solve"

Facts: Known to be true **Values**: Regarded as desirable

FIG. 5.1. Conflicts matrix (facts/values).

image and font to accommodate a user who moves further away, object-oriented software could allow for the document, rather than the application, to become the primary focus, and interlinked information objects could allow manipulations of associations between data. There will be a need to establish standards for advanced data interchange and system integration, agents will be needed to anticipate users' needs and computers will be able to write computer code. These types of things were predicted by Apple in 1987 with their futuristic Knowledge Navigator. Such a device is still seen to be some way off.

Bob Deller, federal program manager at Input, Inc., a Vienna, VA, USA based market research firm said that "Multimedia systems promise to deliver to the mind what it wants and needs — total immersion in sensory data. Without full-brain accommodation, humanity cannot progress to its full potential as an information processing participant" (Deller 1993).

Lania Rivarmonte sees multimedia used in several GIS applications: three-dimensional modelling, vector/raster draping, drive/walk-throughs, fly-overs, video conferencing, video logging and animation. She sees the future of multimedia being in the areas of interactive video, holographic data storage, the use of Photo CD and multi-function drives (Rivarmonte 1992).

The blending of information from both image (computer graphics/digital map) and numbers (data and metadata) is needed for what Berry (1993) calls creative computation. He goes on to say, "....what's holding us back? Two things come to mind: 1) the complex nature of spatial problems and 2) the inhuman nature of GISs".

In many decisions descriptive information is not enough. Figure 5.1, adapted from Berry and his description of a decision support system (DSS), illustrates sociology's enduring quandary of "fact/value conflicts".

Science and technology most frequently are employed to "solve" problems. This approach is only valid if the problem is *computational*, with agreement on facts and values. Other conflicts use technology to communicate ("verify, persuade, inspire") various perspectives. If used creatively, a DSS, incorporating all that multimedia has to offer, can transform information into an understanding of the complex nature of spatial problems (Berry 1993).

Map producers and map users alike should revel in what multimedia offers. Multimedia is the matchmaker between the logical world of computers and the abstract world of video. Cartographers, already attuned to dealing with multimedia, are well placed to exploit the medium. The challenge for map designers and producers is to use multimedia as a new tool for mapping. Mapping packages/programs can be assembled by applying skills which cartographers already have to multimedia.

References

Advanced Imaging (1993) "Intel/Microsoft Video", *Advanced Imaging*, January, p. 10.

AimTech Corporation (1992), "ICON Author", product literature.

Andrews, S. K. and D. W. Tilton (1993) "How multimedia and hypermedia are changing the look of maps", in *Proceedings, Auto Carto 11*, Minneapolis, pp. 348–366.

Anson, L. F. (1993) "Fractal image compression", *Byte*, October, p. 195.
Armour, F. J. (1992) "Using hypermedia in a MAU model based spatial decision support system (SDSS)", in *1992 URISA Proceedings*, pp. 74–86.
AudioVisual International (1993a) "Multimedia set to boom", *AudioVisual International*, March, p. 12.
AudioVisual International (1993b) "Multimedia development mushrooms monthly", *AudioVisual International*, July, p. 8.
Barlaz, E. C. (1993) "Objectives analysis chops cost of property revaluation photo imaging system", *GIS World*, May, p. 50.
Barlaz, E. C. and J. Gottsegan (1992) "Implementing photographic imaging systems in local government", in *1992 URISA Proceedings*, pp. 221–228.
Berry, J. K. (1993) "Distinguishing data from information and understanding", *GIS World*, October, pp. 22-24.
Byte (1993) *Byte*, February, pp. 128–131.
Cartwright, W. E. (1990) "Atlases on optical storage mediums: comparisons between video atlas usage and conventional atlases", *Society of Cartographers SUC*, Portsmouth, UK, September 1990.
Cartwright, W. E. and S. Miller (1992), "Multimedia and Cartography: an innovative interactive mapping medium", *Proceedings from the 1st Australian Mapping and Charting Conference*, Adelaide, Australia, September 1992.
Davis, A. W. (1993) "Digital imaging and audio meet on your desk: DSPs and multimedia", *Advanced Imaging*, April, pp. 24–30.
Deller, R. (1993) "Multimedia: giving the mind what it wants", *Fed Micro PreView*, March, p. 18.
DiBiase, D., A. M. MacEachren, J. Krygier, C. Reeves and A. Brenner (1991) "Animated cartographic visualization in earth system science", *Proceedings of the 15th International Cartographic Association Conference, Bournemouth*, UK, 23 September–1 October 1991, pp. 223–232.
Dickens, D. (1992) "The interactive multimedia format ready for exploitation today", *Conference Proceedings, ITTE 1992, Information Technology for Training and Education*, Brisbane, Queensland, Australia, September/October 1992, pp. 12–16.
Dornier/Deutche Aerospace and Mairs Geographischer (1993) Promotional brochure and demonstration, *International Cartographic Association Conference*, Cologne, Germany, 3–6 May 1993.
Friedman, R. (1993) "Walt Disney knew a good interface", *Byte*, November, p. 48.
GIS World (1993a) "Newslink", *GIS World*, October, p. 10.
GIS World (1993b) "Multimedia public information system introduced", *GIS World*, July, p. 11.
Green, H. (1993) "Imaging standards: which will impact document imaging", *Advanced Imaging*, September, pp. 81–83.
Gooding, K. and D. Forrest (1990) "An examination of the difference between the interpretation of screen based and printed maps", *The Cartographic Journal*, Vol. 27, pp. 15–19.
Halfhill, T. R. (1993a) "Video compression standards vie for acceptance", *Byte*, September, p. 72.
Halfhill, T. R (1993b) "Sony's minidisc for data: future floppy?", *Byte*, November, p. 32.
Hartigan, J. M. (1993) "Closing the 'computer people' vs 'video people' mindset gap", *Advanced Imaging*, January, pp. 69–70.
Hedberg, J. G. (1993) "Eight keywords for interactive multimedia", *Audio Visual International, August*, pp. 28–29.
Hedberg, J. G. and B. Harper (1992) "Creating interface metaphors for interactive multimedia", *Proceedings of the International Interactive Multimedia Symposium*, Perth, Australia, January 1992, pp. 219–226.
IBM (1992a) "DV-I technology — A universe of multimedia creativity and power", *Multimedia Solutions*, Vol. 2, No. 5, pp. 22–23.
IBM (1992b) "The digital future of multimedia", *Multimedia Solutions*, Vol. 2, No. 2, pp. 3–9.
IBM (1992c) "Sony multimedia CD-ROM player", *Multimedia Solutions*, Vol. 2, No. 5, pp. 1–6.
IBM (1992d) "PhotoMotion", *Multimedia Solutions*, Vol. 2, No. 5, pp. 24–25.
IBM (1992e) "IBM LinkWay Live", *Multimedia Solutions*, Vol. 2, No. 5, pp. 20–21.
IBM (1992f) "TouchVision", *Multimedia Solutions*, Vol. 2. No. 2, p. 26.

IBM (1992g) "New World", *Multimedia Solutions*, Vol. 2, No. 2, p. 28.

IBM (1993h) "Information theatre 24 hours a day", *Multimedia Solutions*, Vol. 3, No. 2, p. 11.

Kindleberger, C. (1993) "Multimedia — the next big wave", *URISA*, Vol. 5, No. 1, pp. 122–133.

McNamara, S. (1992) "Multimedia, BUT...(Boffins using Thingummybobs or Beginners Understanding Technology", *Proceedings of the International Interactive Multimedia Symposium*, Perth, Australia, January 1992, pp. 613–619.

Mounsey, H. (1988) "Cartography and interactive video: developments and applications in Britain", *Technical Papers, 7th Australian Cartographic Conference*, Sydney, Australia, August 1988, pp. 189–196.

Nelson, L. R. (1992) "Developing interactive digitised audio courseware on Amiga, Macintosh, and PC platforms: a comparison of common support facilities available", *Proceedings of the International Interactive Multimedia Symposium*, Perth, Australia, January 1992, pp. 483–492.

Niblack, W. and M. Flickner "Find me the pictures that look like this: IBM's image query project", *Advanced Imaging*, April, pp. 32–35.

Pournelle, J. (1993) "The state of multimedia", *Byte*, October, pp. 217–234.

Reinhardt, A. (1993) "Video conquers the desktop", *Byte*, September, pp. 64–80.

Richardson, L. and P. Pemberton (1992) "Interactive CD — right there on your TV!" *Conference Proceedings, ITTE 1992, Information Technology for Training and Education, Brisbane*, Queensland, Australia, September/October 1992, pp. 495–501.

Rivarmonte, L. (1992) "A multimedia primer for GIS", *1992 URISA Proceedings*, pp. 188–199.

Sekioka, K., W. Taniguro, T. Minamisawa and K. Soma (1992) "Pavement management implementation with video-based GIS", *1992 URISA Proceedings*, pp. 38–47.

Shandle, J. (1990) "Who will dominate the desktop in the '90's. IBM and Apple rev their technology engines as the multimedia age begins", *Electronics*, February, pp. 48–50.

Shapiro, E. (1991) "Now, CDs emit sights as well as sound", *The New York Times*, May 12, p. 5.

Shaw, N. A. and G. J. Standfield (1992) "Compact disc interactive (CD-I) — a multi-media system of the future", *Proceedings of the International Interactive Multimedia Symposium*, Perth, Australia, January 1992, pp. 409–414.

Shiffer, M. J. (1993) "Implementing multimedia collaborative planning technologies", *1993 URISA Proceedings*, pp. 86–97.

The Age (1992) "A quiet revolution at home and work", 10/10/92, pp. 37–38.

The Australian (1993) "CDs will never be the same again", *The Australian*, February 9, p. 25.

Udell, J. (1993) "Start the presses", *Byte*, February, pp. 116–134.

Uhler, G., D. Flanagan and K. J. Schoenfein (1993) "Compressed video for real multimedia products: cross-platform", *Advanced Imaging*, May, pp. 45–46.

CHAPTER 6

Visualization Software Tools

TERRY A. SLOCUM*

Department of Geography

University of Kansas

Lawrence, KA 66045, USA

Introduction

The objective of this chapter is to provide an overview of software tools that can assist in the visualization of spatial data. The tools range from traditional programming languages such as C to multimedia packages such as Macromind Director to packages expressly intended for scientific visualization such as the Application Visualization System (AVS). In selecting tools to be reviewed, I restricted myself to those which most clearly met the definition of visualization provided by MacEachren in Chapter 1; thus the tools had to be capable of private–exploratory analysis, revealing unknowns and high interaction with the spatial data. As a result, common mapping packages such as MapViewer and AtlasPro were ruled out, as were general design packages such as Aldus Freehand and Corel DRAW. Multimedia tools designed for more public exploration or presentation are covered in the chapter by Cartwright.

It is common to divide those packages expressly intended for exploratory visualization into data flow and non-data flow groups. Data flow packages contain separate modules to import, manipulate and display data that can be linked in a data flow graph. Non-data flow packages contain numerous tools for visualization, but they lack the linking and thus the interactive capability of the data flow packages (Lucas *et al.* 1992; Williams *et al.* 1992). (For those unfamiliar with the data flow concept, Rhyne provides a thorough discussion in this chapter.) Non-data flow software reviewed in this chapter include Advanced Visualizer, Interactive Data Language (IDL) and SpyGlass, whereas data flow software reviewed include Application Visualization System (AVS), IBM's Data Explorer (DX) and Khoros.

*e-mail: slocum@ukanvm.cc.ukans.edu

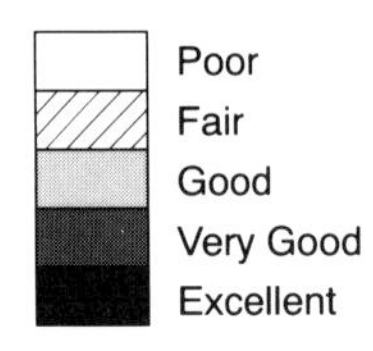

AV - Advanced Visualizer
IDL - Interactive Data Language
SG - Spyglass
AVS - Advanced Visualization System
DX - IBM Data Explorer
K - Khoros

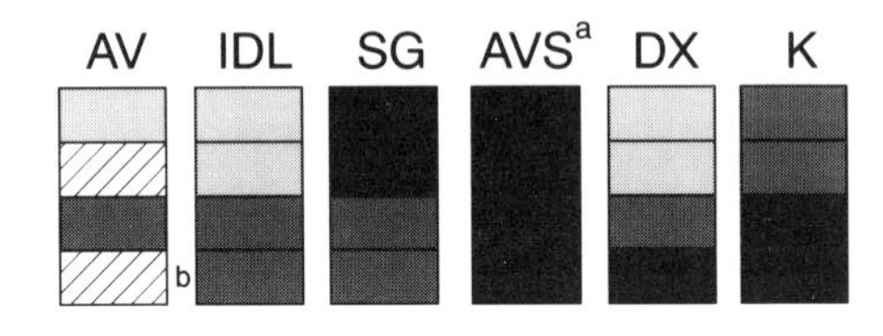

Documentation
Ease of Learning
Ease of Use
Product Support

[a]For a more detailed discussion, see text by Rhyne in this chapter.

[b]based on Australian and limited US experience

FIG. 6.1. Ratings of "true" visualization packages on general software assessment categories.

To make the reviews as consistent as possible, I sent each contributor an e-mail message asking him or her to provide software specifications, hardware and software requirements, and ratings on five-point scales for both general software assessment and geographic visualization (Figs 6.1 and 6.2) and text describing the tool.* Although the figures for general software assessment and geographic visualization can only be used as a rough guide because reviewers did not contrast the packages with one another, the criteria shown in the figures did focus the reviewers on the same set of issues. These criteria also may serve as a basis for future reviews of visualization software. Reviewers of traditional programming and multimedia approaches chose not to rate such approaches on the visualization scale; ratings for general software assessment in these cases are given with the text of the review.

In examining the overall set of reviews, it is natural to ask what general comments can be made about visualization tools at this point in time. On the plus side, we must be impressed by the range of tools that is now available, especially those that are expressly intended for visualization. Given access to the appropriate hardware and software and the time to develop a working knowledge of the visualization package, researchers can now create a broad range of visualizations. Although this represents a considerable advancement from as recently as five years ago, some problems are still apparent.

One problem is that data flow tools, although capable of creating the most flexible and powerful visualizations, are relatively difficult to use for the typical research scientist. A solution to this problem is to use these tools to create a variety of point and click applications, as several contributors have suggested is possible. Good examples of point and click applications that can be purchased include SpyGlass (reviewed here) and PV-Wave's Point & Click (Dungan 1993).

---

*To obtain an example of the e-mail message that was sent, contact Professor Slocum via e-mail.

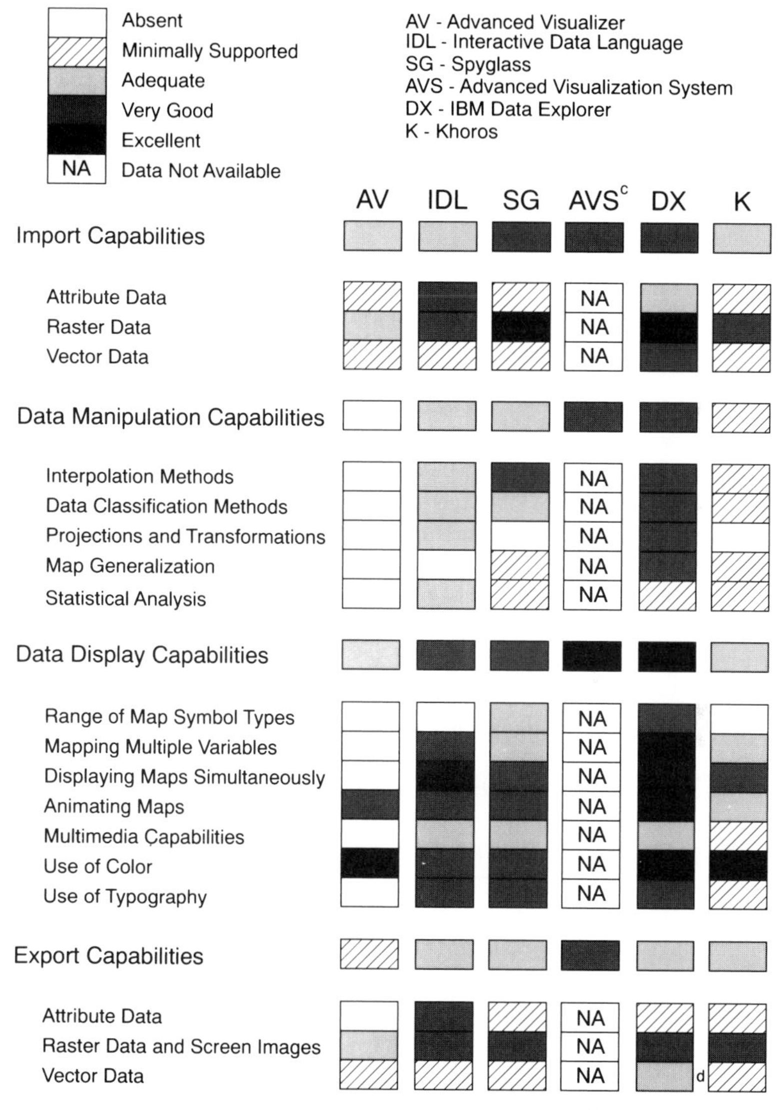

[c] Rhyne has rated AVS only in the very broad categories. See her text for more information.

[d] Very good within DX format; minimally supported for other formats

Fig. 6.2. Ratings of "true" visualization packages on geographic visualization categories.

A second problem with data flow tools is that they were not developed with geographical applications *per se* in mind. For example, Rhyne indicates that AVS "...does not readily support standard geographic manipulation capabilities."

A third problem is that the data flow tools tend to be expensive and require sophisticated graphics workstations. Thus at present such tools are not available to the typical geography teacher or student. Hopefully, as the technology changes and hardware costs decline, this problem will disappear.

For those wishing to develop their own point and click applications, the cheapest alternative (in terms of hardware and software) to data flow tools is to make use of traditional programming languages and associated graphics primitives (as described by Egbert and Slocum). The trade-off in this case is that considerably more programming time will be required and skill in computer programming is essential. A potential middle ground is the use of a program such as Director, which provides some already developed tools and a scripting language for ease in programming.

In addition to considering references mentioned by reviewers, readers should consider the following for the respective software: IDL (Dungan 1993; Kruse *et al.* 1993), SpyGlass (Armstrong 1994), DX (Lucas *et al.* 1992) and Khoros (Krohn and Clark 1992; Wampler 1992; Clark 1993b).* For software not reviewed here, readers should consider: aPE (Dyer 1990), FAST (Bancroft *et al.* 1990), NGS (Treinish 1989), SGI's Explorer (Clark 1992, 1993a) and UNIRAS (Jern 1989). Other overview material includes Prawel *et al.* (1990), Williams *et al.* (1992), Ribarsky *et al.* (1992), Treinish *et al.* (1989) and several articles in the April 1993 issue of *Byte.*

Using C and MetaWINDOW Graphics Primitives to Develop Software for Exploring Spatial Data

STEPHEN L. EGBERT AND TERRY A. SLOCUM

Department of Geography
University of Kansas
Lawrence, KA 66045, USA
e-mail: slocum@ukanvm.cc.ukans.edu

General Software Assessment Ratings

Documentation: Borland C++ (good), MetaWINDOW (very good).
Ease of learning: Borland C++ (see discussion), MetaWINDOW (very good).
Ease of use: Borland C++ (very good), MetaWINDOW (very good).
Product support: Borland C++ (very good), MetaWINDOW (very good).

---

*The articles appearing in *Information Systems Developments* are published by the USGS. For information on obtaining reprints of these publications, readers should contact the *Developments* staff at 703–648–7149.

Our Application and Overview of the Tools

We have used C and MetaWINDOW to develop ExploreMap, software that enables users to explore a database typically associated with choropleth maps (Egbert and Slocum 1992). In ExploreMap users can perform a wide variety of functions, including display map classes in sequence, display any combination of classes, highlight subsets of the data and compare the effect of different numbers of classes and methods of classification (available in version 2.0). Related work in spatial data exploration has also been undertaken by Ferreira and Wiggins (1990), Haslett *et al.* (1991), MacDougall (1992), Monmonier (1992) and Tang (1992).

We began developing ExploreMap about five years ago. At that time our main reasons for using C and MetaWINDOW were that they were standard tools of professional graphics programmers, relatively inexpensive, IBM/PC compatible), and required no royalties for program distribution. Initially, we used Microsoft C for compatibility with the graphics and technical libraries we were using in some of our formalized test procedures for ExploreMap. Although Microsoft C proved adequate in most respects, we switched to Borland C++ 3.1 because we felt that its stand-alone debugger provides superior capabilities. Specifically, the Turbo Debugger allows for virtual memory debugging of large programs by using the available extended memory and it enables reliable screen swapping between the graphics output from ExploreMap and the debugger information screen — this is essential for debugging graphics programs.

Although we have switched to Borland C++, ExploreMap is still written in standard C code, rather than C++. As C++ is a superset of C, the Borland compiler is able to handle C code with the setting of a few options. We have elected not to rate Borland C++ for ease of learning (above) because we feel that both C and C++ have fairly steep and long learning curves, regardless of the compiler used. The Borland programming environment itself is easy to use and it facilitates the automation of large projects. However, for learning C or C++, a novice programmer would be well advised to acquire additional reference books, such as those produced by the Waite Group.

One of the most unattractive features of C compilers, including Borland's, is that they allow the programmer to make egregious errors without a word of warning or complaint. For example, it is possible to write beyond the allocated memory of an array, producing memory corruption errors that are difficult to trace. For this reason, most veteran C programmers avail themselves of specialized debugging tools in addition to the debuggers packaged with the compiler. We currently use PC-lint (Gimpel Software, 3207 Hogarth Lane, Collegeville, PA 19426, USA) and MemCheck (StratosWare Corp., Suite 1500, 1756 Plymouth Road, Ann Arbor, MI 48105, USA). Similar tools are available from a variety of other software companies. Although C can be a frustrating language to work with, its power and flexibility and its adoption as a programming standard make it a potent tool for those needing to produce complex graphics programs.

MetaWINDOW is a popular graphics library that includes numerous functions for screen drawing and painting, bitmap manipulation and object querying. Used by a number of professional developers, it features an event queue system that enables user input from multiple devices, and it supports numerous graphics displays in a large variety of resolutions. Its use is sufficiently widespread that it has spawned a number of add-on libraries that employ MetaWINDOW drawing functions to create state of the art "look and feel" user interfaces, including pop-up windows, button bars, sliders and other graphical objects. Autumn Hill Software, WNDX, Island Systems and Oakland Group/Liant are among the companies providing add-on products. MetaWINDOW provides over 20 sample programs that cover most of the functions in its library, giving the programmer a good picture of how to implement specific functions. MetaWindow Plus offers extended features and the past policy of charging royalties for the distribution of software published with it is no longer applicable.

In comparison with software expressly intended for visualization (e.g. IBM Data Explorer), the primary advantages of our approach are the low initial cost for hardware and software and the ability to tailor the program for specific requirements. The disadvantage is that a great deal of time and effort are required to produce relatively efficient, bug-free code in a graphics programming environment. It is probably true that an attractive, intuitive graphical user interface (GUI) is an essential part of any visualization program. However, programming an effective GUI adds considerably to the size and complexity of a program (possibly half of ExploreMap's 10,000+ lines of C code is devoted to the graphical interface). Even if a software developer uses one of the add-on tools to facilitate interface programming, the task is still complex — Autumn Hill's Menuet library, for example, comes with over 1100 pages of documentation.

One alternative to our approach is to program for the Microsoft Windows environment. Programming for Windows, which in the past has been daunting to program developers, has been made easier by the introduction of several tools that partially automate the graphical interface part of Windows programming. These include Asymetrix's ToolBook, Microsoft's Visual Basic and C/C++ compilers that offer Windows automation tools, including Borland, Microsoft and Symantec. Given the popularity and the standardized nature of the Windows environment, developers of visualization programs may wish to consider one of these tools rather than building a new graphical environment from the ground up.

Multimedia Authoring in Macromind Director

CHRISTOPHER R. WEBER

The DSW Group
1775 The Exchange Suite 640
Atlanta, GA 30339, USA
e-mail: Compuserv (73552,1035)

General Software Assessment Ratings

Documentation (Very Good)

Installation of the software is simple and direct, as outlined in the Getting Started manual. I have installed the software several times without incident. Director's user manuals are divided into systematic subtopics which provide both novices and experienced users with manageable chunks of information in tutorial format. Illustrations are comprehensible, though their screen-captured/bitmapped format makes them a bit fuzzy at first glance. The text is clearly written, but often employs a syntax derived from the software's scripting language, Lingo. Users with minimum programming experience can manage these sections easily. The documentation is generally complete excluding higher level functions and external code and hardware linking through XCMDs (external commands) and XFCNs (external functions). At this level, the user is expected to intuitively decode tutorial examples (make presumptions concerning the result of Lingo commands) or have previous Macintosh system level experience.

Ease of Learning (Good)

Director's graphic interface and graphic programming language make basic animation simple and intuitive. The tutorials are direct and clearly written. It is helpful to have prior knowledge of computer graphic concepts to understand why some functions, such as color palette cycling, perform as they do. On-screen menus are of typical Macintosh design, though keeping track of their appearance and disappearance is disconcerting until one realizes they are linked to the active window in use. Higher level functions and scripting are not as accessible. The manuals too often intermix script functions when presenting examples, or rely upon intuitive interpretation of code examples.

Ease of Use (Very Good)

The tutorials and graphical programming interface make Director easy to use for most animations. Novice users readily grasp its score/cast/stage metaphor. It is possible to create simple but effective visualizations without writing a single line of code.

Lingo allows higher level users to access looping and hypermedia functions, as well as to control the characteristics of bitmap cast members. Basic programming experience allows users to jump right into Lingo's functionality. However, the language lacks enumerated data typing and the ability to handle real numbers and arrays. (These abilities can be faked with much round about converting from strings to integers.) The numerous predefined functions make Lingo extremely powerful, but at the same time the user will be puzzled by the hierarchy of script/score/cast member functionality. Experimentation seems to be the only solution.

Product Support (Fair)

Technical support is the cost of a long-distance phone call and not Director's best feature. Users also can access Compuserve's "MacroMedia Forum" for electronic discussion. The support staff was bewildered by my use of the software to create real-time animations from ASCII data files embedded as cast members. Memory management problems (which are Director's major drawback) are never the software's problem, according to the staff. Because of memory management problems, I recommend that users "Save as" often, especially when using the software's bitmap paint functions.

Director as a Tool for Geographic Visualization

Director appears to have been designed originally as a business and education presentation package. None the less, its multimedia/hypermedia capabilities make it an extremely powerful prototyping and presentation tool for geographers.

It is primarily a raster image package. The software offers raster "in-betweening" and many automated animation capabilities. No vector manipulations are possible. It does have non-raster text and button functionality. Attributes attached to images (beyond palette and dimensions) are not importable; the software does not import database files. All attributes of the raster images in Director including size, position, palette and presentation are controllable through Lingo. Many ink effects are available for image processing. For instance, a user could make use of the Add Pin or Subtract Pin ink effects to overlay two raster maps and discern areas of overlap for various color-coded cells. Map symbols and map appearance are amenable to spatio-temporal display.

Because Director functions under Multifinder, geographic data analysis can take place externally. As an example, I've created surface and isopleth maps using Golden Software's Surfer and then screen-captured and pasted images to Director. Image conversions and incorporation in Director from other geographical information systems (GIS) is likewise possible. Cut and paste under Multifinder is restricted only by machine memory. The software imports images, sound, text and other Director movies, and exports PICT images. Import of images and their

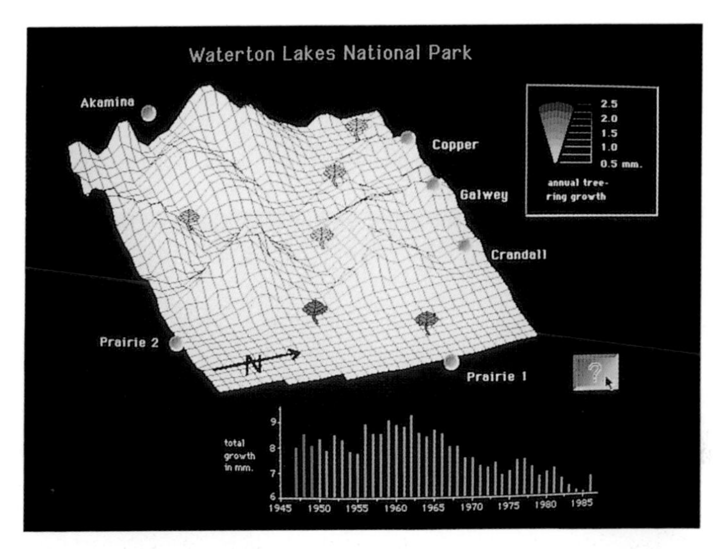

FIG. 6.3. Cartographic animation (Weber, SUNY, Buffalo, 1993) produced in MacroMind Director depicting radial growth in tree populations. Leaf icon color (annual growth) and size (cumulative growth) are determined in real time from embedded data files.

FIG. 6.4. Simulation of a proposed lake created by filling an existing open-cut coal mine with water. The land surface is created by texture mapping a mosaic of aerial photographs onto a digital terrain model. The power station was also created using the Advanced Visualizer software. Atmospheric haze has been used to increase the sense of depth.

FIG. 6.5. Mesh diagram of a digital terrain model of North Fork Elk Creek, Montana, a 17 km^2 watershed used in ecosystem modeling experiments by NASA Ames researchers. Color draped over the mesh corresponds to the six hillslopes used to partition the area. Three IDL procedures were used to generate the picture: SURFACE, TRIANGULATE and POLYFILL.

palettes is simple. Lingo allows for access to external numeric data, and execution of external code and hardware. Unfortunately, those wishing to perform geographic analysis on Director's output will find the task difficult because attribute information is not included in the output file.

I have used Director to create tutorial flip-art animations describing the functions of GIS and interactive flip-art animations of yearly surface temperature maps for the US created outside the system (Weber 1991; Weber and Buttenfield 1993). Currently, I am completing a real-time animation of tree growth data in an interactive hypermedia format (Fig. 6.3). I have also used the software to perform subject testing for cartographic research into sonification, utilizing moveable bitmap "sound coins" which sounded tones when clicked on (Weber 1993). Subject responses demonstrated through coin placement were easily recorded to external data files.

Overall, Director is a valuable animation, hypermedia and prototyping tool. Because it functions in a multi-tasking environment, GIS functions can be performed externally before compilation, or (theoretically) in real-time controlled by the scripting language. It is a powerful animation package whose functionality is affordable when compared with the four-figure prices of high-end authoring packages [such as Authorware Pro ($4000), Quest from Allen Communications (between $2700 and $4000) and MediaScript from Net Work Technology (between $700 and $4000); see Kindelberger 1993]. For anyone new to animation and hypermedia, Director's ease of production and friendly interface make it fun to work with.

Using Wavefront Technology's Advanced Visualizer Software to Visualize Environmental Change and Other Data

IAN BISHOP

Centre for Geographic Information Systems and Modelling
The University of Melbourne
Parkville, Victoria 3052, Australia
e-mail: ian_bishop@mac.unimelb.edu.au

Applications and Overview of Advanced Visualizer

I have used Advanced Visualizer (AV) in a variety of ways. In most cases this has been to visualize change within the natural environment. I have sought to present this change in as natural and realistic a manner as possible given the available technology. There are a number of advantages in using this "natural scene paradigm" (Robertson 1991) and I have argued these elsewhere (Bishop, 1994). My visualization projects to date have included change arising from modeled environmental processes (e.g. the spread of mountain pine beetle; Bishop and Flaherty 1990) and planned change (e.g. construction of a highway bypass; Bishop 1992). Figure 6.4 illustrates the visualization process in a mine redevelopment project. In each case I have based the visualization on accurate spatial data derived from GIS or CAD systems while maximizing visual realism. In general, this has been achieved by texture mapping images of the real world onto the terrain and built structure models derived from GIS and CAD — these models are called "objects" in AV nomenclature.

The process of texture mapping is sometimes called draping in GIS software. It involves fitting a two-dimensional image in raster format onto a three-dimensional surface. In GIS-based draping this surface is nearly always the land surface in the form of a digital terrain model (DTM). In a package such as AV a texture map can be applied to any surface including buildings, forests and individual trees.

Advanced Visualizer has facilitated these processes through its capacity to: (1) work with very large numbers of polygons (this is limited only by system RAM, but I have worked with objects described by over 150,000 polygons) and high resolution texture maps (with 64 MB RAM available, I only have managed to work at 2048 x 2048 resolution, but up to 4096 x 4096 resolution is possible); (2) work with many texture maps at once and to use bump, transparency and reflection maps to achieve special realism effects; and (3) support animation of any elements of the visualization; all of these elements can be moved independently or hierarchically linked.

The flexibility of AV as a data visualization tool is shown by a project in which data, rather than landscape appearance, was visualized. A graduate student used AV to animate census data based on local government areas (enumeration units). Cylinders representing the numbers of migrants of different countries of origin were placed in each government area. An appropriate perspective view was chosen and the cylinders made to lengthen and shorten according to the statistics over a 60-year period.

The non-proprietary nature of AV files also provides great flexibility in using the software. Nearly all files can be stored in ASCII format, making it very easy to write conversion programs for import and export. For example, programs developed in-house have allowed transfer of three-dimensional models and mapped data from Intergraph ".dgn", AutoCAD DXF, ArcInfo ASCII/grid and macGIS formats into AV. These programs will also take coverage data such as land use and use this to allocate colors within the AV object files. Linkage to UNIX shell scripts for complex operations is also simple. Some AV commands include explicit options to call scripts and pre- or post-execution operations. This allows the user, for example, to copy a different file into an animation package for each rendered frame. I have used this ability to map a texture of moving water onto a river surface.

In a more sophisticated use of texture mapping, I have used the software to drape either airborne video imagery or aerial photography onto terrain models. In the case of the mountain pine beetle simulation, a series of texture maps based on different stages of tree disease and defoliation were created. The texture for any particular DTM polygon was based on a pest spread simulation model. C programs were used to run the models and to define texture requirements within the object files. The whole process was controlled by shell scripts.

Advanced Visualizer is clearly not a GIS, hence its low ratings on the GIS-oriented aspects of Fig. 6.2. It does not have built-in functions for statistical analysis or map projection. It is a modelling, animation and rendering package. It performs all these functions well, but linkage to GIS data is a matter of user initiative. Many of the capabilities of AV are of limited use in the visualization of spatial data (e.g. multiple light sources, morphing, Boolean object merging). Some of the functions for which I have used AV are becoming more readily available in major GIS packages. For example, several packages can drape raster images over terrain surfaces, and Intergraph's ModelView supports many of the same animation and rendering capabilities as AV.

As AV has been used in this context primarily as a rendering tool, there is a potential problem that the viewer of the visualization is given no opportunity to explore the data. I have plans to overcome this problem using videodisk technology. Provided the exploratory wishes of the viewer can be anticipated, we can render appropriate sequences and put these onto video disk. With a suitable computer interface to the videodisk, the viewer can access the rendered frames in any chosen sequence or at any chosen rate.

Using IDL for Visualizing Results of Spatio-temporal Simulations of Forest Carbon Flux

JENNIFER DUNGAN

Ecosystem Science and Technology Branch
NASA Ames Research Center
Moffett Field, CA 94035-1000, USA
e-mail: dungan@gaia.arc.nasa.gov

Overview of IDL and Applications

Research System's Interactive Data Language (IDL) can be run interactively to analyze data, create from two- to five-dimensional graphics or build new applications. It includes libraries of high-level procedures to support mapping, image processing, statistics, surface and volume visualization and animation. The product was designed for scientists and promotes the visual investigation of data through fast, easy to use algorithms. It could be used to extend the analytical capabilities of commercially available GIS, many of which are limited in their abilities to extract data from conventional map displays; IDL allows examination of the underlying data as histograms, time series or other types of plots.

Interactive Data Language is a tool one level above a language such as Fortran or C that allows users to write their own procedures and functions. With the ability to create variables of virtually any type, and with the standard and user library functions and procedures, the building blocks for performing spatial analysis on points, lines, areas and surfaces exist, especially using a raster data model. It is possible to call IDL code from Fortran or C programs, or to call Fortran or C programs with the IDL command language. IDL includes functions and procedures for manipulating widgets, the elements of the X Window System, for building new interactive applications.

In IDL, two- or three-dimensional arrays are used for image processing. There are about 50 image processing functions in the standard library, including resizing, convolution, dilation, erosion, fast Fourier transform, histogram equalization, rotation, edge detection and polynomial warping given control points. The vector mapping capabilities of the package are much more limited than those for image processing. There is a CONTOUR function for creating contour maps, and because the user has control over the color tables and can use the line drawing and text functions (including PostScript fonts) for annotation, good quality maps can be produced; however, procedures for interpolation from irregularly spaced data are primitive, so users may want to replace them with their own or external interpolation programs.

There is a set of map functions which is primarily designed for displaying global or continental maps. The MAP_SET function allows the user to select one of 12 map projections, including azimuthal, cylindrical and pseudocylindrical projections, as well as the center, boundaries and polar rotation of the map. MAP_CONTINENTS is then used to draw continental boundaries from the WDBII, with the option of drawing parallels and meridians with the MAP_GRID function. Data can be displayed over these boundaries in the form of contours (with the CONTOUR function) or images (with the special purpose MAP_IMAGE function).

I have used IDL regularly for displaying output from a forest ecosystem simulation model, FOREST-BGC (described by Nemani *et al.* 1993). The simulation code exports ASCII files in which each record contains attribute information for a given spatial unit for a given day of the year. The IDL procedures I have written allow viewing the output as scatterplots, time series plots, choropleth maps, shaded surface perspective views and as animations through time. These views are created directly from the simulation output file. Therefore they do not entail the creation of separate plot or raster files with their concomitant disk usage and file management problems. Figure 6.5 shows a mesh diagram of a DEM of one of the small study areas; the diagram is colored according to hillslope partitions used to organize spatial information input to the model.

I also have used IDL for comparing quasi-point data with pixel data, processing time-dependent spectral data, exploratory spatial data analysis with regularly and irregularly spaced data and examining data from imaging spectrometers. I would recommend IDL to the scientist or engineer who has some programming experience, but would rather be spending more time analyzing data than looking at code.

Using SpyGlass to Animate Simulated Forest Change

MARC P. ARMSTRONG, AMY J. RUGGLES AND DEMETRIUS-KLEANTHIS D. ROKOS

Department of Geography
The University of Iowa
Iowa City, IA 52242, USA
e-mail: armstrng@umaxc.weeg.uiowa.edu

Overview of SpyGlass and our Application

Spyglass is a powerful, general purpose raster-based visualization tool that can be used in a variety of imaging, surfacing, three-dimensional and animation applications. The data import capabilities of the software are extensive. In the Transform module, for example, numerical and graphical data can be read in a variety of formats, including: two- and three-dimensional HDF data sets and images, PICT, TIFF, GIF and PICS animation files. Data can be imported from MatLab files, X-Window screen dumps (XWD) and in matrix or row–column formatted ASCII text or binary files. Color manipulation is very well supported. Output can be formatted in a number of ways and it can be directed to the screen, a variety of printers and to videotape. Specific cartographic output capabilities are not provided.

A macro language, available in the Transform module, provides the user with a rich set of image manipulation capabilities. A set of mathematical functions (e.g. log) is provided, along with a set of matrix handling routines (e.g. transpose). In addition, users can access high level image manipulation functions through the macro interface. Thus, it is possible to apply fast Fourier transforms and kernel functions to an image using this macro facility.

The objective of our research was to visualize how forest stands are affected by fragmentation and barriers in the landscape during periods of climate change (Malanson *et al.* 1993). For this purpose, we used a modified version of the JABOWA-FORET model to simulate processes of establishment, growth and death of forest trees in a grid cell framework. Three factors affecting forest dynamics and structure systematically varied in the model included: climatic range, dispersal ability of species and the spatial structure of the landscape. SpyGlass assisted us in the visualization of the results of these experiments. We ran each simulation for 1000 years and wrote the results to disk at the end of each 10-year period. Thus, each animation sequence consisted of 100 frames that were evenly spaced in the temporal dimension.

The simulation model ran on a UNIX workstation. We transferred ASCII files generated by the model to the Macintosh and then imported the files into SpyGlass to create compatible files that we used to render the cellular simulation results. We designed a color table based on varying hue and intensity to show the total basal area for each cell at each 10 year period of the simulation. Each of these images served as a frame in the animation. We also designed a legend and title which remain fixed in each frame, and a dynamic frame-counter to keep track of the progress of the animation.

In general, the software performed the tasks it was designed to perform very well. The main strengths of the tool in our application were: (1) it provided a capability to custom-design a color scheme; and (2) it enabled us to capture images, animate them and then direct the animation to videotape. The main weakness was the lack of direct support for specific cartographic features such as georeferencing and symbolization. It should also be noted that SpyGlass does not handle vector formatted data very well.

The Application Visualization System (AVS) as a Tool for Visualizing Geographic Data Sets

THERESA MARIE RHYNE

Martin Marietta Technical Services
US Environmental Protection Agency's
Scientific Visualization Center
Research Triangle Park, NC 27709, USA
e-mail: trhyne@vislab.epa.gov

General Software Assessment

Documentation

Advanced Visual Systems provides eight volumes of documentation designed to support new users as well as software developers. The documentation is clearly written with good illustrations. Installation instructions are written in a step by step manner and are in a separate handout from the bound documentation volumes. As copyright protection of the Application Visualization System (AVS) is based on license manager software, installation of the AVS requires contacting Advanced Visual Systems Inc. with appropriate hardware and software identification information. Although awkward, this process is helpful in establishing a product support relationship.

Ease of Learning

Advanced Visual Systems recommend that new users begin with the Users' Guide and work through the Tutorial Guide (Advanced Visual Systems 1992). Unfortunately, for new users who have not worked extensively on UNIX workstation platforms, this quantity of documentation can be overwhelming.

Ease of Use

For very general users of scientific visualization tools (i.e. those not involved with visual programming), the process of connecting AVS modules into a network (or pipeline) can be confusing. As a result, Advanced Visual Systems provide a software development environment in which computer graphics programmers can build point and click interfaces to customized AVS applications.

Product Support

Advanced Visual Systems provides a well defined product support plan. Upon purchase of the AVS software, the first year of support is included in the overall product. At the end of the 12-month period, Advanced Visual Systems automatically bills the customer to encourage re-enrollment in the AVSupport Plan. The AVSupport Plan includes: major AVS software versions; new platform support; telephone hotline support; electronic mail assistance; software updates; on-line updated database; and AVS news.

In addition to Advanced Visual Systems Inc., there is also an International AVS Center (IAC) located at the North Carolina Supercomputing Center. The IAC provides many services for AVS users around the world, including free user-contributed AVS modules via ftp.

Geographic Visualization Categories

Although AVS is a general purpose scientific visualization tool, AVS modules and applications can be written which support geographic visualization efforts. There are many third party vendors writing code which will interface their GIS with AVS; in this respect many public domain modules are available from the International IAC. C, Fortran or C++ can be integrated into AVS as a module simply by adding a "wrapper" around the user's subroutine. Some general concepts related to work underway to support geographic visualization activities are given below.

Import Capabilities

From a geographic visualization perspective, the AVS does not directly address the import of attribute, raster or vector data. Rather, there are two choices for importing data into the AVS: (a) use an existing AVS import module or (b) write a customized new import module. One problem is that it is up to the user to translate between cartography terminology and AVS terminology. For two-dimensional data (e.g. DTMs and multi-spectral images), read field input modules are well suited for importing the data set. The AVS Data Interchange Application (ADIA) is provided to deal with other input formats (Advanced Visual Systems 1992).

At the US EPA's Scientific Visualization Center, we have found a number of circumstances where a customized AVS based data importing module needed to be developed. These cases usually centered around large computational environmental science models which were executed on compute engines such as a Cray YMP (Rhyne 1992b).

Data Manipulation Capabilities

AVS 5.0 does not readily support standard geographic manipulation capabilities, but a strong development environment is available for creating customized modules. It is strongly recommended that AVS users keep in touch with the activities of the IAC to benefit from the development efforts of AVS users around the world (Gelberg and Myerson 1992).

Data Display Capabilities

As a general purpose visualization tool, AVS is excellent for multimedia and animation. The AVS/Animator component of the software is a key-frame system which can interpolate any parameter in an AVS network, thus allowing for sophisticated animations such as isosurfaces across a geographic domain (Advanced Visual Systems 1992).

The AVS supports the creation of modules which interface to standard multi media tools such as Apple's QuickTime capabilities. Public domain modules have been developed which support multivariate mapping. AVS networks can be built which support the simultaneous display of separate maps. The generate colormap module in AVS provides for the creation of color palettes, although extensive point and click control is required to generate customized discrete color maps. AVS supports text labeling capabilities in its visual displays and the quality of the type fonts depends on the hardware platform used rather than on AVS.

Export Capabilities

As mentioned previously, AVS is a general purpose scientific visualization tool, so the export of geographic visualization parameters such as attribute, raster or vector data is not addressed directly. AVS does provide data output modules that can convert AVS images to computer graphics metafiles or PostScript files. Public domain AVS data output modules which convert AVS files to various data output formats for printing still images and for recording to videotape are available from the IAC. At the US EPA Scientific Visualization Center, we were required to write a customized AVS output module which supported recording animation sequences to an optical disk with customized software code. It was possible to link an AVS data output module to this previously developed software code (Gelberg and Myerson 1992).

General Overview of AVS and its Applications

Visualization Toolkits: a Basic Definition

The newest programming environments for supporting the visualization of scientific data focus on data flow-based visual languages. These visual languages provide a graphical interface to support the import, manipulation and display of data. These programming environments are called visualization toolkits. One of the first visualization toolkits created was the Application Visualization System (AVS). With such toolkits, researchers program the software not by typing, but by dragging iconic representations of operations onto a drawing grid and connecting them by means of pipelines. This approach has the advantage of minimizing the researcher's need to cope with the cumbersome UNIX and C shell programming environments, as well as providing an interactive approach to viewing pictorial representations of data (Rhyne 1992a).

In visualization toolkits, software components are connected together to create a visualization. The software components, called modules, implement specific functions such as filtering, mapping and rendering. "Filtering" takes modeled or collected data and "filters" it into another form which is more informative and less massive. Examples of filtering operations include computing derived quantities such as the gradient of an input scalar field, deriving a flow line from a velocity field, or extracting a portion of a computational model data set. The next step "maps" the filtered or newly derived data into geometric primitives such as points, lines or polygons. Once a set of geometric primitives is chosen and calculated, the geometric data are "rendered" into pictures. At this stage, the researcher chooses coloring, placement, illumination and surface properties for the visualized image (Upson *et al.* 1989).

Visualization toolkits allow the software modules for filtering, mapping and rendering to be combined into executable flow networks (Fig. 6.6). To do this, the researcher selects modules from menus and places the icons representing the modules in a diagram. Each module appears as a box and connections between the modules are drawn as lines. Once the structure of the application has been established, the visualization toolkits can execute the network and display the resulting computer image.

Each image that is produced is a visual representation of the scene defined by the modules. The researcher can interact with the image by moving or changing lights, by modifying surfaces, by rotating, shifting or resizing objects, or by changing the point of view and angle of view.

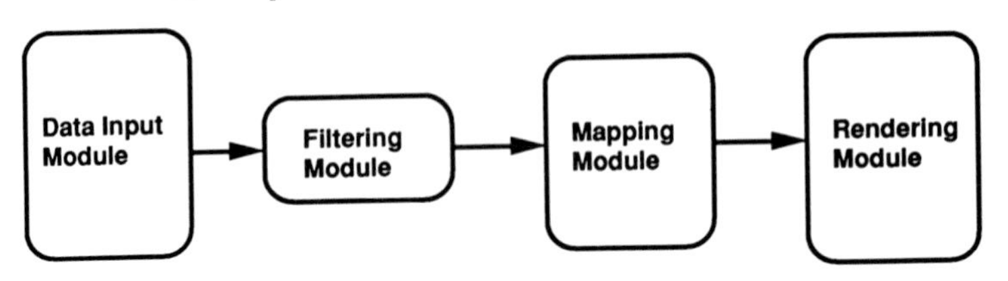

FIG. 6.6. General diagram of visualization toolkit approach.

Applicability to Geographic Visualization

The US EPA has used the AVS to examine environmental research data sets which pertain to air quality, water quality and subsurface contamination studies. Visualization training sessions, using AVS, have been developed and are currently provided at the National Environmental Supercomputing Center (NESC) in Bay City, Michigan. In general, we have found that EPA researchers and policy analysts are not immediately comfortable with building their own AVS pipelines or networks. As a result, visualization specialists have used the AVS development environment to build "turn-key" AVS applications which support the specific requirements associated with data input, visual display and data output requirements of EPA efforts. This has included the building of point and click user interfaces (Fig. 6.7) (Rhyne 1993; Chall and Pearson 1993).

The AVS modules can also link to other software tools or packages which have functions not immediately available in AVS. For example, it is possible to link with Khoros modules (see below), which are more robust in supporting image processing functions (Myerson and Reid 1992). It is also possible to build AVS networks which cross multiple UNIX platforms. At the EPA we have written AVS modules for the Cray YMP which execute photochemical air quality computational models. The Cray-based execution modules can be incorporated into an AVS network which is actually residing on another UNIX (e.g. Silicon Graphics, SUN, Digital, IBM or Data General AViiON) workstation. This provides for "remote module execution" of the computational model (Chall and Pearson 1993).

Simply stated, AVS is a powerful software tool which can also function as an umbrella for supporting metacomputing activities associated with geographic data visualization.

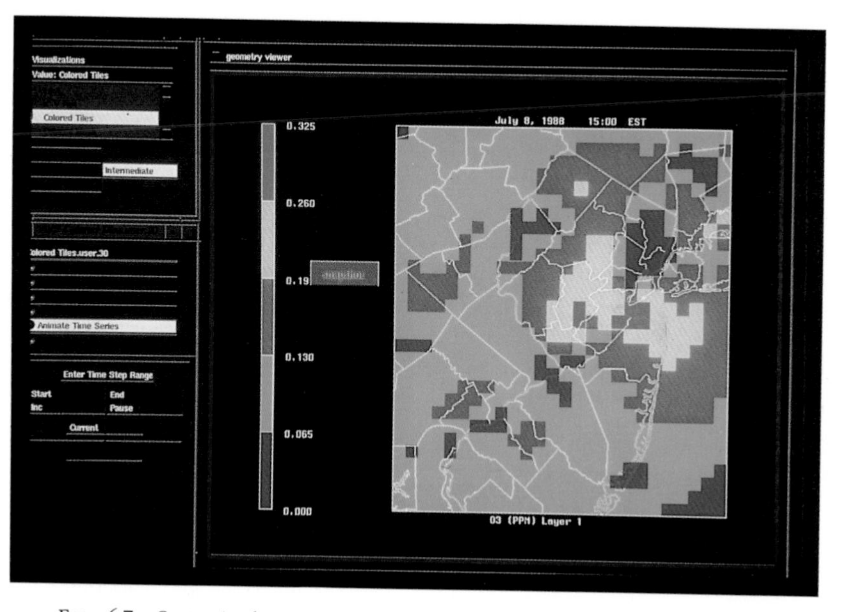

FIG. 6.7. Customized user interface of AVS-based Environmental Sciences Visualization System (developed at the US EPA Scientific Visualization Center, Programmers: Todd Plessel and Kathy Pearson).

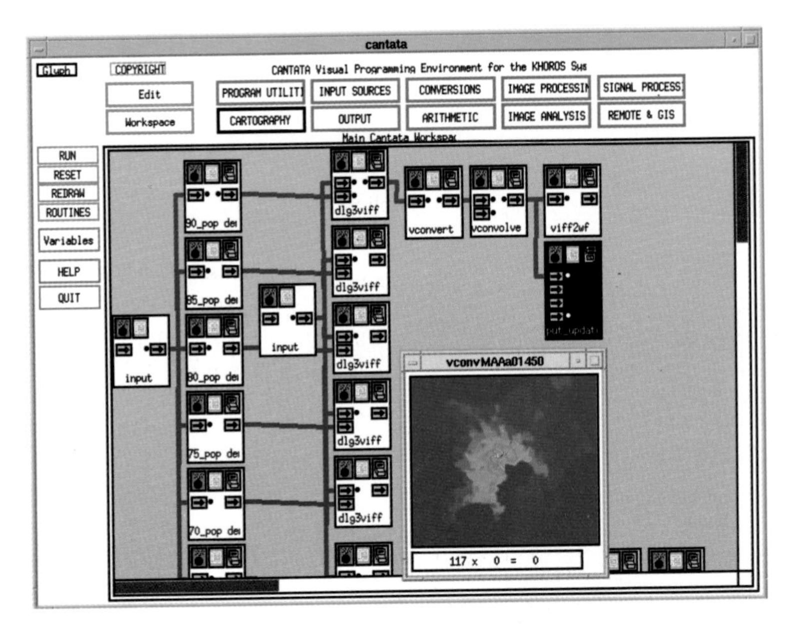

FIG. 6.8. Khoros visual programming environment.

Evaluation of IBM's Data Explorer

LOEY KNAPP AND JOHN CARRON

Department of Geography
University of Colorado
Boulder, Colorado, USA
e-mail: knappl@bldfvm9.vnet.ibm.com

Overview of Data Explorer and Applications

Our assessment considers: (1) the extent to which Data Explorer (DX) is moving beyond the toolkit level; and (2) its applicability to the geographic arena. Our conclusions are very positive from both perspectives. The visual programming interface of DX, which allows the user to work with modules rather than code, is a major step towards providing non-programmers with the capability to develop advanced, flexible applications. Recent enhancements to the product, including the availability of a wide range of map projections and import functions for DLG, DEM and image data from Cornell University, strongly contribute to its value for geographical analysis. Of primary interest is IBM's announcement in August 1993 of intent to develop an integrated analysis and visualization platform together with ESRI's GIS product ArcInfo.

We used DX to examine ways in which visualization techniques could be applied to water resources analysis in the Gunnison River Basin (GRB) of Colorado, specifically the assessment of moisture conditions. Time series data involved in analyzing moisture conditions in the GRB were a mixture of station observations of precipitation and temperature and the two-dimensional gridded output of two models, the RHEA-CSU orographic precipitation model and a moisture index model. The entire time series included 10 years of monthly values. Previous to this study ArcInfo had been used for display purposes, but there was interest in examining the benefit of more interactive display capabilities and time series animation. Some of the key features of the visualization techniques included the following:

1. Animation of time series data, both single maps through time and the simultaneous display of multiple maps through time (e.g. three resolutions of the Rhea model).
2. Two and a half dimensional perspective views of model output.
3. Statistical diagrams including histograms, scatterplots and line graphs displayed simultaneously with spatial data (the diagrams could be applied to single time steps or the entire time series).
4. Simultaneous display of multiple variables by associating one variable with the size of glyphs and another variable with color (one example involved using arrows to represent temperature data where the color of the arrow indicated

the mean daily temperature and the length of the arrow indicated the range from minimum to maximum values; another example involved using color to indicate water quality readings along the river and a flexible line width to indicate flow levels).

5. Use of data probes which are like digital "stickpins" that can be inserted into two- and three-dimensional data sets; these probes can be queried for data values at a specific point or linked together to provide cross-sectional displays of the data values between the probes.
6. Interactive controls allowing the user to define glyph type, colors, data sets, animation sequence, data probe placement, camera angle and relief displacement.

The strengths of DX lie in its flexibility, object-oriented approach, range of interactive tools, two-tier user interface and open architecture. A wide range of *n*-dimensional data structures can be imported allowing the mechanism to display, for example, two-dimensional surface data together with three-dimensional subsurface or atmospheric data. The data can be grouped in a variety of ways including time series or location series which can then be easily animated. Display flexibility is provided by an extensive array of interactors which control series sequencing, color, camera angle, panning, zooming and other user-specified options. The majority of the object-oriented modules work with a variety of data structures, each recognizing the dimensionality of the incoming data and treating it appropriately.

The two-tiered user interface allows both novice users and advanced designers to work with the software, the first group interacting primarily with predefined programs and the second group interacting with the visual programming interface to develop programs at varying levels of complexity. Finally, user-written modules can easily be incorporated into the product providing the requisite environment for fulfilling specific application needs.

The primary weaknesses of DX are the complex documentation, the difficulty of learning the system and importing data (which are due, in part, to a large range of options), the lack of sophisticated statistical functions and the limitations on interactive data selection. The documentation is written at a fairly sophisticated level which makes it more appropriate for advanced users. It provides comprehensive functional descriptions but could benefit from a user's guide with task-oriented examples. The video tapes which come with the software provide the best source of information for new users. Benefits from the robust support for a wide variety of data structures is negated by the difficulty the new user has in defining their data in a DX-readable format. This problem is being addressed through new import modules, the agreement with ESRI and a new data prompter which assists in defining some of the simpler file structures. Display flexibility also mandates user comprehension and manipulation of many variables for rendering and screen arrangement.

The conclusion of the GRB study was that the visualization techniques provided by DX were a substantial enhancement to the analysis of the data. Integration of scientific visualization and GIS is important given the extent of the geographical analysis which is currently being performed by GIS software. The fact that IBM is actively pursuing a partnership with ESRI to fulfill this requirement should make DX even more appealing for geographical applications.

Using Khoros to View and Edit Animated Census Data of Tokyo and Surrounding Prefectures

DONNA OKAZAKI

Department of Geography
2424 Maile Way, Porteus Hall Room 445
University of Hawaii at Manoa
Honolulu, HA 96822, USA
e-mail: donna@uhunix.uhcc.hawaii.edu

Introduction

Khoros is an integrated software development system for data processing and visualization (Rasure and Young 1992). It comes with many image processing, remote sensing and a few GIS modules which are well suited for working with raster or gridded data. I used Khoros as the backbone of a project which involved transforming census data and base maps into a cartographic animation on videotape. The data were aggregated to the level of cities, towns and villages for five prefectures of Japan centered on Tokyo.

Overview of Khoros and my Application

Khoros is a general purpose visualization tool which grew out of the engineering disciplines, so it does not provide modules specific to cartographic visualization. This causes it to be rated poorly in this area (for the non-programmer). C language programmers can write modules to suit their own needs. A fairly active user community provides support over the Internet. The Khoros group runs an integrated mailing list/news group and is very responsive to user queries. The manuals are readable and training classes for the public will be provided after Version 2.0 is released. On-site consulting is available to consortium members.

Khoros offers users different levels of interaction, ranging from point and click applications to visual programming to CASE tools for the software developer. The system comes with hundreds of modules which can be run as stand-alone programs or within a visual programming environment. Modules are pulled down from menus and appear as icons which can be linked together in a data flow network as shown in Fig. 6.8. Many more modules, often bundled as toolkits, can be obtained from other users over the Internet.

The goal of my project was to create a thematic animation of Japanese census data with the objective of being able to view and explore the urban development of Tokyo in time–space. The process of creating the animation was broken into four stages: (1) digitizing the x–y spatial component from base maps; (2) adding a thematic variable as z to make a three-dimensional surface model and creating texture maps for overlay;* (3) creating and rendering the video frames; and (4) editing and composing the frames into the final video product. No one program was sufficient to perform all these tasks. A combination of tools, with Khoros as the backbone, enabled the transition from paper maps and statistical tables to video tape. The work flow and software and hardware tools used are shown in Fig. 6.9.

Khoros provided utilities to create a module which inputs DLG formatted files, and outputs a raster thematic map. In this case, population density was used as the value to fill in polygons which represented the boundaries of cities, towns and villages. Image processing modules were used to smooth the boundaries between adjacent polygons, so when viewed in three-dimensions, the surface appeared continuous instead of disjointed. This made it easier to observe surface trends and small anomalies. Khoros modules created a custom colormap and attached it to the data. Other modules displayed the results on the screen and output them on a color laser printer. One more module needed to be coded to convert the thematic map into a format for export to the WaveFront three-dimensional animation system. The resulting frames were imported back into Khoros to splice in text and the legend inset. Khoros also generated the dissolves between title frames and between an introductory Landsat image and the first data frame.

Khoros is a powerful tool which can help geographers compensate for deficiencies in other software, but it has shortcomings as a stand-alone package. It does not provide data structures to support digitizing polygon or network topology;

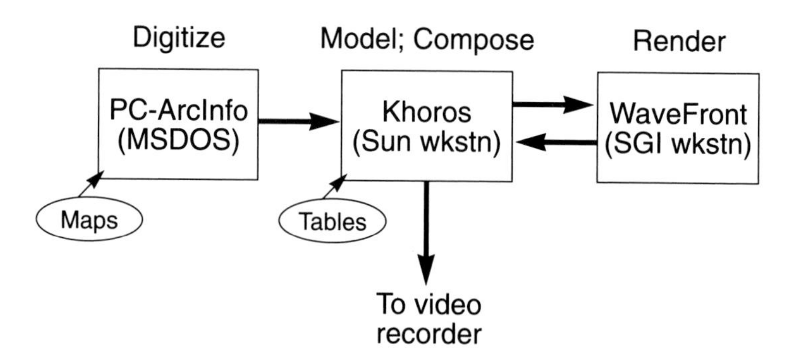

FIG. 6.9. Project work flow.

*In the context of animation, texture map refers to any image which is draped over an object, such as a three-dimensional surface, giving it colour and/or the appearance of texture.

nor does it have a digitizing module. It also does not have the features of a full-blown animation system. However, it was able to bridge between various other software, making it possible to create, view and edit key components of my animation.

The release of Khoros Version 2 will overcome data structure limitations by providing an object-oriented, abstract data model which will include three-dimensional geometry data services (Young and Rasure 1992). It will then be up to the user community to build the necessary modules to make it usable as a true cartographic visualization tool, one that could help us examine and explore, from unique perspectives, the data we use to represent the world.

References

Advanced Visual Systems (1992) *AVS User's Guide*, Release 4, May 1992.

Armstrong, M. (1994) "Spyglass", *Cartography and Geographic Information Systems*, Vol. 21, No. 1, pp. 52–54.

Bancroft, G. V., F. J. Merritt, T. C. Plessel, P. G. Kelaita, R. K. McCabe and A. Globus (1990) "FAST: a multi-processed environment for visualization of computational fluid dynamics", *Proceedings Visualization '90*, San Fransisco, pp. 14–27.

Bishop, I. D. (1992) "Data integration for visualisation: application to decision support", *Proceedings AURISA '92, Australian Urban and Regional Information Systems Association, Gold Coast*, pp. 74–80.

Bishop, I. D. (1994) "The role of visual realism in communicating and understanding spatial change and process" in Unwin, D. and H. Hearnshaw (eds.), *Visualization in Geographic Information Systems*, John Wiley & Sons, Chichester, London, pp. 60–64 and plates 1 and 2.

Bishop, I. D. and E. Flaherty (1990) "Using video imagery as texture maps for model driven visual simulation", *Proceedings Resource Technology '90*, Washington, DC, pp. 58–67.

Chall, S. and K. Pearson. (1993) "UAMWORLD: prototyping a distributed environment for air quality modeling and visualization in AVS", *AVS Network News*, Vol. 2, Issue 2, pp. 26–30.

Clark, B. (1992) "Explorer: a data visualization product", *Information Systems Developments*, US Geological Survey, Vol. 9, Issue 11, pp. 234–237.

Clark, B. (1993a) "Distributing Explorer's processing", *Information Systems Developments*, US Geological Survey, Vol. 10, Issue 4, pp. 105–107.

Clark, B. (1993b) "Khoros update", *Information Systems Developments*, US Geological Survey, Vol. 10, Issue 2, pp. 18–22.

Dungan, J. (1993) "PV-Wave CL and IDL", *Cartography and Geographic Information Systems*, Vol. 20, No. 2, pp. 126–129.

Dyer, D. S. (1990) "A dataflow toolkit for visualization", *IEEE Computer Graphics and Applications*, Vol. 10, No. 4, pp. 60–69.

Egbert, S. L. and T. A. Slocum (1992) "EXPLOREMAP: an exploration system for choropleth maps", *Annals, Association of American Geographers*, Vol. 82, No. 2, pp. 275–288.

Ferreira, J. Jr and L. L. Wiggins (1990) "The density dial: a visualization tool for thematic mapping", *Geo Info Systems*, Vol. 1, pp. 69–71.

Gelberg, L. and T. Myerson (1992) "AVS Tips and Tricks", *AVS Network News*, Vol. 1, Issue 3, pp. 14–18.

Haslett, J., R. Bradley, P. Craig, A. Unwin and G. Wills (1991) "Dynamic graphics for exploring spatial data with application to locating global and local anomalies", *The American Statistician*, Vol. 45, No. 3, pp. 234–242.

Jern, M. (1989) "Visualisation of scientific data", *Computer Graphics 89*, Blenheim Online Publications, London, pp. 79–103.

Kindelberger C. (1993) "Multimedia — the next big wave", *URISA Journal*, Vol. 5, No. 1, pp. 121–133.

Krohn, M. D. and B. Clark. (1992) "Khoros — A distributed image processing system for networked workstations", *Information Systems Developments*, US Geological Survey, Vol. 9, Issue 6, pp. 144–153.

Kruse, F. A., A. B. Lefkoff, J. W. Boardman, K. B. Heidebrecht, A. T. Shapiro, P. J. Barloon and A. F. H. Goetz (1993) "The spectral image processing system (SIPS) — interactive visualization and analysis of imaging spectrometer data", *Remote Sensing of Environment*, Vol. 144, pp. 145–163.

Lucas, B., G. D. Abram, N. S. Collins, D. A. Epstein, D. L. Gresh and K. P. McAuliffe (1992) "An architecture for a scientific visualization system", *Proceedings Visualization '92*, Boston, pp. 107–114.

MacDougall, E. B. (1992) "Exploratory analysis, dynamic statistical visualization, and geographic information systems", *Cartography and Geographic Information Systems*, Vol. 19, No. 4, pp. 237–246.

Malanson, G. P., M. P. Armstrong and D. A. Bennett. "Fragmented forest response to climatic warming and disturbance", *Proceedings of the 2nd International Conference on Integrating Geographic Information Systems and Environmental Modeling*, Breckenridge, Colorado, in press.

Monmonier, M. (1992) "Authoring graphic scripts: experiences and principles", *Cartography and Geographic Information Systems*, Vol. 19, No. 4, pp. 247–260.

Myerson, T. and D. Reid (1992) "New Khoros Modules for AVS", *AVS Network News*, Vol. 1, Issue 3, pp. 6–10.

Nemani, R., S. W. Running, L. Band and D. Peterson (1993) "Regional hydroecological simulation system: an illustration of the integration of ecosystem models in a GIS", in Goodchild, M. F., B. O. Banks and L. T. Steyaert (eds.), *Environmental Modeling with GIS*, Oxford University Press, New York, pp. 296–304.

Prawal, D., K. Bedford, J. Barros and D. Barton (1990) "Tools for visual data analysis — user experiences", *Proceedings Visualization '90*, San Fransisco, pp. 391–392.

Rasure, J. and M. Young (1992). "An open environment for image processing and software development", *1992 SPIE/IS&T Symposium on Electronic Imaging, SPIE Proceedings*, Vol. 1659, pp. 300–310.

Rhyne, T. (1992a) "Evaluating scientific visualization toolkits at the U.S. EPA", *ASPRS/ACSM/RT92 Convention — Mapping and Monitoring Global Change (3–8 August 1992)*, Washington D.C., Vol. 5, pp. 254–258.

Rhyne, T. (1992b) "Using AVS at EPA's Scientific Visualization Center", *AVS Network News*, Vol. 1, Issue 3, p. 13.

Rhyne, T. (1993) "Developing an AVS based training program for environmental researchers at the U.S. EPA", *Proceedings of the 2nd Annual International AVS User Group Conference*, Orlando, paper #2.

Ribarsky, W., B. Brown, T. Myerson, R. Feldmann, S. Smith and L. Treinish (1992) "Object-oriented, dataflow visualization systems — a paradigm shift?", *Proceedings Visualization '92*, Boston, pp. 384–388.

Robertson, P. K. (1991) "A methodology for choosing data representations", *IEEE Computer Graphics and Applications*, May, 56–67.

Tang, Q. (1992) "From description to analysis: an electronic atlas for spatial data exploration", *ASPRS/ACSM/RT 92 Convention — "Mapping and Monitoring Global Change (3–8 August 1992)"*, Vol. 3, pp. 455–463.

Treinish, L. A. (1989) "An interactive, discipline-independent data visualization system", *Computers in Physics*, Vol. 3, No. 4, pp. 55–64.

Treinish, L. A., J. D. Foley, W. J. Campbell, R. B. Haber and R. F. Gurwitz (1989) Panel on "Effective software systems for scientific data visualization", *Computer Graphics*, Vol. 23, No. 5, pp. 111–136.

Upson, C., T. Faulhaber Jr, D. Kamins, D. Laidlaw, D. Schlegel, J. Vroom, R. Gurwitz and A. Van Dam (1989) "The Application Visualization System: a computational environment for scientific visualization", *IEEE Computer Graphics and Applications*, Vol. 9, No. 4, pp. 30–42.

Wampler, S. (1992) "Extending the Khoros visualization system", *Information Systems Developments*, US Geological Survey, Vol. 9, Issue 11, pp. 238–241.

Weber, C. R. (1991) "A cartographic animation of average yearly surface temperatures for the 48 contiguous United States: 1897–1986", *NCGIA Technical Paper Series*, University of California, Santa Barbara (software also available).

Weber, C. R. (1993) "Sonic enhancement of map information: experiments using harmonic intervals", *Unpublished Dissertation*, State University of New York at Buffalo, Department of Geography.

Weber, C. R. and B. P. Buttenfield (1993) "A cartographic animation of average yearly surface temperatures for the 48 contiguous United States: 1897–1986", *Cartography and Geographic Information Systems*, Vol. 20, No. 3, pp. 141–150.

Williams, C., J. Rasure and C. Hansen (1992) "The state of the art of visual languages for visualization", *Proceedings Visualization '92*, Boston, pp. 202–209.

Young, M. and J. Rasure (1992) "An open environment for heterogeneous distributed computing", *1992 Xhibition Proceedings*, pp. 159–170.

Appendix: Product Information

Borland C++ and MetaWINDOW

Software Specifications

Borland C++ 3.1. Borland International, 1800 Green Hills Road, PO Box 660001, Scotts Valley, CA 95067-0001, USA. Phone: 1-800-331-0877. List Price: $495 ($795 with Application Frameworks); substantial discounts are available from dealers and mail order outlets.

MetaWINDOW/Standard 4.2A. Metagraphics, 269 Mount Hermon Road, PO Box 66779, Scotts Valley, CA 95066-9982, USA. Phone: 1-800-332-1550. List price: $249 (Version 4.3B); also available from mail order outlets specializing in programming tools.

Hardware Requirements

Borland C++ 3.1: IBM PC family, including XT, AT, PS/2, 386, 486; hard disk; one floppy drive; any 80-column monitor. [Although this is the required minimum configuration, as a practical matter we recommend at least a 386-based PC with a fast (18 ms or less), high-capacity hard disk, a math coprocessor, a mouse and a VGA monitor.] Hard disk space requirements vary according to installation options selected and the physical format of the hard disk. Our installation occupies approximately 30 MB.

MetaWINDOW 4.2A: As MetaWINDOW is an add-on programming library, its hardware requirements are the same as for the supported compiler. A full MetaWINDOW installation, including sample programs, occupies less than 1 MB.

Software Requirements

Borland C++ 3.1: DOS 2.0 or higher. (We recommend DOS 5.0 or higher. If Borland C++ is to be used for compiling programs for Microsoft Windows, then Windows 3.0 or higher is also required.)

MetaWINDOW 4.2A: Same DOS requirement as the supported compiler.

Macromind Director

Software Specifications
Version 3.1. Macromedia Inc., 410 Townsend St., Suite 408, San Francisco CA 94107, USA. Phone: 415-442-0200. List price: $1100.

Hardware Requirements
Any Apple, Macintosh Plus, Classic, SE, Portable, LC, Macintosh II, or newer computer (e.g. Centris or Quadra series); 14 MB of hard disk space for full installation; 2 MB RAM minimum (4 MB or more recommended).

Software Requirements
System 6.05 or later (System 6.07 or later for sound capabilities).

Advanced Visualizer

Software Specifications
Version 3.0, Wavefront Technologies,
530 East Montecito Street, Santa Barbara, CA 93103, USA. Phone: 805-962-8117. List price: pricing depends on the type of graphics hardware and whether a single or multiple processors are used. Pricing begins at $18,000. Educational pricing begins at $9000.

Hardware Requirements
Advanced Visualizer runs on the following platforms: Silicon Graphics [All models except Challenge (Power) L, XL]; IBM (RS6000 'Sabine', Gt4X, GTO); SUN (SPARC GS, GX); Hewlett-Packard [7xx CRX (PwrShd.), CRX24, CRX24Z, CRX48Z].

IDL

Software Specifications
Version 3.1 (May 1993). Research Systems, Inc., 777 29th Street, Suite 302, Boulder CO 80303, USA. Phone: 313-786-9900. List price: $3750 on workstation platforms, $1500 on IBM-compatible personal computers and Apple Macintosh; educational discounts available. Exchange of technical information about IDL can be obtained via Usenet news group comp.lang.idl-pvwave.

Hardware and Software Requirements
IDL runs on the following platforms and operating systems: Convex C2 and C3 (ConvexOS 10.0.5); Data General Aviion (DG/UX 5.4.1 and later); DEC ALPHA (OSF1 1.2 and OpenVMS AXP1.0); DOS-based personal computers running Microsoft Windows 3.1 or later; HP 9000 (HP-UX 8.0 on Series 300, 400 and 700); IBM 6000 (AIX 3.2); MIPS (Risc/OS 4.52B); Risc Ultrix (Ultrix 4.2); SGI (IRIX 4.0); Sun 3 (SunOS 4.1.1); Sun 4 - sparc (SunOS 4.1.0 and SunOS 5.1 - Solaris 2.1); VAX (VMS 5.1 and up); Apple Macintosh and PowerPC running System 7.0 or higher.

SpyGlass

Software Specifications

Transform Version 2.11, Format Version 1.11, Dicer Version 1.12. Spyglass, Inc., PO Box 6388, Champaign, IL 61820, USA. Phone: 217-355-6000. List prices: $595 for Transform/Viewer, $695 for Dicer and $235 for Format.

Hardware Requirements

Macintosh computers with 68020 CPU or better, 256-color, and 4 MB of RAM; a hard disk is recommended. A floating-point unit is recommended for Transform and Dicer. UNIX versions are also available.

Software Requirements

System Software 6.0.5 or later; System 7 is required to use AppleEvent-based scripting, MathLink or Balloon Help. 32-Bit QuickDraw is required for System 6.0.

Application Visualization System (AVS)

Software Specifications

Release 5.0 (March, 1993). Advanced Visual Systems, 300 Fifth Avenue, Waltham, MA 02154, USA. Phone: 617-890-4300. Internet e-mail: info@avs.com. List price: $6500 CPU-locked, $8000 floating for single copy (discounts available for education, US Government and volume purchasers).

Hardware Requirements

AVS is available on many UNIX workstation platforms, including Data General, Digital, Hewlett Packard, IBM, Silicon Graphics and Sun Microsystems. It is also available for the Evans & Sutherland Freedom and Kubota Pacific Denali graphics subsystems, and Cray, NEC and Fujitsu supercomputers. Third party versions of AVS are available directly from Convex, Intel, Meiko, Set Technologies and Thinking Machines.

Although AVS makes full use of three-dimensional graphics hardware, Rhyne's review focuses on its performance on a Data General AViiON workstation, which requires AVS to use software rendering to produce graphic displays. The AViiON workstation requires a minimum of 16 MB of memory and 50 MB of disk storage to install AVS from the release tapes.

Software Requirements

AVS 5.0 for Data General AViiON workstations requires release 5.4.1 or later of the DG/UX operating system. AVS requires that the X Window System and TCP/IP be running on the Aviion. Thus, at least level 3 of the DG/UX system software is required. AVS will not work in single user mode. A minimum system swap space of twice the amount of main memory or 64 MB (whichever is greater) is recommended.

IBM Visualization Data Explorer (DX)

Software Specifications

Version 2.0. IBM, 8 Skyline Drive, Hawthorne, NY 10532, USA. Phone: 1-800-388-9820. For information on DX modules available through Internet, telnet to info.tc.cornell.edu. Information about DX is available via Usenet news group comp.graphics.data-explorer.

Hardware/Software Requirements

Memory: 32 MB minimum recommended (interactive playback of animated time series will require additional memory). Disk: 30 MB minimum. 8-bit color XWindow support (12- and 24-bit color displays and three-dimensional graphics hardware is supported if installed). Data Explorer runs on IBM Risc Systems and Power Visualization Systems, Sun, SGI, HP and DG.

Khoros

Software Specifications

Version 1.0.5 (August 1993). The Khoros Group, Khoral Research Inc., 4212 Courtney NE, Albuquerque, NM 87108, USA. E-mail: khoros-request@chama.eece.unm.edu. Cost is free if acquired from various FTP sites including ftp.eece.unm.edu (129.24.24.10); code is located in the directory /pub/khoros [answers to frequently asked questions (FAQs) are in the directory /pub/khoros/release]. By mail the cost is $250 US/Canada, $350 International; send e-mail to khoros-request@chama.eece.unm.edu for an order form. Optional consortium membership fee is $50,000/yr, and optional affiliate membership fee is $5000/yr. Consortium members receive free training and proportionately greater influence on software development. All members receive alpha and beta copies of new software.

Current information and help can be obtained through Internet news (comp.soft-sys.khoros) and the Khoros mailing list (subscribe by sending mail to khoros-request@chama.eece.unm.edu).

Hardware and Software Requirements

With respect to the UNIX workstation, binaries available from the Khoros Group include: DEC MIPS (Ultrix 4.2), SUN4, Sparc (SunOS 4.1.3), SUN3 (SunOS 4.1.1), SGI (OS 4.0), IBM RS/6000 (AIX 3.2), HP 9000, Intel 486 (Interactive Unix 3.2.2). UNIX user compilations include: SUN4 (Solaris 2.2), Convex (ConvexOS 8.1, 9.1), Cray (Unicos 5.*, 6.0, 6.1), DG Avion, IBM R/S 6000 (AIX v3.x), Intel PC's with Linux, Alliant, Apollo, Fujitsu, Intergraph, Sequent, Sony.

75-150 MB disk, depending on how binaries are linked and whether sample data, sources and manuals are kept on-line. X Windows version X11R4 or X11R5.

CHAPTER 7

Color Use Guidelines for Mapping and Visualization

CYNTHIA A. BREWER*

Department of Geography
San Diego State University
San Diego, CA 92182, USA

Introduction

When color is used "appropriately" on a map, the organization of the perceptual dimensions of color corresponds to the logical organization in the mapped data. A color scheme typology is presented here that matches a comprehensive listing of the ways in which data are organized to corresponding organizations of hue and lightness, with secondary attention to saturation. Appropriate use of color for data display allows interrelationships and patterns within data to be easily visualized. The careless use of color will obscure these patterns. Careful use of color on maps is particularly important in interactive and animated map contexts, where the map reader must attend to changing patterns on the maps and has little time to look back and forth to a map legend, if one is offered at all. A spontaneously understood color scheme is a great asset in these contexts.

Recommendations on the use of color have a long tradition in cartography (Robinson 1952), and Bertin's (1981) frequently cited work describes the use of a set of visual variables for map symbolization that includes hue and lightness. Guidelines in cartography texts are generally limited to instructions to use hue for qualitative differences on maps and to use lightness for quantitative or ordered differences (Robinson *et al.* 1984; Campbell 1991; Muehrcke and Muehrcke 1992; Dent 1993; Monmonier 1993). The recommendations in this chapter extend these

* Present Address: Department of Geography, The Pennsylvania State University, University Park, PA 16802, USA.

basic guidelines to a more detailed categorization of data conceptualizations, corresponding color scheme types and their combinations. The recommendations are based on my personal experience, cartographic convention and the writings and graphics of others. They have not yet been thoroughly tested.

This discussion of color schemes is limited to the use of color for representing data associated directly with map areas. The guidelines offered apply to many types of thematic maps: both classed and continuous–tone choropleth maps, filled isoline maps and qualitative areal extent maps. The guidelines may also be extended to maps with colored point and line symbols and to the other numerous graphic forms for which color is used to represent data. The use of lightness for relief shading and advances in surface rendering, such as transparency and gloss effects, used to produce realistic three-dimensional forms are beyond the scope of this chapter. Shading and other rendering techniques, however, reduce the variation in color available for thematic data mapped onto modeled surfaces. Brewer and Marlow (1993) describe a scheme for terrain visualization (shown on the cover of this book) that combines shaded relief, aspect, and slope representations.

Terminology

In addition to recommendations on color use, a focus of this chapter is to establish unambiguous vocabulary that is easy to use during spoken explanation. Many disciplines, other than cartography and geography, use color to visualize data: engi-

TABLE 7.1

Terminology from Diverse Sources. Alternatives to, Variations on and Subsets of Qualitative *and* Quantitative *that have been Used to Describe Color Schemes and Data Organizations. Some Disciplines Class Ordinal Data with Categorical Rather than Quantitative Data*

Qualitative	Quantitative
Nominal	Semi-quantitative
Naming	Ordered, ordering
Categorical	Ranked
Classification	Ordinal
Selective	Interval/ratio
Denotative	Scalar
Identity	Continuous, continua
Discrete	Progressive
Mutually exclusive	Ordered progression
Different, differential	Monotonically increasing
Different in:	Magnitude change
Kind	Differences in amount
Type	Single-sequence
Name	Unipolar
Category	Double-ended
Class	Bipolar
Binary	Balance

TABLE 7.2

Color Scheme Types of Olson and Mersey. Figure References (at right) Provide Examples for these Previously Proposed Scheme Types

Olson's schemes (1987, and personal communications in 1993)	
• Qualitative	Fig. 7.3a
• Single-sequence monochrome	Fig. 7.3c and 7.4a
• Single-sequence polychrome	Fig. 7.4b and c
• Shading scheme	(relief shading)
• Double-ended bipolar	Fig. 7.5b and c
• Double-ended balance	Fig. 7.7b
• Qualitative/quantitative	Fig. 7.6b
• Two-variable quantitative	Fig. 7.6c and 7.7a
• Three-variable proportions	Fig. 7.7c
• Multi-variable counts	(multi-color dot map)
Mersey's series for one-variable choropleth maps (1990)	
• Hue series	
• Double-ended	Fig. 7.5b and c
• Spectral	Fig. 7.5a
• Hue value	Fig. 7.4b
• PMS value	Fig. 7.4a
• Black and white	Fig. 7.3c

neering, medicine, statistics, computer science, graphic design and psychology, for example. Members of this diverse group are unlikely to know or care about the nuances of color theory or cartographic conventions in terminology, so recommendations about color use that are accompanied by opaque language will be ignored. Cartographers do have a substantial store of knowledge and expertise to offer the visualization community. Use of readily understood terms is critical if cartographic recommendations are to be welcomed in diverse fields that share similar graphic challenges.

Hue, lightness and saturation are the three perceptual dimensions of color used in these recommendations. Hue is the aspect of color described by the color names we use such as red, yellow and green. Dominant wavelength is the physical correlate of hue. Color is sometimes used synonymously with hue, but color more accurately describes the combination of all three perceptual dimensions. Many different terms are used for the lightness dimension in the work listed in the references and bibliography: value, darkness, brightness, luminance, intensity. Value is a frequently used term and is often clarified with lightness as its definition. Use of value is awkward, however, because simultaneous discussion of data values and color values becomes confusing.

The third dimension, saturation, can be thought of as the amount of hue in a color. For example, reds of constant lightness can range from grayish to pure red. Alternative terms for saturation are purity, chroma, colorfulness and intensity (avoid intensity, which is used in a confusing manner for both lightness and saturation). Of these choices, saturation is the most commonly used term in the varied disciplines represented in the references and bibliography. In addition, I feel that

TABLE 7.3
One-variable Data Types and Color Schemes

Data conceptualization and scheme type (emphasis of data display)	Perceptual characteristics of scheme	
	Hue	Lightness
Qualitative	Hue steps (not ordered)	Similar in lightness
Binary (special case of qualitative)	Neutrals, one hue or one hue step	Single lightness step
Sequential	Neutrals, one hue or hue transition	Single sequence of lightness steps
Diverging	Two hues, one hue and neutrals, or two hue transitions	Two diverging sequences of lightness steps

saturation is more readily understood than chroma by readers who are not familiar with color jargon because the meaning of saturation is akin to the lay use of the word. White, grays and black are neutral colors that have no saturation or hue.

Continuing on the topic of terminology, Table 7.1 lists the terms used to describe color schemes and data characteristics in a wide range of fields. The publications from which this varied vocabulary has been gleaned are listed in both the references and bibliography. Some of these terms will be unfamiliar, and one purpose of Table 7.1 is to demonstrate that terms common to one discipline are unfamiliar to others. Sets of terms for color scheme types that have been suggested are listed in Table 7.2.

The schemes in Table 7.2 provide a good start, but there are inconsistencies within and omissions from the collections. Mersey's (1990) one-variable scheme types combine perceptual dimensions with specifics of production. Pantone Matching System (PMS) ink use is irrelevant to the representation of data in varied media. Olson's augmentation of her 1987 set* is more thorough than Mersey's, but there are gaps such as the combination of two qualitative variables. Olson does not elaborate on the combinations of, in her language, single-sequence and double-ended schemes within her two-variable quantitative scheme type. I also move her balance scheme to a two-variable category. These differences are elaborated below. I omit her multi-variable count (dot) maps and relief shading schemes from my typology because I limit my schemes to color areas directly associated with areal data, as described in the Introduction.

*My scheme types are greatly influenced by Olson through the classes I have taken with her and the research we have worked on together. I am building on concepts that were initially hers.

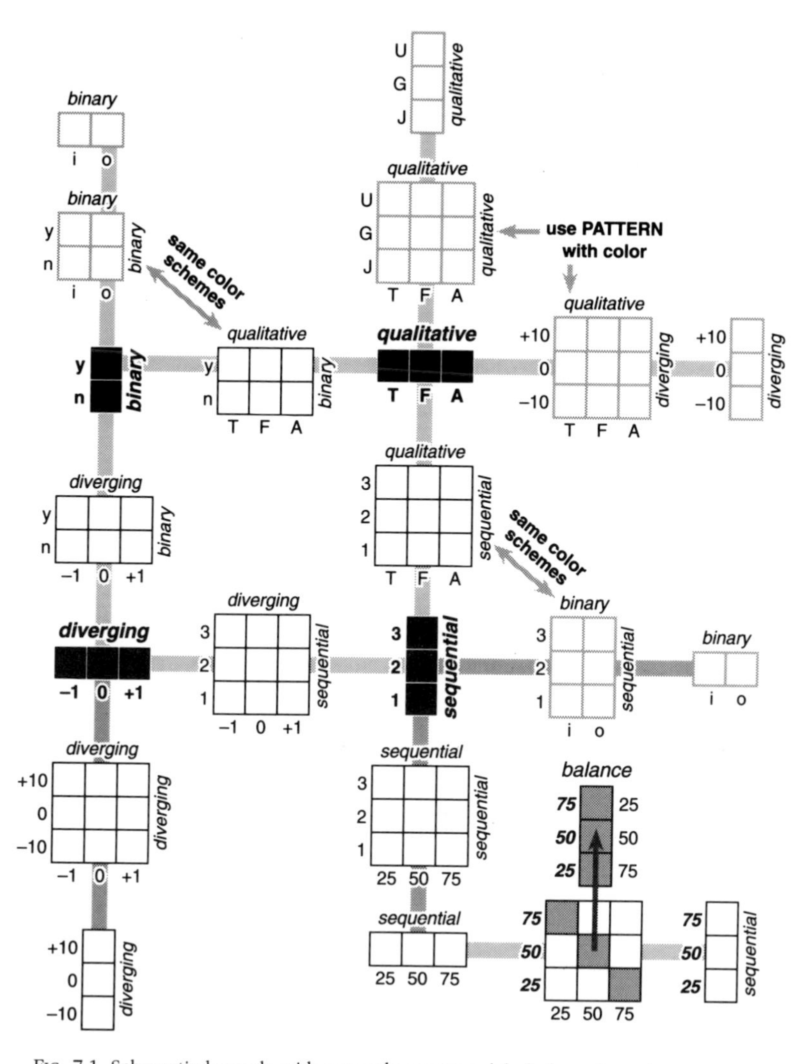

FIG. 7.1. Schematic legends with example category labels for color scheme types. One-variable scheme types are shown with black-filled legends. Binary schemes are a special case of qualitative schemes. Sequential and diverging are different conceptualizations of quantitative data that have corresponding scheme types. The figure shows all two-variable combinations of these four basic scheme types. Combination schemes described in the chapter are shown with black outlines. Combination schemes shown with gray outlines are omitted from the color scheme guidelines because they are either redundant or should be represented by a combination of color and pattern rather than color alone. The gray-filled balance legend is a special case of the sequential/sequential scheme type.

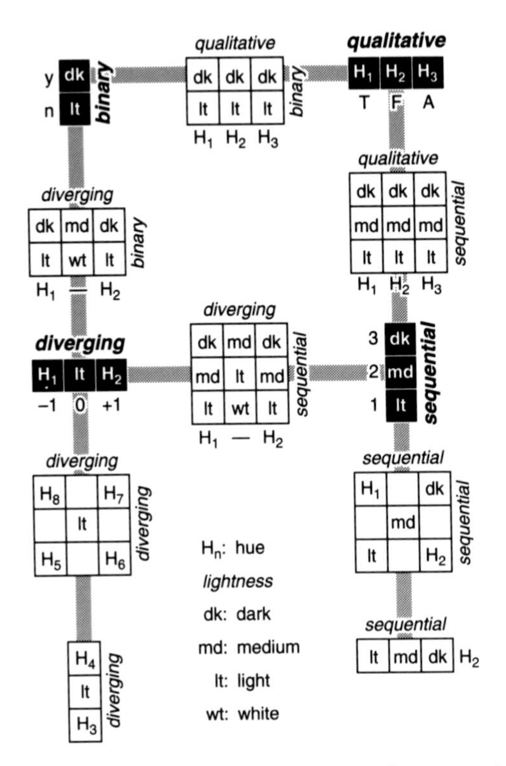

FIG. 7.2. Schematic legends with color use guidelines for one- and two-variable color schemes.

One-variable Color Schemes

I have selected four labels for the basic color scheme types: qualitative, binary, sequential and diverging. These four terms are concise and comprehensive and they should be readily understood in varied fields. They also have the advantage that they are all single-word adjectives, which makes them easier to use in the composite labels for two-variable schemes (diverging/sequential is simpler than double-ended/single-sequence). Table 7.3 summarizes the perceptual characteristics of the four basic scheme types that are linked with four data conceptualizations. This section is devoted to a detailed explanation of the information summarized in Table 7.3.

The four basic color scheme types are represented by black schematic legends in Figs 7.1 and 7.2. Figure 7.1 emphasizes categories of data organization and Fig. 7.2 shows color orderings for the schemes that are discussed in detail. These two figures summarize the differences between scheme types and illustrate the logic of the two-variable combinations that are discussed later.

The color figures (Figs 7.3–7.8) provide examples of individual schemes, as well as combination schemes. These color maps were constructed using an initial set of

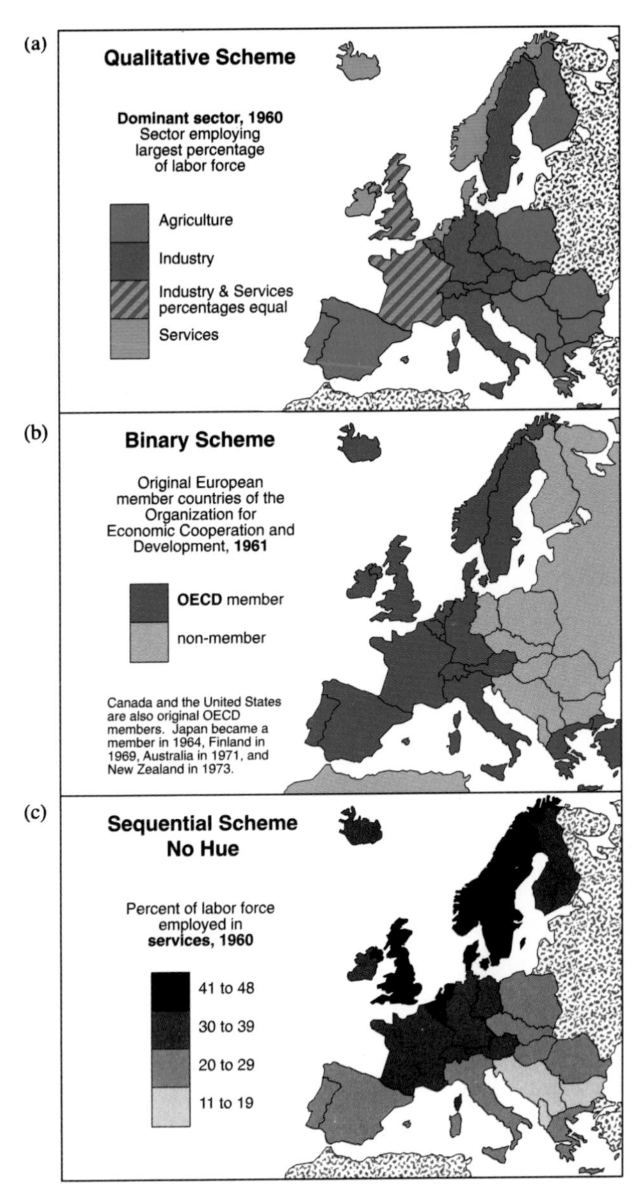

FIG. 7.3. One-variable schemes. (a) Qualitative scheme with three hues of similar lightness. (b) Binary scheme with one hue and lightness step. (c) Sequential scheme with lightness steps of neutral grays (data are classed by 10% increments of employment and legend colors are labeled with extreme data values in class).

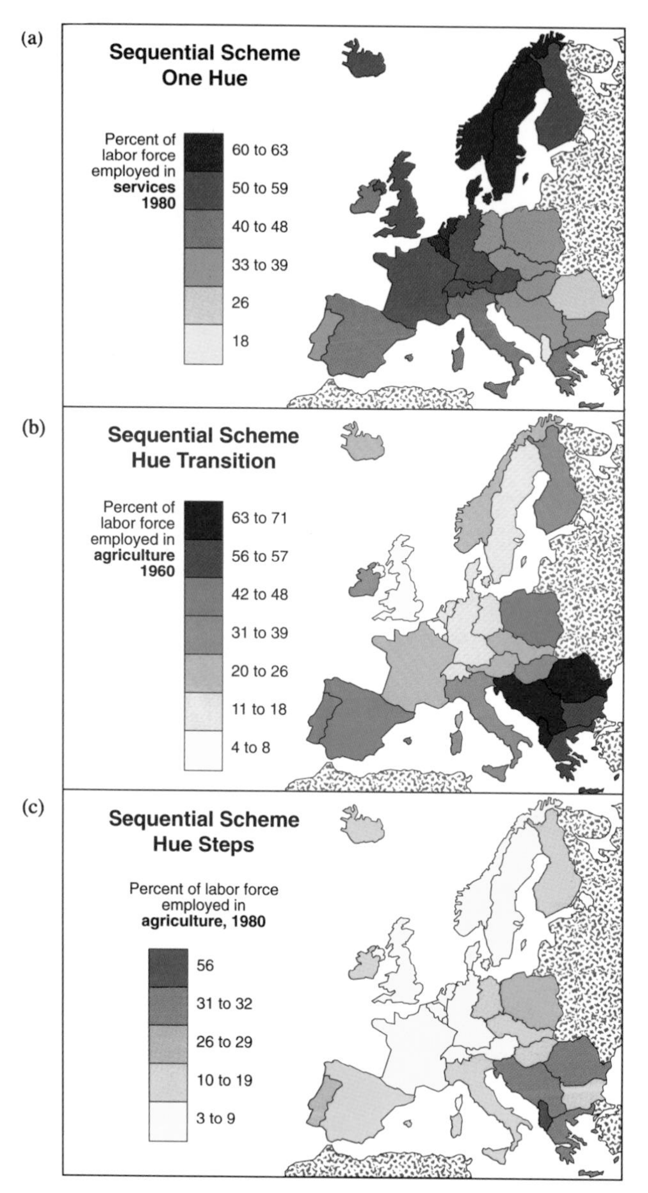

FIG. 7.4. One-variable sequential schemes. (a) Lightness steps of a single hue. (b) Lightness steps with a part-spectral transition in hue. (c) Lightness steps with hue steps that progress through all spectral hues but begin in the middle of the spectrum with low values represented by light yellow. Data are classed by 10% increments of employment and legend colors are labeled with extreme data values in each class.

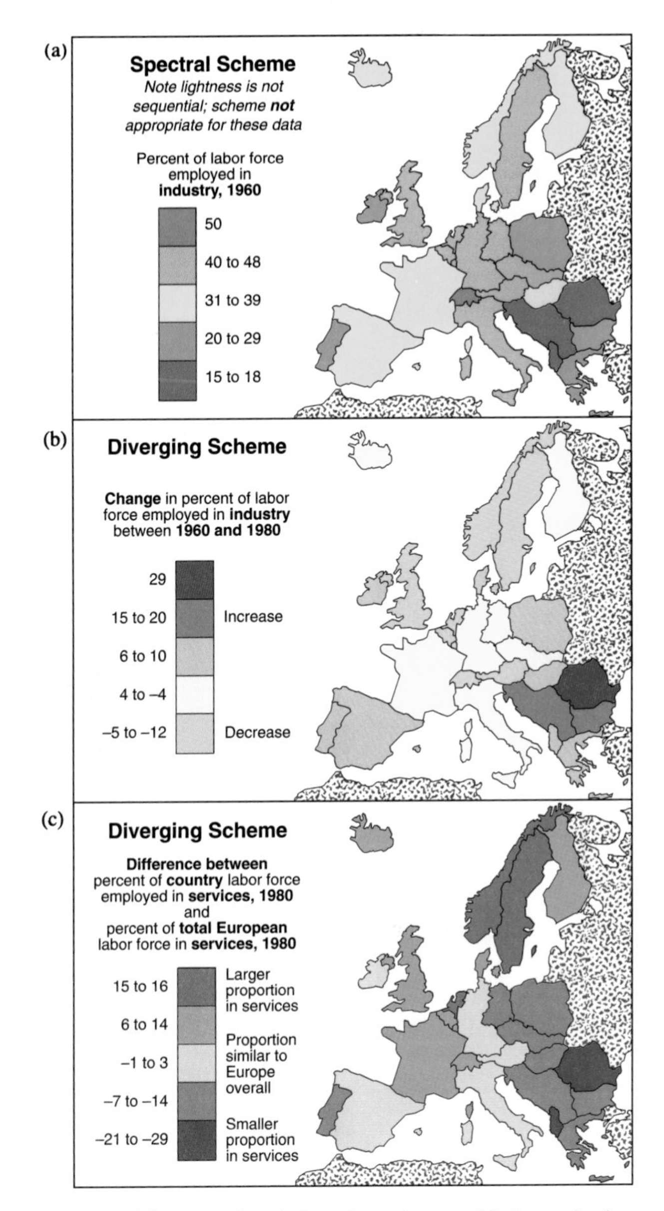

FIG. 7.5. One-variable spectral and diverging schemes. (a) Spectral scheme with both hue and lightness steps (poor choice because large hue steps interfere with groupings above and below the midpoint of a diverging scheme, and colors not appropriate as a sequential scheme because lightness steps are not arranged in a single sequence). (b) Two hues differentiate increase from decrease and lightness steps within each hue diverge from the lightest color that signifies minimal change. (c) Two hues with lightness steps diverging from the midpoint. Data are classed by 10% increments of employment and legend colors are labeled with extreme data values in each class.

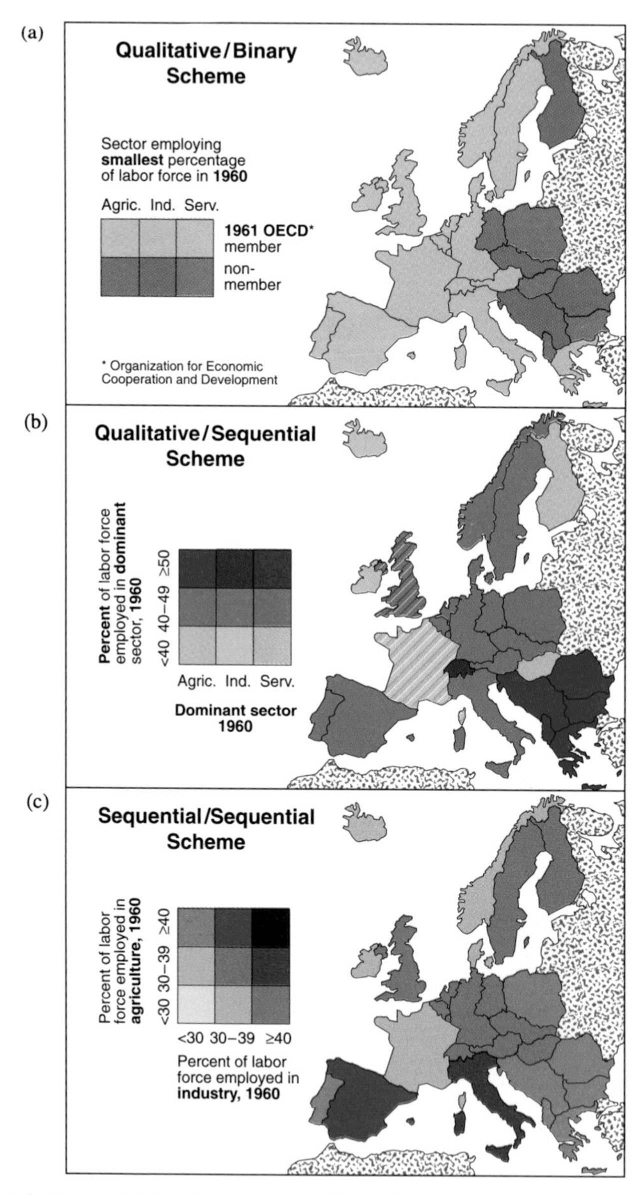

FIG. 7.6. Two-variable schemes. (a) Different hues for the qualitative variable crossed with a lightness step for the binary variable. (b) Different hues for the qualitative variable crossed with lightness steps for the sequential variable. (c) Sequential/sequential scheme with cross of lightness steps of two complementary hues with mixtures producing a neutral diagonal.

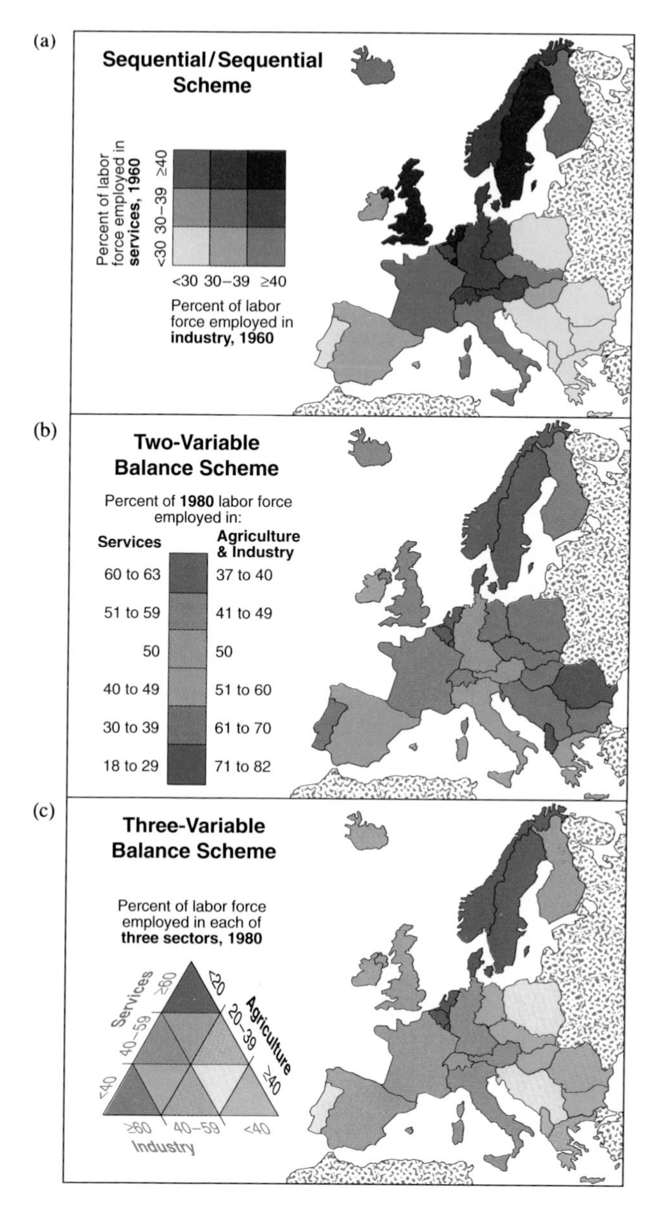

FIG. 7.7. Combinations of sequential schemes. (a) Sequential/sequential scheme with cross of lightness steps of two hues, resulting in transitional hue mixtures. (b) Two balanced quantitative variables represented by transition between two hues, with similar lightness throughout the scheme (colors similar to top-left to lower-right diagonal of Fig. 7.7a). (c) Mixtures of lightness steps of three hues used for three balanced quantitative variables.

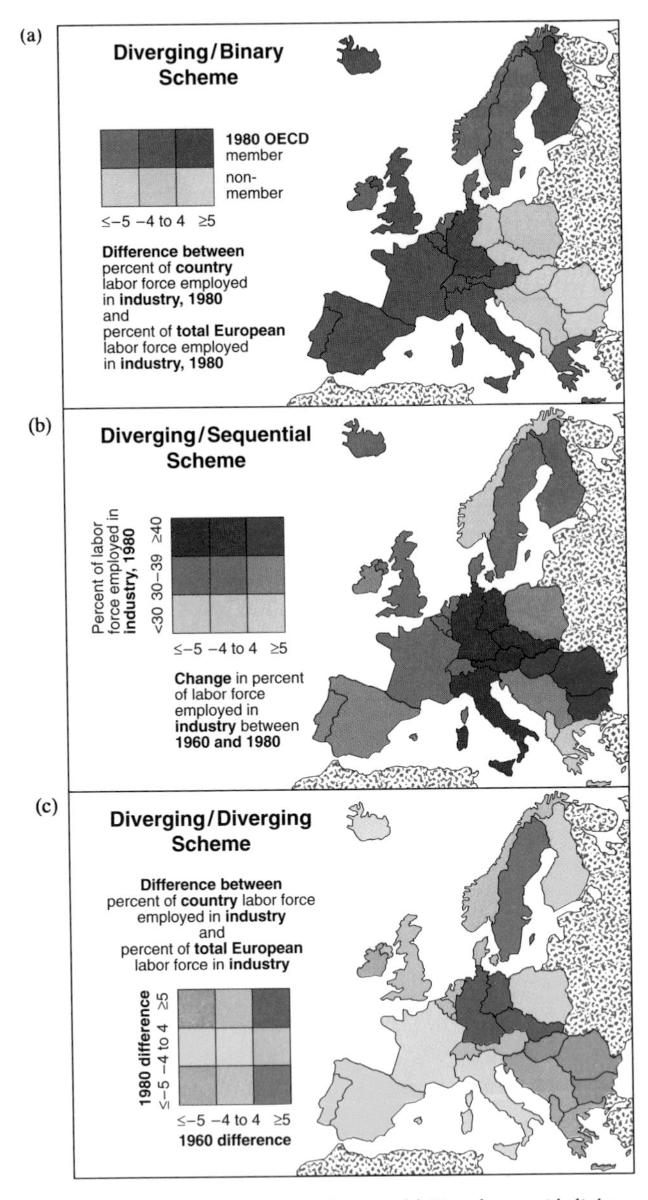

FIG. 7.8. Combinations with diverging schemes. (a) Two hues with lightness steps diverging from the midpoint of the quantitative variable crossed with a greater lightness step for the binary variable. (b) Two hues with lightness steps for the diverging variable crossed with greater lightness steps for the sequential variable. (c) Different hue at each corner with hue transitions for lighter midpoints is a logical cross of two diverging schemes.

seven variables: percentage of the labor force employed in agriculture, industry and services for countries of Europe in 1960 and 1980 (World Resources Institute 1990). Membership in the OECD (Organization for Economic Co-operation and Development 1990) is the binary variable chosen to complement the European employment data. These maps were constructed specifically for this chapter in the hope that their common theme and well known region will allow distinctions in data organizations and scheme types to be readily understood by an audience of diverse experience. These examples happen to be choropleth maps, but the color guidelines are intended for application to any type of map or graphic on which color symbolizes data.

Qualitative Schemes

Qualitative is the most commonly used term from the surveyed literature for nominal differences or differences in kind, and I feel it is the most unambiguous of the choices (Table 7.1). Nominal is a less widely used candidate common in cartography, but in lay contexts nominal also means insignificant or approximate and in statistical contexts it has a more precise implication as a measurement scale than is appropriate for the wide range of categorizations that may be mapped (Stevens's 1946 report summarizes the original definition of nominal as a measurement scale and also describes the frustrating process of coming to a consensus on terminology). Differences in hue should be used to represent qualitative differences in a data display. Green, blue and magenta hues of similar lightness and saturation are used for each of the qualitatively different employment categories shown in Fig. 7.3a (economic sector that employs the largest percentage of the labor force by country). Land use and land cover are other qualitative variables that are frequently mapped.

The lightness of the hues used for qualitative categories should be similar but not equal. Large differences in lightness or saturation between hues connote differences in importance and may inappropriately draw attention to categories that are not greater in significance than others on a qualitative map. Small lightness differences between hues, however, are essential for colors to be easily differentiated. The portion of the visual system responsible for distinguishing hues has poor edge and shape detection (Livingstone and Hubel 1988). Thus small differences in lightness allow area shapes and hue boundaries to be identified more easily, particularly on displays in which lines do not bound colored areas on the map. For example, it would be difficult to literally focus on the interwoven colors in France and the United Kingdom in Fig. 7.3a (where two sectors employ equal percentages of the labor force) if they were of identical lightness.

When dealing with numerous qualitative categories, it may be necessary to use large differences in lightness within and between hues to differentiate all map colors once they appear together on the map (also consider reducing the number of mapped categories or using pattern with color). Apply knowledge of the data to assign lightness differences wisely. Assign the lightest, darkest and most saturated colors to categories that warrant emphasis on the map. Categories that appear infrequently or as very small areas on a map (such as a lone wildlife sanctuary, a polluted dump site or narrow streams) will also benefit from greater lightness contrast or greater saturation.

Ordinal relationships may also be embedded within a qualitative categorization. For example, fallow and planted fields may be interpreted as intensities of agricultural activity. Express these differences with lightness steps of the same or similar hues, such as lighter and darker greens. These lightness differences are appropriate, although the map is considered to be generally qualitative and the majority of color differences are hue differences. Likewise, categories of greater similarity are appropriately represented by colors closer on the hue circle (for example, orange and red are adjacent hues of related appearance, whereas red and cyan are complementary or opposite hues with high contrast). Although the three-category qualitative map of Fig. 7.3a provides a fairly straightforward example, qualitative maps can be challenging problems that invite intelligent adjustments in lightness and saturation among their many hues.

Binary Schemes

I have isolated binary color schemes as a special case of schemes appropriate for qualitative data (Figs 7.1 and 7.2). Binary variables are nominal differences that are divided into only two categories, such as yes/no, present/absent, private/public or inside/outside. The primary perceptual difference between the two categories of a binary scheme may be a lightness step, unlike the use of hue for multi-valued qualitative variables. The lightness step choosen for a binary variable might be between, for example, gray and white (neutrals), red and white, light blue and darker blue, or light blue and darker green. No hue, one hue or two hues may be used, and the primary difference may be lightness contrast.

Figure 7.3b shows binary membership status in the OECD for European countries. One reason for isolating binary schemes as a separate category is that two is the greatest number of qualitative categories that may be effectively symbolized with color on two-variable maps that combine qualitative with diverging data or with a second qualitative variable (discussed in a later section). Another reason to identify binary as a special case is the explicit appropriateness of using lightness for this qualitative distinction. The map maker decides which category is more significant given the purpose of the map, and the more important of the two data categories should be darker. If neither binary category is of greater importance, treat the scheme as qualitative with only a slight lightness difference between two hues.

Sequential Schemes

Sequential data classes are logically arranged from high to low and this stepped sequence of categories should be represented by sequential lightness steps.

Usually, low data values are represented by light colors and high values are represented by dark colors. The association may be reversed so that light colors represent high data values when the overall display area is dark and high data values are emphasized by light colors that have maximum contrast within the display (this reversal is common to maps of remotely sensed data). Be aware that this reversal will contradict the expectations of a large proportion of the map audience (McGranaghan 1989) and should be made clear with an obvious legend. Regardless of whether high or low data values are lightest, it is imperative that the remaining colors form a lightness sequence that corresponds with the order of the data categories. Figures 7.3c, 7.4a, 7.4b and 7.4c present sequential schemes for quantitative sequences of high to low percentage employment.

The simplest sequential scheme (Fig. 7.3c) progresses from dark to light gray (neutral colors with no hue). A black to white sequence may also be used if a greater contrast range is desired to accommodate more classes. This choice, however, has the disadvantages that black lines are obscured and white areas with no data (such as water bodies in Fig. 7.3c) will be colored the same as the lowest class.

Figure 7.4a presents an example scheme of one hue, orange, that varies in lightness. In my teaching, I have found that the most common difficulty novices have in designing lightness sequences comes from beginning with a fully saturated hue in the lightest and lowest data category. A saturated hue is not usually a light color and thus it limits the contrast range available for the remaining categories. Conversely, novices may place a fully saturated hue in the darkest and highest data category and, again, they are restricted by limited lightness contrast between white and the saturated hue. In both cases the resulting maps are prone to failure because colors will not be differentiable once they appear together on the map.

Most hues are available only as desaturated colors when they are light and dark (exceptions are the high saturations available for light yellow and for dark purple–blues). Therefore, large lightness steps between light and dark colors are usually accompanied by a saturation progression from low saturation light colors to higher saturation at mid-lightness and back to low saturation for darker categories (Fig. 7.4a). I do not recommend depending on saturation alone to distinguish more than two (maybe three) categories because of the limited contrast range available with saturation (MacEachren 1992 describes the intuitive appropriateness of using saturation to represent data uncertainty). Similarly, do not ignore saturation when selecting colors. If saturation changes are systematic throughout sequential schemes and if high saturation colors do not overemphasize unimportant categories, this third color dimension is helpful in increasing contrast between categories.

More than one hue may be used in a sequential scheme, but it is important that hue differences are subordinate to lightness differences between categories. The light to dark progression should dominate the look of the map with the accompanying transition in hues increasing contrast between categories. Figure 7.4b presents seven categories with sequential lightness steps and with a transition from light yellow to dark purple–blue through greens. The hue differences are small and the transition moves through only part of the color circle or spectrum. Another useful transition is from yellow through orange to dark red. These are the longest hue ranges that I recommend for sequential schemes. With careful attention to lightness, sequential schemes may be constructed that use the entire color circle from light yellow through green, blue, purple, red and brown (dark orange) or from light yellow through orange, red, purple, blue and dark green (Fig. 7.4c).

Do not use the saturated spectrum as a sequential scheme. This advice is not new; Bertin discusses examples that demonstrate problems with spectral schemes in his 1977 writings (translation 1981). The opinion is also not unique to cartographers. For example, Livingstone works in physiology and also criticizes spectral schemes (her critique appeared in the popular media; Grady 1993). The order of hues associated with the visible electromagnetic spectrum begins with red and ends with purple–blue (Fig. 7.5a), but it is extremely difficult to produce a successful sequential scheme that moves through the spectrum from red to purple–blue in a sequential manner from light to dark. The difficulty with this scheme is that saturated yellow (at the middle of the spectrum) is an intrinsically light hue and a darkened yellow is by necessity desaturated. Thus sequential lightness steps force a desaturated hue to the middle of the scheme and it is very difficult to create transitions to this murky yellow that provide colors resembling the saturated spectrum.

In the past, the use of spectral schemes was dictated by hardware limitations. With the current availability of eight-bit color and better, computing technology that is restricted to a limited set of spectral colors is not appropriate for producing complex data visualizations and map displays. Instead of spectral schemes, use simple transitions between a few adjacent hues to effectively enliven a lightness sequence, increase contrast between categories and ensure that colors on the map fall in an obvious perceptual order.

Diverging Schemes

The emphasis of a quantitative data display may be progressions outward from a critical midpoint of the data range. Example data that are appropriately represented with diverging lightness steps are deviations above and below a mean, median, or zero point. For example, the most effective way to map residuals from a regression model is to use one hue for positive residuals, use another hue for negative residuals and to darken these hues for larger positive and negative residuals. Weber and Buttenfield (1993) make good use of a red–gray–blue diverging scheme in their animation of 90 years of deviations from mean US surface temperatures. I recommend the term diverging for this type of scheme instead of double-ended, which has been used in the cartography literature. In addition to the advantage of being a single word, the term diverging emphasizes the importance of critical values within the data range, rather than the extremes at the ends of the range.

Figures 7.5b and c both present example data that are mapped with diverging schemes. Figure 7.5b shows changes in percentage industrial employment between

1960 and 1980. Note that the lightest class, a pale yellow, straddles zero change. This class is not at the midpoint of the mapped data range. The importance of this class is its role as the break between categories of increase and decrease in industrial employment. Likewise, Fig. 7.5c emphasizes the difference in services employment between each country and the entire European labor force. The overall pattern of high to low data values is the same as that of Fig. 7.4a, which shows percentage services employment for the same year with a sequential scheme. The difference is that the comparison value for Europe as a whole (47%) is not given graphic emphasis with the sequential scheme.

Positive and negative values are not a prerequisite for the application of a diverging scheme. Examples of data that do not diverge from a zero point are voting results (50% being the critical value with two parties) and median incomes above and below the poverty level (or above and below another significant income level). Likewise, the existence of a zero value in a data range may be an artifact of the measurement scale. Zero may not be conceptually important and, thus, does not warrant emphasis within a diverging scheme. For example, 0° Fahrenheit is probably not a critical break point on a temperature map, but the freezing point (at 32°F) may deserve emphasis with a diverging scheme. The decision of whether or not to present quantitative data as sequential or diverging is often subjective and depends on the attributes of the mapped data to be emphasized in explorations or communicated to an audience. In an interactive environment, alternating between sequential and diverging schemes representing the same data can reveal different aspects of the data and aid data investigation.

Understanding diverging schemes as two sequential schemes running end to end encourages greater freedom in their construction. Pair a sequence of neutrals with a hue sequence (such as dark gray to white and white to dark blue), pair two hue sequences (Fig. 7.5c), or pair two hue transitions (such as green to yellow and yellow to orange). There is considerable freedom in the lightness sequences you may pair, but they must converge systematically on a shared light hue or light neutral (it is not mandatory that the critical value has its own class; it may be at the break between two classes and be between lightness sequences). In diverging schemes, classes similar in absolute value above and below the critical value should have similar lightness and saturation levels so they are perceived as similar in magnitude.

Another look at the example spectral scheme in Fig. 7.5a shows that its lightness characteristics are akin to those of a diverging scheme with two wide-ranging and stepped hue progressions. When a spectral scheme is used inappropriately as a sequential scheme, however, the lightest color (yellow) does not usually mark a critical value of the data range.

Two-variable Color Schemes

Analysts are seldom content to examine one variable at a time, particularly in an interactive exploratory visualization context. Two-variable mapping allows detailed comparisons of distributions. In Chapter 13, for example, Koussoulakou describes her interactive use of composite maps that combine varied data on air pollution, the physical environment and urban settlements. Elsewhere, MacEachren *et al.* (1993) describe the construction of an interactive visualization program that allows changes between different combination schemes to show relationships between water pollution data and measures of data uncertainty. The schematic legends in Fig. 7.1 are aids to understanding how variables are combined for all of the possible pairings of the basic scheme types. The black outline schematic legends in Fig. 7.2 provide color use guidelines for the following combination schemes for two-variable mapping:

- qualitative/binary
- qualitative/sequential
- sequential/sequential (balance scheme is a special case of sequential/sequential)
- diverging/sequential
- diverging/diverging
- diverging/binary

Qualitative/Binary Schemes

The cross of qualitative and binary schemes produces a set of hues with light and dark versions of each hue that correspond to the value of the binary variable that is simultaneously mapped. Figure 7.6a presents a qualitative/binary scheme (the economic sector employing the smallest percentage of the labor force and country membership in the OECD). A multi-hued vegetation map with darker hues for vegetation on public lands and lighter hues for vegetation on private lands is another example of a qualitative/binary scheme. Increasing the saturation of all hues for the aspect of the binary variable to be emphasized will improve visual emphasis and coherence within the map.

Qualitative/Sequential Schemes

Figure 7.6b presents a qualitative/sequential scheme (dominant employment sector is shown with three hues and the percentage employment in the dominant sectors is shown with sequences of lightness steps within each hue). Note that the same pattern of hues is seen in Fig. 7.3a, but Fig. 7.6b offers added information about the magnitude of sector dominance. Similarly, population percentages of varied dominant ethnic groups or religions are well represented with qualitative/sequential schemes.

Duplicate Schemes

The combination schemes that do not appear in the above list are shown in Fig. 7.1 as schematic legends with gray outlines. The binary/binary combination can be represented with a scheme having the same perceptual characteristics as a qualitative/binary scheme. One binary variable is represented by a difference in lightness and the other is represented by a difference in hue, as a multi-valued qualitative variable would be represented. The binary/sequential combination is represented by lightness steps for the quantitative variable and a difference in hue for the binary qualitative variable, which are the same characteristics used in a qualitative/sequential scheme. To avoid redundant descriptions, separate color use guidelines are not provided for binary/binary and binary/sequential schemes in Fig. 7.2.

Sequential/Sequential Schemes

Of the many two-variable schemes discussed, the combination of two sequential schemes has received the most attention in the literature (Olson 1975; 1981; Wainer and Francolini 1980; Trumbo 1981; Eyton 1984). Figures 7.6c and 7.7a show sequential/sequential schemes that cross percentages of employment in two sectors (agriculture and industry in Fig. 7.6c and services and industry in Fig. 7.7a). Mapping two quantitative variables together emphasizes locations where both variables are high valued, where both are low valued, and where one variable is high while the other is low. In these two example figures, colors along the positive sloping diagonals of the legends (grays in Fig. 7.6c and purple–blues in Fig. 7.7a) mark countries that have similar employment levels in the sectors compared. Alternatively, the two variables may be measured along different scales, such as the education and income variables mapped together by the US Bureau of the Census in the 1970s (Meyer *et al.* 1975). Eyton relates the arrangement of sequential/sequential legends to data distributions within statistical scatter diagrams.

A sequential/sequential scheme may be considered as the logical mix of all combinations of the colors in two sequential schemes. Thus the schemes are based on two hues. If the two hues crossed are approximate complements, their mixtures produce a neutral diagonal and desaturated transitional colors. Figure 7.6c shows the combination of orange–yellow and blue lightness sequences that mix to neutrals on the diagonal. If the two hues are subtractive primaries, their mixtures form a third hue (magenta and cyan sequences produce a variety of transitional purple–blues in Fig. 7.7a). In constructing sequential/sequential schemes, be sure to include systematic lightness differences throughout the scheme and do not depend on hue to reveal or impart differences in magnitude. Use hue transitions to designate differences in proportions of the two variables mapped.

Balance Schemes

Eastman (1986) compares balance and bipolar color schemes, and he uses these two terms to describe different scheme types within Olson's (1987) double-ended category (bipolar is equivalent to the diverging scheme category). A balance scheme emphasizes both ends of the sequence with different hues and remaining categories are progressive mixtures of these two hues. Figure 7.7b provides an example of a balance scheme: services employment is compared with the sum of agricultural and industrial employment and these variables always sum to 100%. Thus a rate of 60% employment in services is balanced by 40% combined employment in agriculture and industry. The example Eastman offers is percentage English speakers balanced by percentage French speakers, with the two components adding to 100% in a region where these are the only languages spoken.

A balance scheme is used to compare two quantitative variables and I consider it to be a special case of the sequential/sequential scheme in which the only classes that occur on the map are from the diagonal of the legend (note the three gray-filled legend boxes in the example sequential/sequential scheme in the lower right of Fig. 7.1). A class centered on 75% will occur only with a 25% class because the values sum to 100%. Cartographers often do not include classes in a legend for which values do not appear on the map. Thus with balanced variables it is reasonable to show only the classes from the diagonal of a corresponding sequential/sequential legend and arrange the classes in a column, a form common for one-variable maps. Compare the colors from the top-left to lower-right diagonal of the legend of Fig. 7.7a to the colors arranged in a column in Fig. 7.7b. Similar mixtures of magenta and cyan are used with some added lightness contrast to improve color differences with the increased number of classes.

Understanding the balance scheme as a special case of a sequential/sequential scheme should aid in its design. However, I recommend choosing to emphasize the high end of one of the balanced variables and using a standard sequential scheme to obtain better lightness contrast between categories. Alternatively, if the evenly balanced midpoint is a critical value in the classification, use a standard diverging scheme, which will place a much stronger visual emphasis on the midpoint than does the mixture hue of the balance scheme.

The only three-variable map included in this chapter is the special case of balanced proportions of three variables that sum to 100% (Fig. 7.7c shows agricultural, industrial and services employment percentages). In the example, the three variables are represented by subtractive primaries and their mixtures represent varied proportions of employment in the three sectors. Other variables that are appropriate for this sort of representation are voting results in a three-party system or proportions of sand, silt and clay that combine to characterize soil types (graphs of this form are sometimes called ternary diagrams). Variables that are not constrained to sum to 100%, or an otherwise defined whole, will produce category combinations that do not fall within the triangular legend. The potential perceptual organization appropriate for mapping three quantitative variables that are not balanced results in a cube-shaped legend that uses a volume from three-dimensional color space. The resulting diversity of colors takes us beyond the goal of ready interpretation sought by recommending the careful matching of perceptual and data organizations.

Diverging/Binary and Diverging/Sequential Schemes

Diverging/binary and diverging/sequential schemes have the same perceptual characteristics (they remain as distinct schemes in Figs 7.1 and 7.2 because the first is a combination of quantitative and qualitative variables and the second is a combination of two quantitative variables). Figure 7.8a and b are examples of diverging/binary and diverging/sequential schemes, respectively. Figure 7.8a shows differences from the European labor force in industrial employment for countries that are further distinguished by OECD membership. Figure 7.8b shows changes in industrial employment between 1960 and 1980 crossed with percentage industrial employment in 1980 (adding knowledge of long-term growth or decline in the sector to information about absolute employment rates).

The success of the schemes in Fig 7.8a and b hinges on the large contrast range available in the lightness dimension. In each map, large lightness steps are used for the binary or sequential variable. Smaller lightness steps that are bolstered by a change in hue represent the diverging component of the scheme within each large lightness step of the comparison variable. Compare Fig. 7.8a and b with the schematic guidelines in Fig. 7.2 and note the variation in lightness within each larger lightness step. This variation in lightness and the use of a desaturated midpoint hue distinguish these schemes from the qualitative/binary and qualitative/sequential schemes (Fig. 7.6a and b). Note that hue is also used differently between the sequential/sequential schemes (Figs 7.6c and 7.7a) and the diverging/sequential scheme (Fig. 7.8b). Rather than positioning hues at opposite corners, hues are constant within columns of the Fig. 7.8b legend. In addition, no transitional hues are present in the three by three diverging/sequential legend.

Diverging/Diverging Schemes

The diverging/diverging scheme is the only two-variable scheme that departs from the idea of a direct overlay of the component one-variable schemes. Compare the schematic diverging/diverging legends of Figs 7.1 and 7.2 to the color example in Fig. 7.8c (the map compares industrial employment differences between countries and Europe overall for 1960 and for 1980). Different moderately dark hues are placed at each of the four corners of the legend to represent categories that are extremes for both variables. A very light or white color is placed at the center of the legend, creating an appropriately light color for the class that contains the critical value or midpoint of both variables. The remaining colors are lighter than the corners because they contain the midpoint of one of the two variables and are transitional hues that lie between adjacent corner hues. The color circle or spectrum is essentially stretched around the perimeter of the legend and lightness adjusted in response to critical values within the data ranges of both variables. Other example data appropriate for a diverging/diverging scheme are two quantitative variables classed by standard deviations above and below the means.

Qualitative/Qualitative and Qualitative/Diverging Schemes

The diverging/diverging scheme must depart from the logic of crossing two one-variable schemes because it requires that hue differences represent both variables. Likewise, qualitative/qualitative and qualitative/diverging schemes (gray outline legends in the upper right of Fig. 7.1) require changes in hue in perpendicular directions as their component schemes intersect. Here we reach the limits of what can be presented with color in a logically organized manner. If you have data that fall into these categories of two-variable schemes, I suggest using a pattern difference for one of the pair of variables that you wish to compare.

Summary

I have presented a comprehensive set of color scheme types and corresponding guidelines for the use of hue and lightness for each scheme (Fig. 7.2):

- qualitative
- binary
- sequential
- diverging
- qualitative/binary
- qualitative/sequential
- sequential/sequential
- diverging/sequential
- diverging/diverging
- diverging/binary

The schemes are matched with parallel conceptualizations of data. Examining data with different schemes may reveal different characteristics of distributions and their interrelationships. Software that allows interactive switching between scheme types will facilitate accurate and thorough understanding through data visualization.

Random or perceptually ill-fit assignments of colors to combinations of two variables remains a possible "solution" to mapping problems. This solution, however, will mask the interrelationships that the map-maker should be attempting to illuminate by mapping variables together. Better results would be produced by comparing maps of the individual variables (displayed with suitable schemes). The characteristics of both distributions will be rendered indecipherable by failing to organize hue, lightness and saturation in a way that corresponds with logical orderings within the mapped variables. A disorderly jumble of colors produces a map that is little more than a spatially arranged look-up table. The goal of this chapter is to help you do better than that by using color with skill.

Acknowledgements

Discussions with Judy Olson, David DiBiase and Alan MacEachren had important influences on the development of this typology of color schemes. Richard Becker's commentary on an earlier version of the material (presented at the GIS/LIS'92 conference in San José, California) helped to clarify the importance of the midpoint in diverging schemes. Cheryl Rogers assisted with the literature search.

References

Bertin, J. (1981) *Graphics and Graphic Information Processing* (Berg, W. J. and P. Scott translators; original copyright 1977), Walter de Gruyter, New York.

Brewer, C. A. and K. A. Marlow (1993) "Color representation of slope and aspect simultaneously", *Proceedings, Auto-Carto 11*, Minneapolis, pp. 328–337.

Campbell, J. (1991) *Introductory Cartography*, 2nd edn, Wm C. Brown, Dubuque.

Dent, B. D. (1993) *Cartography: Thematic Map Design*, 3rd edn, Wm C. Brown, Dubuque.

Eastman, J. R. (1986) "Opponent process theory and syntax for qualitative relationships in quantitative series", *The American Cartographer*, Vol. 13, No. 4, pp. 324–333.

Eyton, J. R. (1984) "Complementary-color, two-variable maps", *Annals of the Association of American Geographers*, Vol. 74, No. 3, pp. 477–490 (and folded supplement).

Grady, D. (1993) "The vision thing: mainly in the brain", *Discover*, Vol. 14, No. 6, pp. 57–66.

Livingstone, M. and D. Hubel (1988) "Segregation of form, color, movement, and depth: anatomy, physiology, and perception", *Science*, Vol. 240, pp. 740–749.

MacEachren, A. M. (1992) "Visualizing uncertain information", *Cartographic Perspectives*, Vol. 13, pp. 10–19.

MacEachren, A. M., D. Howard, M. Von Wyss, D. Askov and T. Taormino (1993) "Visualizing the health of Chesapeake Bay: an uncertain endeavor", *GIS/LIS'93 Proceedings*, Minneapolis, Vol. 1, pp. 449–458.

McGranaghan, M. (1989) "Ordering choropleth map symbols: the effect of background", *The American Cartographer*, Vol. 16, No. 4, pp. 279–285.

Mersey, J. E. (1990) "Colour and thematic map design: the role of colour scheme and map complexity in choropleth map communication", *Cartographica*, Vol. 27, No. 3, Monograph 41.

Meyer, A. M., F. R. Broome and R. H. Schweitzer Jr (1975) "Color statistical mapping by the U.S. Bureau of the Census", *The American Cartographer*, Vol. 2, No. 2, pp. 100–117.

Monmonier, M. (1993) *Mapping It Out: Expository Cartography for the Humanities and Social Sciences*, University of Chicago Press, Chicago.

Muehrcke, P. C. and J. O. Muehrcke. (1992) *Map Use: Reading, Analysis, and Interpretation*, 3rd edn, JP Publications, Madison.

Olson, J. M. (1975) "The organization of color on two-variable maps", *Proceedings, Auto-Carto 2*, Reston, UA, pp. 289–294, 251, 264–266.

Olson, J. M. (1981) "Spectrally encoded two-variable maps", *Annals of the American Association of Geographers*, Vol. 71, No. 2, pp. 259–275.

Olson, J. M. (1987) "Color and the computer in cartography", in Durrett, H. J. (ed.), *Color and the Computer*, Academic Press, Boston, pp. 205–219.

Organization for Economic Co-operation and Development (OECD) (1990) *Labour Force Statistics 1968–1988*, Department of Econonomics and Statistics, OECD, Paris.

Robinson, A. (1952) *The Look of Maps: an Examination of Cartographic Design*, The University of Wisconsin Press, Madison.

Robinson, A. H., R. D. Sale, J. L. Morrison and P. C. Muehrcke (1984) *Elements of Cartography*, 5th edn, Wiley, New York.

Stevens, S. S. (1946) "On the theory of scales of measurement", *Science*, Vol. 103, No. 2684, pp. 677–680.

Trumbo, B. E. (1981) "A theory for coloring bivariate statistical maps", *The American Statistician*, Vol. 35, No. 4, pp. 220–226.

Wainer, H. and C. M. Francolini (1980) "An empirical inquiry concerning human understanding of two-variable color maps", *The American Statistician*, Vol. 34, No. 2, pp. 81–93.

Weber, C. R. and B. P. Buttenfield (1993) "A cartographic animation of average yearly surface temperatures for the 48 contiguous United States: 1897–1986", *Cartography and Geographic Information Systems*, Vol. 20, No. 3, pp. 141–150.

World Resources Institute (1990) *World Resources 1990–1*, Oxford University Press, New York.

Bibliography

Carr, D. B., W. L. Nicholson, R. J. Littlefield and D. L. Hall. (1986) "Interactive color display methods for multivariate data" in Wegman, E. J. and D. J. DePriest (eds.), *Statistical Image Processing and Graphics*, Marcel Dekker, New York, pp. 215–250.

Castner, H. W. (1980) "Printed color charts: some thoughts on their construction and use in map design", *ACSM-ASP Technical Papers*, pp. 370–378.

Coughran Jr, W. M. and E. Grosse (1990) "Techniques for scientific animation", in Farrell, E. J. (ed.), *Extracting Meaning from Complex Data: Processing, Display, Interaction*, Vol. 1259, The International Society of Optical Engineering (SPIE), Santa Clara, pp. 72–79.

Cuff, D. J. (1972) "Value versus chroma in color schemes on quantitative maps", *The Canadian Cartographer*, Vol. 9, No. 2, pp. 134–140.

Cuff, D. J. (1973) "Colour on temperature maps", *The Cartographic Journal*, Vol. 10, No. 1, pp. 17–21.

Cuff, D. J. (1974a) "Perception of color sequences on maps of atmospheric pressure", *The Professional Geographer*, Vol. 26, No. 2, pp. 166–171.

Cuff, D. J. (1974b) "Impending conflict in color guidelines for maps of statistical surfaces", *The Canadian Cartographer*, Vol. 11, No. 1, pp. 54–58.

de Valk, J. P. J., W. J. M. Epping and A. Heringa (1985) "Colour representation of biomedical data", *Medical and Biological Engineering and Computing*, Vol. 23, No. 7, pp. 343–351.

Farrell, E. J. (1983) "Color display and interactive interpretation of three-dimensional data", *IBM Journal of Research and Development*, Vol. 27, No. 4, pp. 356–366.

Farrell, E. J. (1987) "Visual interpretation of complex data", *IBM Systems Journal*, Vol. 26, No. 2, pp. 174–200.

Gilmartin, P. P. (1988) "The design of choropleth shadings for maps on 2- and 4-bit color graphics monitors", *Cartographica*, Vol. 25, No. 4, pp. 1–10.

Gilmartin, P. P. and E. Shelton (1989) "Choropleth maps on high resolution CRT's. The effects of number of classes and hue on communication", *Cartographica*, Vol. 26, No. 2, pp. 40–52.

Graedel, T. E. and R. McGill. (1982) "Graphical presentation of results from scientific computer models", *Science*, Vol. 215, No. 5, pp. 1191–1198.

Hopkin, V. D. (1992) "Issues in color application", in Widdel, H. and D. L. Post (eds.), *Color in Electronic Displays*, Plenum Press, New York, pp. 191–207.

Horton, W. (1991) *Illustrating Computer Documentation: the Art of Presenting Information Graphically on Paper and Online*, Wiley, New York.

Levkowitz, H. and G. T. Herman (1986) "Color in multidimensional multiparameter medical imaging", *Color Research and Application*, Vol. 11(suppl.), pp. S15–S20.

Moellering, H. and A. J. Kimerling (1990) "A new digital slope-aspect display process", *Cartography and Geographic Information Systems*, Vol. 17, No. 2, pp. 151–159.

Muller, J. C. and Z. Wang (1990) "A knowledge based system for cartographic symbol design", *The Cartographic Journal*, Vol. 27, No. 1, pp. 24–30.

Rheingans, P. (1992) "Color, change, and control for quanititative data display", *Proceedings of Visualization '92*, IEEE Computer Society Technical Committee on Computer Graphics, Boston, pp. 252–259.

Rice, J. F. (1991) "Ten rules for color coding", *Infomation Display*, Vol. 7, No. 3, pp. 12–14.

Robertson, P. K. (1991) "A methodology for choosing data representations", *IEEE Computer Graphics and Applications*, Vol. 11, No. 3, pp. 56–67.

Robertson, P. K. and J. F. O'Callaghan (1986) "The generation of color sequences for univariate and bivariate mapping", *IEEE Computer Graphics and Applications*, Vol. 6, No. 2, pp. 24–32.

Robinson, A. H. (1967) "Psychological aspects of color in cartography", *International Yearbook of Cartography*, Vol. 7, pp. 50–61.

Taylor, J. M. and G. M. Murch (1986) "The effective use of color in visual displays: text and graphics applications", *Color Research and Application*, Vol. 11(suppl.), pp. S3–S10.

Thorell, L. G. and W. J. Smith (1990) *Using Computer Color Effectively: an Illustrated Reference*, Prentice Hall, Englewood Cliffs.

Travis, D. (1991) *Effective Color Displays: Theory and Practice*, Academic Press, New York.

Ware, C. (1988) "Color sequences for univariate maps: theory, experiments, and principles", *IEEE Computer Graphics and Applications*, Vol. 8, No. 5, pp. 41–49.

Ware, C. and J. C. Beatty (1986) "Using colour to display structures in multidimensional discrete data", *Color Research and Application*, Vol. 11(suppl.), pp. S11–S14."

Ware, C. and J. C. Beatty (1988) "Using color dimensions to display data dimensions", *Human Factors*, Vol. 30, No. 2, pp. 127–142.

Weibel, R. and B. P. Buttenfield (1992) "Improvement of GIS graphics for analysis and decision-making", *International Journal of Geograpical Information Systems*, Vol. 6, No. 3, pp. 223–245.

CHAPTER 8

SOUND AND GEOGRAPHIC VISUALIZATION

JOHN B. KRYGIER*

Department of Geography
The Pennsylvania State University
302 Walker Building
University Park, PA 16802, USA

Who the hell wants to hear actors talk?[1]

Introduction

The issue of sound in a book on visualization may at first seem incongruous. There is, however, evidence to support the claim that sound is a viable means of representing and communicating information and can serve as a valuable addition to visual displays. Abstracted two-dimensional space and the visual variables — the traditional purview of cartography — may not always be adequate for meeting the visualization needs of geographers and other researchers interested in complex dynamic and multivariate phenomena. The current generation of computer hardware and software gives cartographers access to a broadened range of design options: three-dimensionality, time (animation), interactivity and sound. Sound — used alone or in tandem with two- or three-dimensional abstract space, the visual variables, time, and interactivity — provides a means of expanding the representational repertoire of cartography and visualization.

This chapter discusses the use of realistic and abstract sound for geographic visualization applications. Examples of how and why sound may be useful are developed and discussed. Uses of sound in geographic visualization include sound as vocal narration, a mimetic symbol, a redundant variable, a means of detecting anomalies, a means of reducing visual distraction, a cue to reordered data, an

*e-mail: jbk5@psuvm.psu.edu

alternative to visual patterns, an alarm or monitor, a means of adding non-visual data dimensions to interactive visual displays and for representing locations in a sound space. The chapter concludes with research issues concerning sound and its use in geographic visualization.

Experiencing and Using Sound to Represent Data

Our sense of vision often seems much more dominant than our sense of hearing. Yet one only has to think about the everyday environment of sound surrounding us to realize that the sonic aspects of space have been undervalued compared with the visual (Ackerman 1990; Tuan 1993). Consider the experience of the visually impaired to appreciate the importance of sound and how it aids in understanding our environment. Also consider that human communication is primarily carried out via speech and that we commonly use audio cues in our day to day lives — from the honk of a car horn to the beep of a computer to the snarl of an angry dog as we approach it in the dark (Baecker and Buxton 1987).

There are several perspectives which can contribute to understanding the use of sound for representing data. Acoustic and psychological perspectives provide insights into the physiological and perceptual possibilities of hearing (Truax 1984; Handel 1989). An environmental or geographical perspective on sound can be used to examine our day to day experience with sound and to explore how such experiential sound can be applied to geographic visualization (Ohlson 1976; Schafer 1977, 1985; Porteous and Mastin 1985; Gaver 1988; Pocock 1989). Understanding how sound and music is used in non-western cultures may inform our understanding of communication with sound (Herzog 1945; Cowan 1948). Knowledge about music composition and perception provides a valuable perspective on the design and implementation of complicated, multivariate sound displays (Deutsch 1982). Many of these different perspectives have coalesced in the cross-disciplinary study of sound as a means of data representation, referred to as sonification, acoustic visualization, auditory display and auditory data representation (Frysinger 1990). Within this context both realistic and abstract uses of sound are considered.

Using Realistic Sounds

Vocal narration is an obvious and important use of realistic sound.[2] Details about the physiological, perceptual and cognitive aspects of speech are well known (Truax 1984; Handel 1989) and film studies offer insights into the nature and application of vocal narration (Stam *et al.* 1992).

Another use of realistic sounds is as mimetic sound icons, or "earcons" (Gaver 1986, 1988, 1989; Blattner *et al.* 1989; Mountfort and Gaver 1990). Earcons are sounds which resemble experiential sound. Gaver, for example, has developed an interface addition for the Macintosh computer which uses earcons. An example of an earcon is a "thunk" sound when a document is successfully dragged into the trash can in the Macintosh computer interface.

Using Abstract Sounds

Abstract sounds can be used as cues to alert or direct the attention of users or can be mapped to actual data. Early experiments by Pollack and Ficks (1954) were successful in revealing the ability of sound to represent multivariate data. Yeung (1980) investigated sound as a means of representing the multivariate data common in chemistry after finding few graphic methods suitable for displaying his data. He designed an experiment in which seven chemical variables were matched with seven variables of sound: two with pitch, one each with loudness, damping, direction, duration and rest (silence between sounds). His test subjects (professional chemists) were able to understand the different patterns of the sound representations and correctly classify the chemicals with a 90% accuracy rate before training and a 98% accuracy rate after training. Yeung's study is important in that it reveals how motivated expert users can easily adapt to complex sonic displays.

Bly ran three discriminant analysis experiments using sound and graphics to represent multivariate, time-varying and logarithmic data (Bly 1982a). In the first experiment she presented subjects with two sets of multivariate data represented with different variables of sound (pitch, volume, duration, attack, waveshape and two harmonics) and asked subjects to classify a third, unknown set of data as being similar to either the first or second original data set. The test subjects were able to successfully classify the sound sets. In a second part of the experiment she tested three groups in a similar manner, but compared the relative accuracy of classification among sound presentation only (64.5%), graphic presentation only (62%) and a combination of sound and graphic presentation (69%). She concluded that sound is a viable means of representing multivariate, time-varying and logarithmic data — especially in tandem with graphic displays.

Mezrich, Frysinger and Slivjanovski confronted the problem of representing multivariable, time series data by looking to sound and dynamic graphics (Mezrich *et al.* 1984). They had little success in finding graphic means to deal with eight-variable time series data. An experiment was performed where subjects were presented with separated static graphs, static graphs stacked atop each other (small multiples), overlaid static graphs and redundant dynamic visual and sound (pitch) graphs. The combination of dynamic visual and sound representation was found to be the most successful of the four methods.

An ongoing project at the University of Massachusetts at Lowell seeks to expand the use of sound for representing multivariate and multidimensional data. The "Exvis" project uses a one-, two- and three-dimensional sound space to represent data (Smith and Williams 1989; Smith *et al.* 1990, 1991; Williams *et al.* 1990). The project is based upon the idea of an icon: "an auditory and graphical unit that represents one record of a database" (Williams *et al.* 1990: 44). The visual attributes of the icon are "stick-figures" which can vary in "length, width, angle, and color" (Williams *et al.* 1990: 45). The sonic attributes of the icons are "pitch, attack rate, decay rate, volume, and depth of frequency modulation" (Williams *et al.* 1990: 45). An experimental Exvis workstation has been set up to run various human factors experiments and initial tests of subjects have been completed. The results reveal that using visual and sonic textures together improves performance.

Two-dimensional sound displays which locate sounds up/down, right/left via stereo technology and three-dimensional sound displays which add front/back to two-dimensional displays are also being developed. A three-dimensional virtual sound environment has been developed at the NASA-Ames Research Center (Wenzel *et al.* 1988a, b, 1990). The ability to locate sound in a multidimensional "sound space" will undoubtedly be important for representing spatial relationships.

Almost all of the above studies and applications which use abstract sound to represent data rely upon a set of basic and distinct elements of sound — pitch, loudness, timbre, etc. These abstract elements can be called "sound variables" (Fig. 8.1). Most of these abstract sound variables naturally represent nominal and ordinal levels of measurement.[3] As such, a "variables" approach, analogous to that developed by Bertin (1983) for visual variables, can serve as a useful heuristic for incorporating sound in geographic visualization displays. This approach can be contrasted with one based on music theory and composition (Weber 1993; Weber and Yuan 1993). Visual map symbolization and design have been approached from many different perspectives — psychophysics, cognitive psychology, Arnheim's art theory and Bertin's semiotics to name but a few — and all have added to our knowledge of cartographic design. The same multiplicity of approaches will undoubtedly underpin our approaches to the use of sound.

Using Abstract Sounds in Geographic Visualization: The Sound Variables

The following discussion reviews a basic set of abstract sound variables — not a complete taxonomy — which is viable for geographic visualization applications. This set of abstract sound variables can be used in tandem with voice narration and mimetic earcons as discussed above. The term "variable" is used loosely and does not imply that the elements of sound are wholly separable from each other. Abstracted elements of sound, like those of vision, interact and affect each other (Lunney and Morrison 1990; Kramer and Ellison 1992). However, abstract sound variables, as with the visual variables, serve to clarify initial design choices and can serve as a viable starting point for incorporating sound into visual displays.

Data display applications using realistic and abstract sounds require a temporal dimension. This is in part due to the need to compare different sounds to glean information from the sounds. For example, the use of relative pitch — comparison with other pitches — is a key factor in using pitch to represent data (Kramer, personal communication 1992). A tone of a certain pitch heard alone means less than when that same tone is heard in comparison with an array of varying pitches. In addition, a temporal dimension is required for certain variables of sound which must vary in some way over time for their character to be identified. Duration, for example, only exists when there is some beginning and end of a sound over time.

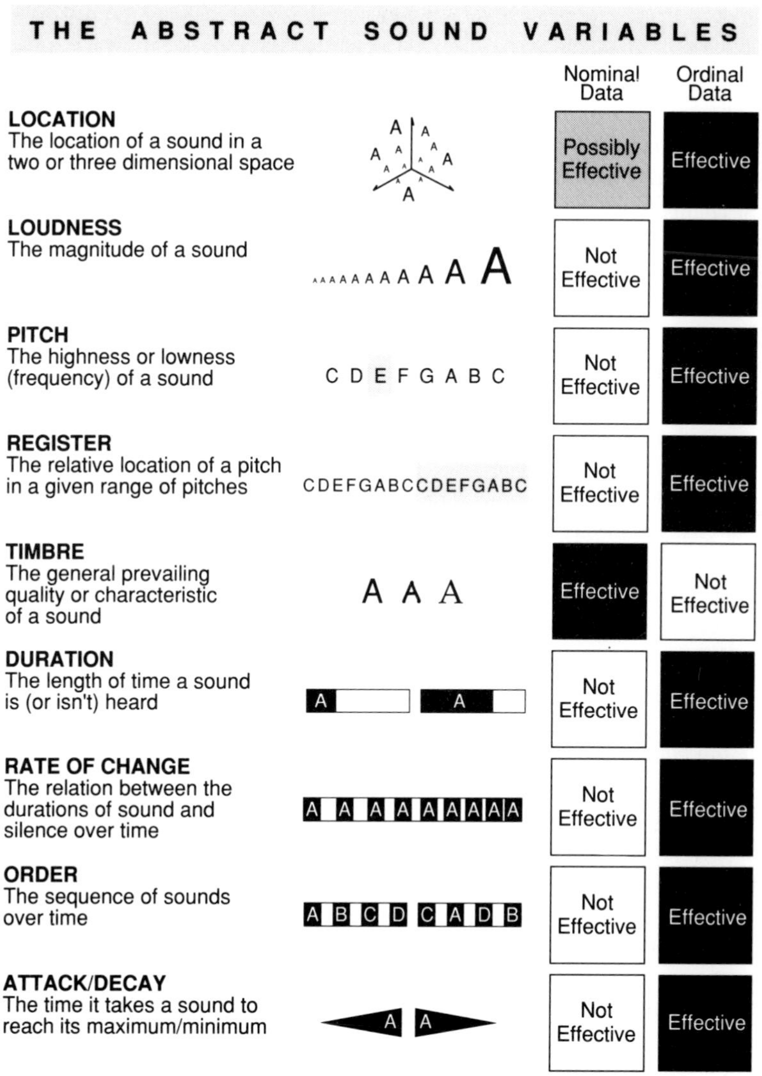

THE ABSTRACT SOUND VARIABLES

Variable	Example	Nominal Data	Ordinal Data
LOCATION The location of a sound in a two or three dimensional space		Possibly Effective	Effective
LOUDNESS The magnitude of a sound	A A A A A A A A A A A A	Not Effective	Effective
PITCH The highness or lowness (frequency) of a sound	C D E F G A B C	Not Effective	Effective
REGISTER The relative location of a pitch in a given range of pitches	CDEFGABCCDEFGABC	Not Effective	Effective
TIMBRE The general prevailing quality or characteristic of a sound	A A A	Effective	Not Effective
DURATION The length of time a sound is (or isn't) heard	A A	Not Effective	Effective
RATE OF CHANGE The relation between the durations of sound and silence over time	A A A A A A A A A	Not Effective	Effective
ORDER The sequence of sounds over time	A B C D C A D B	Not Effective	Effective
ATTACK/DECAY The time it takes a sound to reach its maximum/minimum	A A	Not Effective	Effective

FIG. 8.1. Abstract sound variables.

The Abstract Sound Variables

Location: the Location of a Sound in a Two- or Three-dimensional Sound Space

Location is analogous to location in the two-dimensional plane of the map. A sound variable location requires stereo or three-dimensional sound displays. Two- and three-dimensional sound allows for the mapping of left/right, up/down and (in three-dimensions) forward/backward locations. Location can represent nominal and ordinal data. For example, a two-dimensional stereo sound map could use location to direct attention to a specific area of the graphic map display where the fastest change is occurring in a spatial data set over time.

Loudness: the Magnitude of a Sound

Loudness is measured in terms of the decibel (dB) and implies an ordinal difference. The average human can just detect a 1 dB sound, can detect differences in loudness of about 3 dB and can tolerate up to approximately 100 dB (the loudness of a jet taking off). We would like to avoid 100 dB maps. Loudness is inherently ordered and thus seems appropriate for representing ordinal level data. Loudness may be used to imply direction and can be varied over time to represent ordinal changes in data over time (e.g. to alert users to important but infrequently occurring phenomena). It is known that humans usually become unconscious of constant sounds (Buxton 1990: 125). For example, although the hum of a computer's fan becomes inaudible soon after switching it on, even a slight variation in the fan will be instantly noticed. This effect can be used to represent information where a quiet tone represents a steady state and any variation represents change.

Pitch: the Highness or Lowness (Frequency) of a Sound

Pitch is highly distinguishable and is one of the most effective ways of differentiating order with sound. Judgements of pitch will vary from person to person. Western music has traditionally employed a scale of eight octaves comprised of 12 pitches each; extreme pitches, however, are hard to distinguish. On average, individuals can easily distinguish 48 to 60 pitches over at least four or five octaves, and this implies that pitch (divided by octaves) can be used to represent more than a single variable in a sonic display (Yeung 1980: 1121). Mapping with pitch is appropriate for ordinal data. In addition, pitch may imply direction where, for example, an increasing pitch represents upward movement. Tonal sharps and flats can also be used to some effect, possibly to represent a second variable such as variations in data quality. Every twelfth pitch has the same pitch color (chroma) and this may serve to represent nominal or ordinal data (Weber, personal communication 1992). Pitch, then, can represent quantitative data, primarily ordinal. Time can be added to pitch to create a sound graph which tracks ordinal change in data over time.

Register: the Relative Location of a Pitch in a Given Range of Pitches

Register describes the location of a pitch or set of pitches within the range of available pitches. Register is a more general case of pitch, where one can specify a high, medium and low register, each retaining a full set of chromatic pitches.[4] It can add to pitch as a broader ordinal distinction. An application which uses register and pitch is discussed later in this paper.

Timbre: the General Prevailing Quality or Characteristic of a Sound

Timbre describes the character of a sound and is best described by the sound of different instruments: the brassy sound of the trumpet, the warm sound of the cello, the bright sound of the flute. Timbre, then, implies nominal differences (Risset and Wessel 1982; Kramer and Ellison 1992). For example, a brassy sound could be used to represent an urban phenomena, whereas a warm or mellow sound could be used to represent a rural phenomena. Such an example draws attention to the evocative nature of sound.

Duration: the Length of Time a Sound Is (or Isn't) Heard

Duration refers to the length of a single sound (or silence) and can represent some quantity mapped to that duration. Silence must be used in tandem with duration to distinguish the duration of multiple sounds (Yeung 1980: 1122). Duration is naturally ordinal.

Rate of Change: the Relation Between the Durations of Sound and Silence Over Time

Rate of change is primarily a function of the varying (or unvarying) durations of sounds/silences in a series of ordered sounds over time and can represent consistent or inconsistent change in the phenomena being represented.

Order: the Sequence of Sounds Over Time

The order in which sounds are presented over time can be "natural" — such as the progression from a low pitch to a high pitch — and this means that it should be easy to detect general trends (patterns) in data presented with sound variables such as pitch or loudness. The "natural order" of sounds can be manipulated to represent data "disorder" or different orders. For example, if a natural order of sounds (say pitch from low to high) is matched to chronological temporal order, any non-ordered sound will be recognizable as an indication that the data are out of chronological order. An example will be discussed later in this chapter.

Attack/Decay: the Time it Takes a Sound to Reach its Maximum/Minimum

The attack of a sound is the time it takes for a sound to reach a specific level of loudness; the decay is the time it takes to reach quiet. Attack has been found to be much more successful in conveying information than decay (Lunney and Morrison 1990: 144). Attack/decay could be used to represent the spread of a specific data variable in a given unit: for example, pitch may represent an average value for the income in a county and attack/decay the spread of values; a long attack and decay would represent a wide range of incomes in that county. Attack/decay may also be used to represent rates of diffusion or recession.

Thus far this chapter has described the use of realistic sound (vocal narration and mimetic earcons) and the use of abstract sound (summarized as a basic set of abstract sound variables) for representing data. The next section describes a series of geographically oriented applications of sound in geographic visualization.

Sound and Geographic Visualization: Applications

Animation and Sound

Sound is an inherently temporal phenomena. As a result, it is particularly suited to use with map animation. Recent work on cartographic animation has led to the derivation of a set of dynamic variables — duration, rate of change, and order — and some suggestions for their application (DiBiase *et al.* 1992). Sound can be closely linked to the dynamic variables and their applications and may be used to enhance the comprehension of information presented in a dynamic display. In addition, potential uses of the dynamic variables may be suggested by examining temporal issues in sound and music.

At least three distinct kinds of change can be visualized by a map animation. Spatial change, often called a "fly-by", is visualized by changing the observer's viewpoint of some static object. Computerized flight simulators provide an excellent example of visualized spatial change. Voice-over has been used with fly-by applications to provide an explanation of what is being seen (Jet Propulsion Laboratories 1987; DiBiase *et al.* 1991). Vocal narration is an important way of using sound to enhance dynamic geographical visualizations. Mimetic sounds — earcons — can also be used to enhance dynamic geographic visualizations. Thus sound can be used as a mimetic symbol. The sound of fire and wind, for example, has been incorporated into an animation of forest growth to cue the viewer into what is happening in the animation (Krygier 1993). In this case sound serves as a redundant variable with which to enhance certain key events in the dynamic display.

Chronological change, or "time series," may be visualized by mapping chronologically ordered phenomena onto an animated series. A map of the diffusion of AIDS over time and space is an example of visualized chronological change (Gould *et al.* 1991; Kabel 1992). Such spatial and chronological change is intuitive and minimal explanation is needed to make such representations understandable for most users. Sound has been used to add additional information to the chronological display of AIDS data (Krygier 1993). Loudness is used to represent total cases of AIDS for each of the years displayed in the animation. The increasing loudness adds both a dimension of information (increasing number of AIDS cases) as well as a sense of impending disaster. Pitch is also used in the same AIDS animation to represent the percentage increase of new cases for each year. The pitch can be heard "settling down" as the percentage increase drops and steadies in the late 1980s. An anomaly can be heard in 1991 where the animation switches from actual AIDS cases to model predicted AIDS cases. Thus sound can be used to detect anomalies in data.

Initially less intuitive but valuable for expert users of visual displays is a third kind of change which can be visualized with map animation. Attribute change, or "re-expression," is visualized by mapping attribute-ordered phenomena onto an animated series. Such a visualization of change in attribute may be very useful for enhancing or revealing patterns not evident in the original time series. Graphic methods to alert the animation viewer to the fact that the animation is ordered in terms of attribute change have been used. For example, a time-scale can be included at the bottom of the animation and a pointer can indicate the year of each animation scene. The problem with this graphic solution is that the viewer's attention can be focused on the map or on the time bar and not both at the same time. This is obviously a situation where sound may provide a better solution as it is possible to watch the map and listen to it at the same time. Thus sound can be used to replace a distracting visual element on a map display. Pitch is used to replace the time bar in an animation of presidential election landslides (Krygier 1993). The animation is shown in chronological order with pitch mapped to years (increasing pitch = increasing years). This familiarizes the user with the meaning of pitch in this animation. The same data are then re-expressed in terms of an attribute — magnitude of the landslide — and shown again. The fact that the pitches are heard out of order cues the viewer that the visual sequence is out of chronological order. Patterns noticed in the visual or the sonic display, or both, may then be more carefully examined. Sound patterns may be more easily distinguished than visual patterns and are especially valuable for dealing with cyclic temporal data (Weber, personal communication 1993). Such an application of sound is more important as the amount of data being visualized increases and it becomes necessary to isolate the few interesting patterns from the many uninteresting ones.

Interactivity and Sound

Interactive multimedia displays have begun to attract the attention of cartographers (Andrews and Tilton 1993; Armenakis 1993; Buttenfield 1993; DiBiase *et al.* 1993; Huffmann 1993; Shiffer 1993). Voice-over, realistic sounds and abstract sound used as cues and mapped to data can be incorporated into such displays. A prototype interactive display which uses graphics and sound has been developed to display up to four data variables simultaneously (Krygier 1993). The prototype is based on 1990 US census data from Pennsylvania. A choropleth map is used to display the percentage population not in the labor force. A graduated circle map displaying median income is then added to the choropleth map. At this point one could add a third data variable to the display by changing the choropleth map into a bivariate choropleth map, by adding a data variable as a fill for the graduated circles, or by going to a second map. All of these have problems: bivariate maps are difficult to interpret and understand (Olson 1981); the third variable in the fill of the graduated circle will be hard to see in the small circles; and multiple maps may lead to comparison problems. Sound can provide an alternative to these visual methods. The prototype uses a single pitch in three different octaves (register) to display a "drive to work index". The index is either high, medium or low and refers to the relative distance workers have to drive to their places of work. When one points and clicks Pike County with the mouse, a high octave pitch is heard representing a long drive to work. Thus two variables are seen and one is heard. A fourth data variable can be added by using the range of pitches within each of the three octaves. In the case of the prototype, this was done with another high/medium/low index, that relating to the percentage poor in each county. For example, when one points and clicks Pike County a high octave pitch is heard followed by a low pitch within that octave, representing a long drive to work and low rate of poverty. After a short period of using such a "quad-variate" display it becomes relatively easy to extract the four data variables. Such a supposition will, of course, have to be more carefully evaluated, but experience with the prototype suggests that sound is a viable way to add more data dimensions to visual displays.

Sound and Geographic Visualization: Some Research Issues

This chapter has thus far reviewed the various ways that sound can be incorporated into visualization displays. These methods include the use of realistic and abstract sounds. A basic set of abstract sound variables has been defined and illustrated with some geographic examples and applications. Many issues, obviously, remain to be investigated.

Learning and Sonic Legends

If sound maps are to work, then the design of effective sound legends will have to be investigated. Because sound is not a traditional mapping variable, the user of a map display which incorporates sound will have to be acclimatized to the idea of sound as a data presentation method as well as what the sound variables used in the display represent. How best to design a sonic legend is unclear: should it be "all sound" and set up as an interactive tutorial before the use of the display begins, or should it be akin to a traditional map legend, available if and when needed? The idea of sequencing may be useful in helping sound map users to understand the elements of a multivariate sound display (Slocum *et al.* 1990).

Perceptual Issues

A solid body of knowledge exists (primarily in acoustics, psychology and music) detailing the sound perception capabilities of the human physiological system. This knowledge can be used to underpin our understanding of the possibilities and limitations of the sound variables as visual design elements. We must also be aware of the problem of "sonic overload", of barraging the user with too many different variables and dimensions of sound (Blattner *et al.* 1989: 12; O'Connor 1991). Attendants at the Three Mile Island nuclear power plant were addled by more than 60 different auditory warning systems during the plant's 1979 crisis (Buxton 1990: 125).

Cognitive Issues

More difficult issues of identification, problem solving, judgment, remembering and understanding of sound displays await attention. The sequential nature of sound raises questions of knowledge acquisition and memory. There are also questions of how much information people can deal with. A combined visual and sonic display may be one way to deal with the ever increasing complexities that geographers want to approach; such complex displays, however, have few precedents and may be more confusing than enlightening, especially for non-expert users. However, one of the goals of visualization is the construction of representations which can serve the needs of motivated expert users who are dealing with complex data and thus require sophisticated display methods. Evaluations of such methods must consider the capabilities of these users. One promising way to make the sonic display of complex information feasible is to adapt sound structures we are adept at dealing with — primarily those from music — to display design (Weber and Yuan 1993). Musical structures such as rhythm, melody and harmony are consensual, defined elements of music and must be differentiated from arbitrary and abstract sound representations such as those discussed in Yeung's (1980) study (two pitches, loudness, damping, direction, duration and silence). To what degree a familiarity with common musical structures will help people to distinguish and recognize sonic patterns is, however, unclear. Indeed, it should be expected that the sound variables — based on common musical structures or arbitrarily based on duration, rate of change and order — will interact, interfere and affect each other (Kramer and Ellison 1992; Lunney and Morrison 1990).

Location of Sound

The ability to locate sounds in a two- or three-dimensional "sound space", analogous to the two or three dimensions of the map, is an important aspect of sound which is particularly applicable to the display of spatial data. The location of sound can be used in an abstract manner, as a cue to direct attention to a specific area of a visual display, or can be used to represent the actual location of phenomena in a display. Such applications of sound have been investigated (Blauert 1983; Wenzel *et al.* 1987, 1988a, b, 1990; Begault 1990; Smith *et al.* 1990), but not in terms of geographically referenced data. Questions concerning hardware and software requirements (for two- or three-dimensional sound generation) and issues of the human ability to adequately locate sounds in a sound space need to be investigated.

Sound Maps for the Visually Impaired

The use of sound displays have been explored in the context of communicating scientific data to visually impaired students. Lunney and Morrison have used high/low pitches and pitch duration to map out "sound graphs" and have found that visually impaired users are able to comprehend the graphs and understand the patterns with relative ease (Lunney and Morrison 1981; Lunney 1983). Mansur *et al.* (1985) compared tactile graphs to sound graphs (created with continuously varying pitch) and evaluated subjects based upon judgements of line slopes, curve classification, monotonicity, convergence and symmetry. They found comparable accuracy of information communication capabilities between tactile and sound graphs, yet sound graphs were a quicker way of communicating information and were easier to create (Mansur 1984; Mansur *et al.* 1985).

We can speculate about the use of sound as a means of representation for the visually impaired. Although there is a body of cartographic research on tactile maps (Andrews 1988), there is no cartographic research on sound maps for the visually impaired. The nature of the map — with its two graphic dimensions and one or more data variables — complicates the matter and makes sound maps more difficult to create than simple sound graphs. Is there any way to construct spatial representations using a one-dimensional sound? If a high/low pitch is used to represent high/low location, can this (or other similar sonic metaphors) be used to map with one dimension of sound? Or will we have to look to stereo (two dimensions) and three-dimensional sound? If we can create a two- or three-dimensional sound space, how will maps be represented in that space? How finely can the sound variable location be specified? Can both dimensions of the plane and a data variable be represented? How easy is it to comprehend, remember and use a sonic spatial display? The ability to create and locate a sound in two or three dimensions remains a major problem hindering the use of sound for spatial data representation.

A hybrid of tactile materials and sound may prove more useful than either alone: the research carried out by Yeung, Bly and Williams has shown that complex multivariate single dimension sounds can be detected and understood. Thus a map display for the visually impaired could use a tactile display for base map information and could use sound to represent single or multivariate data located at points or areas on the map. The sonic representations could be roughly located in a two- (or three-) dimensional sound space, or they could be selected by an interactive tactile display. These approaches would allow the communication of complex, multivariate data to the visually impaired — something to which tactile maps are not particularly well suited. It would use the tactile display for base locational attributes which may be more difficult to create and interpret with sound.

Sound Maps of Data Uncertainty and Quality

Maps often impose strict points, lines and areas where no strict structures actually exist or where the certainty of their location or magnitude is low. In addition, maps are often compiled from multiple data sources, and these data vary in quality and reliability. Maps tend to be "totalizing" creatures: variations in uncertainty and quality are smoothed over to create an orderly, homogeneous graphic. On one hand, this is why maps are so useful, and it is obvious that maps enable us to deal with our uncertain and messy world by making it look more certain and tidy. Yet it seems important that some sense of the uncertainty or quality of the represented data is available. The cartographer's reflex is to conceive of uncertainty as a statistical surface and to represent it graphically. There is a rich history of graphical presentations of uncertainty — many historical atlases, for example, show the past migration of people in a manner which stresses that what is known about the migration is "fuzzy" and not well established. On the other hand, taken to its logical extreme, a map which visually displays uncertainty may become a blurred mess. The purpose of maps, remember, is to impose order, not to accurately represent chaos. Further, there is only so much visual "headroom" on a display: using visual variables to display uncertainty may have the effect of limiting the display of other data variables. A final problem with visual representations of uncertainty is that it is difficult to model visually the composite uncertainty of two or more map overlays — the realm of multivariate data displays.

An alternative approach to "visualizing" uncertainty takes advantage of sound (Fisher 1994). A "sound map" can be created which underlies the visual map and can be accessed if and when necessary. This sound map may be multivariate: register and pitch could be used to distinguish different layers of information. The click of a mouse at any position on the visual map would cause a specific sound mapped to that specific point, line or area to be heard; dragging the mouse would reveal a variation of sound as the sound-mapped data varied. A variable pitch (low to high) can represent the level of uncertainty. By dragging the mouse — the sonic equivalent of Monmonier's "brushing" (Monmonier 1989) — one begins to move toward a representation of a two-dimensional space, but the effect would be like having a small "window" with which you could only see a small portion of a map at one time. Can people build up sound "image" of the entire sound map from these small glimpses? The creation of a two dimensional sound space would allow a fuller representation of the uncertainty surface. However, if the entire uncertainty surface needs to be known, a visual representation may be more appropriate. Using sound to represent data quality or uncertainty has the advantage of preserving the sharp image of the map while allowing for the extraction of data quality or uncertainty data if and when it is needed or if it passes a predetermined threshold. Sound, in this case, serves as an invisible source of information and may be one solution to the problem of representing the quality and uncertainty of data in an already crowded visual display.

Applicability and Viability

Sound has been only minimally used for data display to date, in part because of the limitations and costs of producing and using sound. Such limitations are rapidly diminishing as computers incorporate sophisticated sound capabilities. The musical instrument device interface (MIDI) standard for sound and music generation in computers is well accepted across all computer platforms and MIDI-compatible software is currently available which can create and manipulate all of the sound variables this paper has discussed. It is possible to incorporate sound into visual displays with commonly available hardware and off the shelf software.[5]

Finally, it seems reasonable to approach the use of sound in visual displays with a critical sense of its viability and value in terms of actual applications which have real (expert and/or motivated and interested) users. Evaluations can be made using traditional quantitative methods as well as qualitative methods such as focus groups (Monmonier and Gluck 1994). Although interesting in and of itself as a possible addition to visual display, it is important to avoid using sound just for the sake of its novelty.

Conclusions

This chapter reviews the possibilities of using sound as a design variable for geographic visualization. It describes how and why sound may be used as vocal narration, a mimetic symbol, a redundant variable, a means of detecting anomalies, a means of reducing visual distraction, a cue to reordered data, an alternative to visual patterns, an alarm or monitor, a means of adding non-visual data dimensions to interactive visual displays and for locating sounds in a "sound space."

In general, the exploration of sound as a visualization design method for geographic visualization is important for two interrelated reasons. It is necessary to explore the ways in which we can take full advantage of human perceptual and cognitive capabilities in our visualization designs. Much of the inspiration behind the surging interest in visualization lies in the desire to exploit the tremendous and often unappreciated visual capabilities of humans to cope with increasing amounts of data about our physical and human worlds. Our sense of hearing, which has until recently been unappreciated as a means of representing data, can be used to expand the representational repertoire of cartographic design. At the same time, it is important to realize that the ideas and phenomena geographers wish to represent may not always be best represented by static, two-dimensional visual displays. Sound offers a way to represent information for map users who lack the sense of vision. Sound, in tandem with time, offers a way to enhance the comprehension of non-chronological uses of time. Sound offers a way to expand the limited possibilities of representing multivariate data with graphics. Sound, in other words, provides us with more choices for representing ideas and phenomena and thus more ways in which to explore and understand the complex physical and human worlds we inhabit.

Acknowledgements

For constructive criticism and ideas, thanks to Sona Andrews, Mark Detweiler, David DiBiase, Roger Downs, Gregory Kramer, Alan MacEachren, Mark Monmonier, David Tilton and Chris Weber.

References

Ackerman, D. (1990) "Hearing" in Ackerman, D. (ed.), *A Natural History of the Senses*, Random House, New York, pp. 173–226.

Ammer, C. (1987) *The Harper Collins Dictionary of Music*, Harper Collins, New York.

Andrew, S. (1988) "Applications of a cartographic communication model to tactual map design", *The American Cartographer*, Vol. 15, No. 2, pp. 183–195.

Andrews, S. and D. Tilton. (1993) "How multimedia and hypermedia are changing the look of maps", *Proceedings, Auto-Carto 11*, Minneapolis, pp. 348–366.

Armenakis, C. (1993) "Hypermedia: an information management approach for geographic data", *Proceedings, GIS/LIS 1993 Annual Conference*, Minneapolis, pp. 19–28.

Baecker, R. and W. Buxton (1987) "The audio channel" in Baecker, R. and W. Buxton (eds.), *Readings in Human–Computer Interaction*, Morgan Kaufman, Los Altos, pp. 393–399.

Begault, D. (1990) "The composition of auditory space: recent developments in headphone music", *Leonardo*, Vol. 23, No. 1, pp. 45–52.

Bertin, J. (1983) *Semiology of Graphics*, University of Wisconsin Press, Madison.

Blattner, M., D. Sumikawa and R. Greenberg (1989) "Earcons and icons: their structure and common design principles", *Human–Computer Interaction*, Vol. 4, No. 4, pp. 11–44.

Blauert, J. (1983) *Spatial Hearing: the Psychophysics of Human Sound Localization*, MIT Press, Cambridge.

Bly, S. (1982a). "Sound and computer information presentation", *Unpublished PhD Thesis*, University of California, Davis.

Bly, S. (1982b) "Presenting information in sound", *Human Factors in Computer Systems, Proceedings*, Gaithersburg, MD, pp. 371–375.

Buttenfield, B. (1993) "Proactive graphics and GIS: prototype tools for query, modeling, and display", *Proceedings, Auto-Carto 11*, Minneapolis, pp. 377–385.

Buxton, W. (1985) "Communicating with sound", *Human Factors in Computer Systems, Proceedings*, San Francisco, pp. 115–119.

Buxton, W. (1989) "Introduction to this special issue on nonspeech audio", *Human–Computer Interaction*, Vol. 4, No. 4, pp. 1–9.

Buxton, W. (1990) "Using our ears: an introduction to the use of nonspeech audio cue" in Farrell, E. (ed.), *Extracting Meaning from Complex Data: Processing, Display, Interaction, Proceedings of the International Society for Optical Engineering*, Vol. 1259, SPIE, Bellingham, pp. 124–127.

Cowan, G. (1948) "Mazateco whistled speech", *Language*, Vol. 24, pp. 280–286.

Deutsch, D. (ed.) (1982) *The Psychology of Music*, Academic Press, New York.

DiBiase, D., A. MacEachren, J. Krygier, C. Reeves and A. Brenner. (1991) "Uses of the temporal dimension in cartographic animation + visual and dynamic variables", *Computer Animated Videotape*, Deasy Geographics Laboratory, Penn State University, University Park.

DiBiase, D., A. MacEachren, J. Krygier and C. Reeves (1992) "Animation and the role of map design in scientific visualization", *Cartography and GIS*, Vol. 19, No. 4, pp. 201–214.

DiBiase, D., C. Reeves, A. MacEachren, J. Krygier, M. von Wyss, J. Sloan and M. Detweiler (1993) "A map interface for exploring multivariate paleoclimate data", *Proceedings, Auto–Carto 11*, Minneapolis, pp. 43–52.

Evans, B. (1990) "Correlating sonic and graphic materials in scientific visualization" in Farrell, E. (ed.), *Extracting Meaning from Complex Data: Processing, Display, Interaction, Proceedings of the International Society for Optical Engineering*, Vol. 1259, SPIE, Bellingham, pp. 154–162.

Farrell, E. (ed.) (1990) *Extracting Meaning from Complex Data: Processing, Display, Interaction, Proceedings of the International Society for Optical Engineering*, Vol. 1259, SPIE, Bellingham.

Farrell, E. (ed.) (1991) *Extracting Meaning from Complex Data: Processing, Display, Interaction II, Proceedings of the International Society for Optical Engineering*, Vol. 1459, SPIE, Bellingham.

Fisher, P. (1994) "Hearing the reliability in classified remotely sensed images", *Cartography and Geographical Information Systems*, Vol. 21, No. 1, pp. 31–36.

Frysinger, S. (1990) "Applied research in auditory data representation" in Farrell, E. (ed.), *Extracting Meaning from Complex Data: Processing, Display, Interaction, Proceedings of the International Society for Optical Engineering*, Vol. 1259, SPIE, Bellingham, pp. 130–139.

Gaver, W. (1986) "Auditory icons: using sound in computer interfaces", *Human–Computer Interaction*, Vol. 2, No. 2, pp. 167–177.

Gaver, W. (1988) "Everyday listening and auditory icons", *Unpublished PhD Thesis*, University of California, San Diego.

Gaver, W. (1989) "The sonic finder: an interface that uses auditory icons", *Human–Computer Interaction*, Vol. 4, No. 4, pp. 67–94.

Gould, P., J. Kabel, W. Gorr and A. Golub. (1991) "AIDS: predicting the next map" *Interfaces*, Vol. 21, No. 3, pp. 80–92.

Handel, S. (1989) *Listening: an Introduction to the Perception of Auditory Events*, MIT Press, Cambridge.

Herzog, G. (1945) "Drum signaling in a West African tribe", *Word*, Vol. 1, pp. 217–238.

Huffmann, N. (1993) "Hyperchina: adventures in hypermapping", *Proceedings, 16th International Cartographic Conference*, Cologne, pp. 26–45.

Jet Propulsion Laboratories (1987) "LA: the movie", *Computer Animated Videotape*, Jet Propulsion Laboratories, Pasadena.

Kabel, J. (1992) "AIDS in the United States," *Computer Animated Videotape*, Deasy Geographics Laboratory, Penn State University, University Park.

Kendall, G. (1991) "Visualization by ear: auditory imagery for scientific visualization and virtual reality", *Computer Music Journal*, Vol. 15, No. 4, pp. 70–73.

Kramer, G. and S. Ellison (1992) "Audification: the use of sound to display multivariate data", unpublished data.

Krygier, J. (1993) "Sound and cartographic design," *Computer Animated Videotape*, Deasy Geographics Laboratory, Penn State University, University Park.

Iverson, W. (1992) "The sound of science", *Computer Graphics World*, January, pp. 54–62.

Lunney, D. (1983) "A microcomputer-based laboratory aid for visually impaired students", *IEEE Micro*, Vol. 3, No. 4, pp. 19–31

Lunney, D. and R. Morrison (1981) "High technology laboratory aids for visually handicapped chemistry students", *Journal of Chemical Education*, Vol. 8, No. 3, pp. 228–231.

Lunney, D. and R. Morrison (1990) "Auditory presentation of experimental data" in Farrell, E. (ed.), *Extracting Meaning from Complex Data: Processing, Display, Interaction, Proceedings of the International Society for Optical Engineering*, Vol. 1259, SPIE, Bellingham, pp. 140–146.

McCormick, B., T. DeFanti and M. Brown (1987) "Visualization in scientific computing", *Computer Graphics*, Vol. 21, No. 6.

Mansur, D. (1984) "Graphs in sound: a numerical data analysis method for the blind", *Lawrence Livermore National Laboratory Report UCRL–53548*.

Mansur, D., M. Blattner and K. Joy (1985) "Sound graphs: a numerical data analysis method for the blind", *Journal of Medical Systems*, Vol. 9, No. 3, pp. 163–174.

Mezrich, J., S. Frysinger and R. Slivjanovski (1984) "Dynamic representation of multivariate time series data", *Journal of the American Statistical Association*, Vol. 79, No. 385, pp. 34–40.

Monmonier, M. (1989) "Geographic brushing: enhancing exploratory analysis of the scatterplot matrix", *Geographical Analysis*, Vol. 21, pp. 81–84.

Monmonier, M. and M. Gluck (1994) "Focus groups for design improvement in dynamic cartography", *Cartography and Geographical Information Systems*, Vol. 21, No. 1, pp. 37–47.

Mountford, S. and W. Gaver (1990). "Talking and listening to computers" in Laurel, B. (ed.), *The Art of Human Computer Interface Design*, Addison-Wesley, Reading, pp. 319–334.

O'Connor, R. (1991) "Workers in close quarters may not be ready for noisy computers", *Centre Daily Times*, Monday, 18 March, State College p. 9E.

Ohlson, B. (1976) "Sound fields and sonic landscapes in rural environments", *Fennia*, Vol. 148, pp. 33–45.

Olson, J. (1981) "Spectrally encoded two-variable maps", *Annals of the Association of American Geographers*, Vol. 71, No. 2, pp. 259–276.

Peterson, I. (1985) "Some labs are alive with the sound of data", *Science News*, Vol. 27 (June 1), pp. 348–350.

Pocock, D. (1989) "Sound and the geographer", *Geography*, Vol. 74, No. 3, pp. 193–200.

Pollack, I. and L. Ficks (1954) "Information of elementary multidimensional auditory displays", *Journal of the Acoustical Society of America*, Vol. 6, pp. 155–158.

Porteous, J. and J. Mastin (1985) "SoundScape". *Journal of Architectural Planning Research*, Vol. 2, pp. 169–186.

Rabenhorst, D. (1990) "Complementary visualization and sonification of multi-dimensional data" in Farrell, E. (ed.), *Extracting Meaning from Complex Data: Processing, Display, Interaction, Proceedings of the International Society for Optical Engineering*, Vol. 1259, SPIE, Bellingham, pp. 147–153.

Risset, J. and D. Wessel (1982) "Exploration of timbre by analysis and synthesis" in Deutsch D. (ed.), *The Psychology of Music*, Academic Press, New York.

Scaletti, C. and A. Craig (1991) "Using sound to extract meaning from complex Data" in Farrell, E. (ed.), *Extracting Meaning from Complex Data: Processing, Display, Interaction, Proceedings of the International Society for Optical Engineering*, Vol. 1259, SPIE, Bellingham, pp. 207–219.

Schafer, R. (1977) *The Tuning of the World*, Knopf, New York.

Schafer, R. (1985) "Acoustic space" in Seamon, D. and R. Mugerauer, (eds.), *Dwelling, Place, and Environment*, Martinus Nijhoff, Dordrecht, pp. 87–98.

Shiffer, M. (1993) "Augmenting geographic information with collaborative multimedia technologies", *Proceedings, Auto–Carto 11*, Minneapolis, pp. 367–376.

Slocum, T., W. Roberson and S. Egbert (1990) "Traditional versus sequenced choropleth maps: an experimental investigation", *Cartographica*, Vol. 27, No. 1, pp. 67–88.

Smith, S. and M. Williams (1989) "The use of sound in an exploratory visualization environment", *Department of Computer Science University of Lowell Technical Report No. R-89-002*, Computer Science Department, University of Massachusetts at Lowell, Lowell.

Smith, S., R. Bergeron and G. Grinstein (1990) "Stereophonic and surface sound generation for exploratory data analysis", *Proceedings of the Association for Computing Machinery Special Interest Group on Computer–Human Interfaces*, pp. 125–132.

Smith, S., G. Grinstein and R. Pickett (1991) "Global geometric, sound, and color controls for iconographic displays of scientific data" in Farrell, E. (ed.), *Extracting Meaning from Complex Data: Processing, Display, Interaction II, Proceedings of the International Society for Optical Engineering*, Vol. 1459, SPIE, Bellingham, pp. 192–206.

Stam, R., R. Burgoyne and S. Flitterman-Lewis (1992) *New Vocabularies in Film Semiotics: Structuralism, Post-Structuralism and Beyond*, Routledge, New York.

Szlichcinski, K. (1979) "The art of describing sounds", *Applied Ergonomics*, Vol. 10, No. 3, pp. 131–138.

Truax, B. (1984) *Acoustic Communication*, Ablex, Norwood.

Tuan, Y. (1993) "Voices, sounds and heavenly music" in *Passing Strange and Wonderful: Aesthetics, Nature, and Culture*, Island Press, Washington, pp. 70–95.

Weber, C. (1993) "Sonic enhancement of map information: experiments using harmonic intervals", *Unpublished PhD Dissertation,* State University of New York at Buffalo, Department of Geography.

Weber, C. and M. Yuan (1993) "A statistical analysis of various adjectives predicting consonance/dissonance and intertonal distance in harmonic intervals", *Technical Papers: ACSM/ASPRS Annual Meeting, New Orleans*, Vol. 1, pp. 391–400.

Wenzel, E., F. Wightman and S. Foster (1987) "Development of a three-dimensional auditory display system" in Yost, W. and G. Gourevitch (eds.), *Directional Hearing*, Springer, New York.

Wenzel, E., F. Wightman and S. Foster (1988a) "Development of a three-dimensional auditory display system", *SIGCHI Bulletin*, Vol. 20, No. 2, pp. 52–157.

Wenzel, E., F. Wightman and S. Foster (1988b) "A virtual display system for conveying three-dimensional acoustic information", *Proceedings of the Human Factors Society*, Vol. 32, pp. 86–90.

Wenzel, E., S. Fisher, P. Stone and S. Foster (1990) "A system for three-dimensional acoustic "visualization" in a virtual environment workstation", *Visualization '90: First IEEE Conference on Visualization*, IEEE Computer Society Press, Washington, pp. 329–337.

Williams, M., S. Smith and G. Pecelli (1990) "Computer–human interface issues in the design of an intelligent workstation for scientific visualization", *SIGCHI Bulletin*, Vol. 21, No. 4, pp. 44–49.

Yeung, E. (1980) "Pattern recognition by audio representation of multivariate analytical data", *Analytical Chemistry*, Vol. 52, No. 7, pp. 1120–1123.

Endnotes

[1] Harry Warner (of Warner Brothers fame) on being confronted with the prospect of the sound movie. Quoted in A. Walker (1979) *The Shattered Silents: How the Talkies Came to Stay*, William Morrow and Co., New York.

[2] This assumes that the content and meaning of the language used in a narration is unproblematical, which is, of course, an oversimplification. The use of vocal narration in visualization displays opens up interesting possibilities for investigating the relations between spoken and visual languages.

[3] I have collapsed interval and ratio levels of measurement into the category of ordinal pending further research on the capacity of sound to represent these finer distinctions.

[4] In music, distinctions between soprano, alto, tenor and bass are more commonly used.

[5] The applications created for this paper were constructed on a Macintosh. Sounds were generated or digitized with MacroMind SoundEdit Pro and were combined with animations in MacroMind Director.

CHAPTER 9

Designing a Visualization User Interface

MIKKO LINDHOLM AND TAPANI SARJAKOSKI

Finnish Geodetic Institute

Department of Cartography and Geoinformatics

Ilmalankatu 1 A, SF-00240 Helsinki, Finland

Introduction

Cartographic user interfaces form a new branch in cartography. Display maps in them inherit many of their properties from paper maps and traditional cartographic knowledge is necessary in building the user interface. There is, however, more to the user interface than just static maps. A computer application can be interactive, which puts a number of additional requirements on the design. With interactive applications the conceptual basis of the interface design must be more explicitly stated. At the same time, interactivity opens new possibilities.

This chapter concentrates on the general user interface design principles of cartographic visualization software. Practical design guidelines and application dependent visualization issues are discussed in the other chapters of this book. This chapter is divided into three parts. Firstly the visualization situation is outlined. In the second part general user interface design principles are discussed. The third part introduces two special techniques for interface construction: hypermedia and user models.

As noted in the Introduction to this book, visualization can be defined narrowly as a private, interactive search for the unknown. This view comes close to the concept of scientific visualization. On the other hand, we could say that all maps are visualizations of geographical data and thus the field of cartographic visualization would be much broader. In the rest of this chapter, however, the word "visualization" is used in the narrower sense, mainly in the context of geographic information systems (GIS). The principles discussed should apply well to all cartographic/geographic user interfaces.

The Visualization Task

Observation

The use of GIS is a means of observation for society as a whole and for the scientific community. Observation is an activity through which our knowledge of the world around us is increased. Visualization is a part of this activity in which we try to make sense of our observations so that they can be added to the body of knowledge we already have. This can be done through finding patterns in the data and generating hypotheses of them.

In Fig. 9.1, an analogy between the observation processes of an individual and an information system is depicted. A human is not only passively monitoring incoming data. He/she is also directing attention depending on the earlier observations and all the time relating this new data to what is already known (Neisser 1980). In the same way, an information system, be it a research community or a planning organization, directs its data collection and the interpretation of the

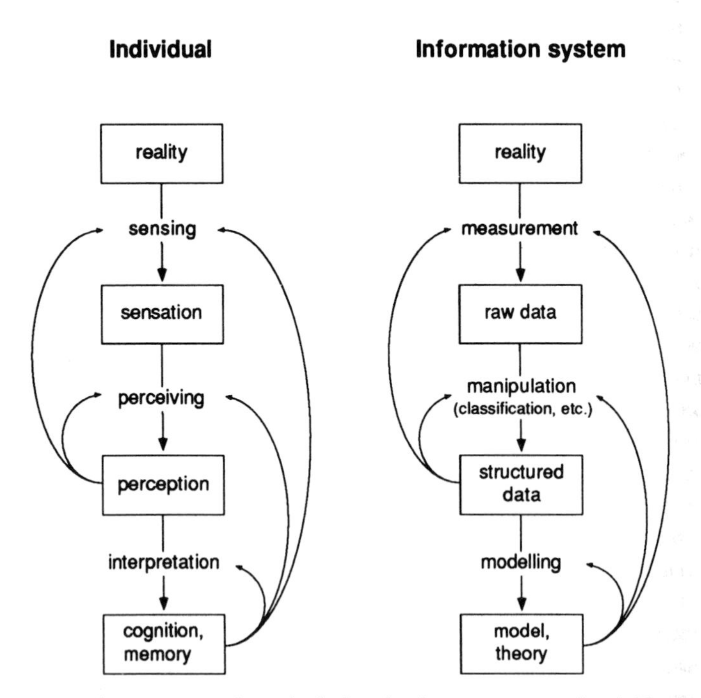

FIG. 9.1. Observation on the indvidual and information system level. The boxes represent information sets and the arrows indicate the flow of information. An information system here may be computerized or not. An example of an information system is the research community. Notice the central role of previous observations in guiding the making of new ones.

measurements on the basis of existing classifications, models and theories. What is called visualization in the context of this book happens in the perceiving and interpretation stages on the individual level and in the manipulation and modelling stages on the information system level. The modelling stage in an information system's observation process includes simulation and forecasting, which are common tools in visualization. They can be thought of as the derivation of new structured observations from the existing models and theories. The corresponding individual process can be called imagination.

In Chapter 3, "Cognitive Issues in Geographic Visualization" by Peterson, human information processing is divided into three memory stores: sensory register, short-term memory and long-term memory. The individual level in Fig. 9.1 is closely related to this. The sensation is stored in the sensory register. The process of perceiving happens in the visuo-spatial scratch pad in the short-term memory and the resulting recognized object is the perception in Fig. 9.1. When this perception is given an interpretation, i.e. when it is linked to the knowledge the observer already has, it can be stored in the long-term memory.

When the researcher or planner performs visualization, the corresponding phases on the individual and information system level merge. This leads to some basic requirements for a visualization system. At the data manipulation stage, the user interface should aid the user in perceiving meaningful structures in the data. At the modelling stage the system should provide easy ways to connect perceptions to the models of reality.

The Visualization Cycle

Figure 9.2 shows a model of the information flows in a visualization situation. The system is here seen as consisting of information sets coded in some language and the translations between these languages. The main parts are the user communicating with the database. The word "database" refers here to the collection of available data, structured or not (see below). The database is created through capturing data about reality. Data manipulation belongs partly to queries and partly to mapping. "Mapping" does not necessarily mean here producing cartographical maps, but rather a function from one language to another. The figure does not represent a physical program structure, but a logical division to identify the different languages we have to deal with.

The human user can receive messages through perception and send messages by physical action. The input language consists of the user's commands, for instance menu selections or written commands. The database receives messages as queries and maps the result of the queries to the output. The output language consists of displays and their elements, such as menus, icons and maps, and possibly of sound. Through input the user can affect both the query and the mapping process (or how the data are to be displayed: the type of thematic map, for instance).

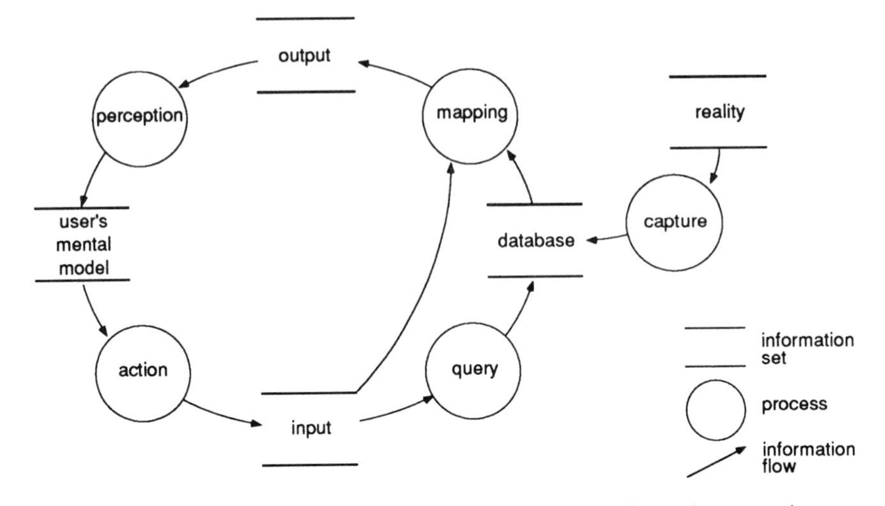

FIG. 9.2. The communication system in a GIS query interface. The rectangles represent information sets coded in some language, the ovals represent processes of translation from one language to another and the arrows indicate the flow of information or messages. The notation is changed from an earlier version (Lindholm and Sarjakoski 1992) to comply to a more standard usage (e.g. DeMarco 1979; Ward *et al.* 1986).

Each message in the model, such as a query or a map display, can be considered on three different levels — syntactic, semantic and pragmatic — basing on semiotic and information theory (Morris 1938; Nauta 1972). A more detailed discussion is found in Lindholm and Sarjakoski (1992).

- The syntactic aspect of information relates to the coding and transmission of the message. The concept was first introduced by Shannon and Weaver (1949) and was later called syntactic information. The amount of syntactic information in a message is inversely related to its frequency of appearing in the communication channel. The size of an optimally coded (compressed) file on a disk is one measure of the syntactic information of its contents
- Semantic information is the general meaning of the message. Its amount in a message depends on how precisely it defines the state of things (Carnap and Bar-Hillel 1952)
- Pragmatic information is the personal meaning of a message to the receiver. If the receiver knew the thing already, the pragmatic information content of the message would be nil. There are no clear theories of pragmatic information. Cherry (1957), Ackoff (1958) and Nauta (1972) among others have dealt with pragmatic information.

The function of each translation in Fig. 9.2 is to convert the messages from the syntax of one language to that of another without losing or altering their semantic contents. The computer operates only on the syntactic level. The semantic and pragmatic values are evaluated only by the user.

User Interface Design

Interaction Styles

The input and output languages have to be closely related to form a consistent and comprehensible system. Different frameworks have evolved for structuring the interaction between the user and the application. Shneiderman (1987) distinguishes five different styles of interaction between humans and computers which are discussed in the following.

Menu Selection

The user is presented with a list of choices from which one or sometimes several alternatives may be selected. A menu may consist of text or of icons, such as a tool palette in a drawing program. The use of menus reduces errors because the user cannot give a command that the application does not understand. They also help the user to remember the available commands. Well defined menus structure decision-making. On the other hand, long menus can be difficult to comprehend and many nested submenus may slow experienced users. These problems can be alleviated to some extent by careful menu layout.

Form Fill-in

Text is typed into the fields of a standardized input display, which looks like a paper form. Form fill-in is well suited to data input, and to query by example database interfaces when dealing with alphanumeric data.

Command and Query Languages

The user types commands to be executed immediately or to be stored for later use. Command languages offer the most flexible means of interaction. Complex operations can be described in a compact form. It is easy to create macros for frequent operations. Command languages are, however, difficult to learn and remember. They often invoke errors. In cartographic or geographic applications, character-based command languages may bring an unnecessary abstraction level between the user and the data. A solution to this might be the use of graphical query and command languages, such as described in Ichikawa *et al.* (1990) and Maiguenaud and Portier (1990).

Natural Language

The user gives the commands in natural language, e.g. English. Spoken input of natural language would be preferable. Owing to the ambiguousness of natural language, there is frequently a need for confirmations and clarifications, which slows down working and annoys users. Furthermore, natural language is cumbersome in describing many spatial concepts, and may thus not be suited to cartographic applications. That's why we have maps in the first place!

Direct Manipulation

The application's concepts have a pictorial representation — icons, for example — which can be manipulated with a pointing device, such as the mouse. Direct manipulation interfaces are easy to learn and remember. They encourage exploration and experimentation. If the pictorial representation is done badly, however, it is misleading and will frustrate users. Often it is possible to distinguish between the manipulation of control items on the display — such as adjusting a scroll bar to change the view — and the manipulation of the data itself — moving an end point of a line feature, for example. Most geographic features have a visual cartographic representation already, so direct manipulation seems natural.

Summary

These interaction styles include the whole of the input language and the subset of the output language, which is needed to present the options and use of the input language. Direct manipulation interfaces may include the whole output language. In practice, the designer usually has to blend different types of interaction depending on the task. A region to zoom in might be selected by directly framing the area from the map display with the mouse. An index of place names in turn would probably be best implemented as a menu.

User Interface Levels

The interaction styles provide only the basic methods for communicating between human and computer. A more general framework is needed to keep the design process together and the user interface clear. Below, the user interface is divided into three levels: conceptual, functional and appearance. The conceptual level consists of the general idea of use and operation of the application. The functional design deals mostly with the input language (i.e. what operations are available and how the user initiates them) and appearance design with the output language (how to present the application and data to the user). This outline is a modification of the division by Foley *et al.* (1990: 391–395) for user interface design. The three levels represent a schedule for design, in the sense that the conceptual level needs to be outlined first, before decisions about the functionality and appearance of the application are made. The functionality, in turn, must be known to some extent before the appearance of the application can be created. Form follows function. In practice, all three levels develop in parallel and mix, but when the application is ready, all three levels should be clearly defined and logically related.

It is difficult to draw a border between user interface design and general application design. The functions incorporated in the interface for example, are also part of the overall application definition. This three-level scheme could, therefore, be used as an aid in the design of the whole application.

Conceptual Level

The basic problem of conceptual design is how to tie the input and output languages into one intuitive and powerful user interface language. Questions of the following type have to be answered: What need does the application meet? How is this goal reached? What is the result of working with the application? Who are the intended users?

The meaning of the conceptual level is to give coherence and consistency to the user interface. The application must have some underlying logic which must be easily grasped by the user. Consistency in the user interface is important. Similar tasks should be performed similarily, using consistent terminology, related tools should look related, the cancel button should always be in the same place, and so on (Shneiderman 1987).

Two main components of the visualization process are the user and the data. They interact in the various ways made possible by the application. The user interface design rests on the assumptions made about the two.

The User. We assume a person doing visualizations on a computer is a professional, often with an academic education. He or she is working with dedicated software on some specialized problem on which he or she is an expert. Planning and research are two prime examples of fields in which visualization is needed. The most important property of an application would probably be the ability to control the data and its presentation easily.

Experts need no introduction to the subject matter and no alleviations to conceptually difficult problems. This is often misunderstood by designers of GIS software, who do not give enough attention to user interface design, thinking that an expert can handle even a complex and difficult program. However, the expert should be able to use his or her energy and abilities for solving the problem and not for figuring out how to use the software. Intuitiveness and ease of use are equally important to experts and to the general public. Design for experts is even more difficult, though, because the power and flexibility of the application should not be sacrificed.

The Data. The data to be processed are of two basic types: (1) raw data from measuring or input devices or (2) structured data stored in a database. A typical source of raw data in geosciences is unclassified satellite imagery. An example of structured data could be a GIS database. The main difference is not in the information contents, but in the level of conceptual organization — i.e. in syntactic structure — and thus also in the mode of access. Structured data have some internal hierarchy and relationships which give meaning to the data items. Raw data are one mass of undifferentiated values on which many different structures or interpretations can be imposed. Structured data can be manipulated on a higher abstraction level and the user interface can be more refined than when dealing with raw data.

Visualization of raw data is performed at the manipulation stage in Fig. 9.1. The goal is to find a meaningful structure for the data. This could be achieved through ordering, filtering and/or classifying. The goal in visualizing the resulting structured data is to find indicators of larger trends or relationships which cannot be measured directly. This is called modelling in Fig. 9.1. Roughly, we could say that the aim of the manipulation and modelling stages is to give a semantic and a pragmatic interpretation to the data set. During manipulation the data items are identified and named. At the modelling stage, their influence and relevance to a given problem is evaluated.

Metaphors for Visualization. Basing the interface design on a metaphor — making the application resemble something more familiar to the user — is a way of controlling the increasing complexity of the applications without sacrificing their power (Carroll *et al.* 1988). The user's abilities and knowledge of the real world exemplar of the metaphor are utilized. A well designed metaphor is thought to guide the user to operate the application more "correctly".

This kind of user interface relies heavily on pictorial representation of the application concepts and they are often implemented in a direct manipulation fashion. It has been pointed out (Gould and McGranaghan 1990) that the common desktop metaphor is not very well suited to cartographic/geographic user interfaces and other metaphors have been proposed (Gould and McGranaghan 1990; Kuhn 1991; Lindholm and Sarjakoski 1993).

Some kind of workshop metaphor might be good for visualization. A person performing visualizations works up the data, twists and bends it, cuts it to pieces and glues them together. The user is a craftsman and the data are the material he or she is working on. The application could make use of a kitchen metaphor. The scissors and the knife tools cut the data set into parts, different data sets can be put togeteher in a "kettle", recipes are found in a statistical cook book, and so on. This way metaphors can also help in maintaining consistency.

Functional Level

On the functional level of the user interface, all the operations the user can perform, their meaning and interaction style are defined. In a cartographic visualization system the basic visualization tasks consist of selection, transformation and presentation of the data. In the visualization cycle in Fig. 9.2, selection belongs to the query process, presentation to mapping and transformation to both.

Selection. The user needs to express the criteria by which certain database items are chosen for transformation and presentation. This can be anything from simply dragging a selection rectangle for zooming in to an extensive database query.

Selection criteria	Suitable interaction style				
Identity or name		ff	cl	nl	
Location			cl		dm
Attribute values		ff	cl		dm
Existence in time		ff	cl	nl	dm
Internal structure (e.g. a linear feature)	ms		cl		dm
Topological connections	ms		cl		dm

The abbreviations are: ms = menu selection; ff = form fill-in; cl = command language; nl = natural language; dm = direct manipulation. The suggested interaction styles are only recommendations and a creative designer could also use other methods.

Transformation. The selected data is transformed — filtered or classified, for example. In other words, the thematic values for the features are calculated. The transformation may be ready-made or defined on the spot.

Transformation type	Suitable interaction style			
Predefined	ms		cl	
Algebraic			cl	
Geometric			cl	dm
Rule-based		ff	cl	

Presentation. Map and diagram displays are produced from the transformed data. To produce a display map several things have to be specified

Map parameter	Suitable interaction style			
Region in view			cl	dm
Visible features	ms		cl	dm
Current theme*				
Scale	ms	ff	cl	dm
Projection	ms		cl	
Map type	ms		cl	
Symbolization, e.g. colours	ms		s	dm

*Essentially the result of selection and transformation.

All these operations should have default values or rules which are easily overridden. In a series of displays it is natural that all the parameters which are unchanged from the previous map will affect the new one. The working session should start with a default map so the user has something to refer to right away.

There are still more design aspects to consider. Is the operation performed interactively (i.e. instantly, such as selecting an area) or as a batch job (such as printing)? Should the operations be performed in a certain sequence or can they run in parallel? Is the application modeless (i.e. all operations are available all the time), or does the task need a structured procedure (some operations might be sensible only in a given context)? Hypermedia (see below) is a way to build structured user interfaces to model complex phenomena.

The Principle of the Least Effort. The user interface can be thought to be everything that is between the user and the data. Thus it is an obstacle for the information flow. This barrier should be lowered as much as possible. All unnecessary commands, mouse clicks and filenames should be removed. Everything the system can infer from the context should be performed automatically. This design approach could be called the principle of least effort. Every operation in the application should be accessed with the least possible effort from the part of user. All the work that can be given to the processor without limiting the user should be removed from the user's burden. Processor time is cheap, but the expert's time is expensive.

Appearance Level

The appearance of the application is the perceptible aspect of it. The visual appearance consists of the graphical and alphanumerical symbology needed to control the operation and present the data. Such matters as the layout of the displays, choice of colours, menu phrasing, the syntax of the command language and the appearance of default maps are defined here. Some important properties of a good user interface appearance are consistency, ease, clear structure and beauty. Practical advice for display design and the choice of symbology for different operations are given, for instance, in Marcus (1992), Helander (1988), Shneiderman (1987) and Durrett (1987). Examples of interface appearance can be found throughout this volume.

Usually a distinction has been made in the display layout between the control items and the data. The data are often shown in a rectangular area — a window — and the control items are in special tool windows or menus. The control items are active; they react to the user's actions, such as mouse clicks or keyboard entry. The window contents, the presentation, is usually passive and is manipulated only via the control items.

In direct manipulation interfaces the presentation itself can be used as a versatile control item. The features on the map can react in various ways when pointed at. Alexander *et al.* (1990) have written about the use of "smart objects" in a user interface. A smart object represents a data item and its visual appearance reflects some of its properties, like the features on a thematic map. For instance, cities could be shown as icons on the display map. The size of the icon would represent the population of the city, its form the main source of income, its colour age, and so on. All changes to these attribute values would be automatically updated on the map. When pointed at, a smart object could open a window containing more detailed information. This approach is closely related to hypermedia, which will be discussed later. This kind of user interface is easier to realize with structured data than with raw data. For a map feature to act as a smart object, it has to be given a semantic interpretation: an identity, a structure and its relation to other features.

In a cartographic application the most important part of the appearance is the display maps. Little experimental data on how they are perceived are yet available [some exceptions are Gilmartin (1988) and Gooding and Forrest (1990)]. At this point, no definitive guidelines for design can be given. Some principles of the use of display maps are studied in Makkonen and Sainio (1991) (also, see Chapters 4, 5 and 12). In the following, some notes gained from experience in designing display maps for a thematic computer atlas of Finland are given.

Because the resolution of the display is much lower than that of a printed product, the display map can deliver much less information per unit area. This, combined with the relatively small size of computer screens (usually less than an A4 sheet) means that display maps have to be much more generalized than paper maps. This calls for a new kind of design strategy. The user's attention can be directed to things that are considered important. Phenomena which used to require complex and difficult symbology can be replaced by showing many maps side by side or in sequence, perhaps using animation.

The display map is a new, dynamic medium. All the different methods of selection, transformation and presentation are available. An interactive map could be made up of smart objects, or perhaps the user could make a cross-section profile along any line on the map. Furthemore, new ways of communication, such as aerial and landscape photographs, video clips, sound environments, spoken guidance, animated maps and drawings can be mixed with the map. After successive zoom-ins the map display could change to an aerial photograph (Rhind *et al.* 1988).

The geometric accuracy of measurement is an important requirement for conventional printed maps. The map reader has to be able to measure locations, directions, distances, areas, slopes and volumes from topographic maps or exact thematic values from the thematic maps. This dates back to the map's earlier role as a storage medium for data. In modern database-driven systems the exact data values can always be made available and there is no need to measure them from the map. The display map is like a mental map in this regard [see Gould and White (1974) for information on mental maps]. It can depart from the measured geometric reality in an arbitrary manner.

In Peterson (1987) human cognition was divided into propositional memory, which stores facts as linguistic statements, and image memory, which contains visuo-spatial representations of phenomena or mental images. Facts in the propositional memory are precise and retrieved very quickly. The image memory is used to store more complex and imprecise perceptions and thoughts, thus giving a general view of the subject. The task of the display maps is to support and direct the formation of the mental images (see Chapters 2 and 3). A database query with a textual output provides propositional knowledge for the user.

In short, the unique properties of a display map are:

- Dynamics — the map can change over time
- Interactivity — the map can react to user's actions
- Modifiability — the user can modify the map
- Multimedia — new communication forms can be added
- Schematic communication — the map need not be accurate

Special Techniques

Hypermedia

Visualizing the properties of geographical features is a fairly straightforward process. Countless different types of thematic maps have been developed for this purpose. Presenting the relationships between features is much more difficult. One method of modelling them is through hypermedia. A hypermedia application is a collection of discrete elements or packets which may contain any kind of information. These nodes are connected to each other with links which organize the information into semantic contructs. Information search within a hypermedia application is called browsing.

A node usually has a visual representation on the display, such as a window, a text or an image field. Links from this node to others are expressed in some way on the display: within or near the node or in some reserved area. They may have the form of a word or an icon. By activating the link, for instance by pointing with the mouse cursor at the link icon, the node to which the link points is brought up.

In recent years many studies of the use of hypermedia in GIS and computer-based map systems have been undertaken. Wallin (1990) introduced the concept of a hypermap. Laurini and Milleret-Rafford (1990) presented a data structure for hypermap applications. Further studies are Camara (1991), Lewis (1991) and Raper (1991). However, in most of these papers maps are seen as single nodes, separate entities connected to each other and other kinds of documents via the links. In our opinion the hypermedia concept should be planted deeper into the structure of the database. The links should be between features, not map sheets. This is where the greatest improvements in information search can be made.

The use of hypermedia requires some internal structure for the data. It is not very easily applied to raw data, but is well suited to visualizing the database structure. In a hypermedia application the relationships in the database can be presented as such in the user interface. The main advantage of hypermedia is, however, the possibility of giving several different organizations to the same information elements. The same set of nodes can be given a different structure when looked at from another aspect. A context can be defined as a set of links which give a certain structure on a set of nodes. There are a huge number of different relationships between the geographic features. Individual features, such as lakes, roads, cities and countries form the nodes and the contexts represent different models and views of the structure of the space. Although individual nodes can be defined at the manipulation stage in Fig. 9.1, contexts are the result of the modelling stage. Contexts seem to be a useful way of reducing the browsing possibilities to a manageable amount. One context at a time is used to browse the nodes. At any time another set of links can be loaded and the context changes.

In Lindholm (1991) two main types of geographical or spatial contexts were defined: vertical and horizontal (Fig. 9.3). Vertical contexts are hierarchies within spatial features. They form tree-like branching structures. There are two kinds of hierarchies: functional and classificatory. The first group is formed of the hierarchies which reflect the functional relationships between features. Typical examples are administrative regionalization (e.g. country — county — municipality — village)

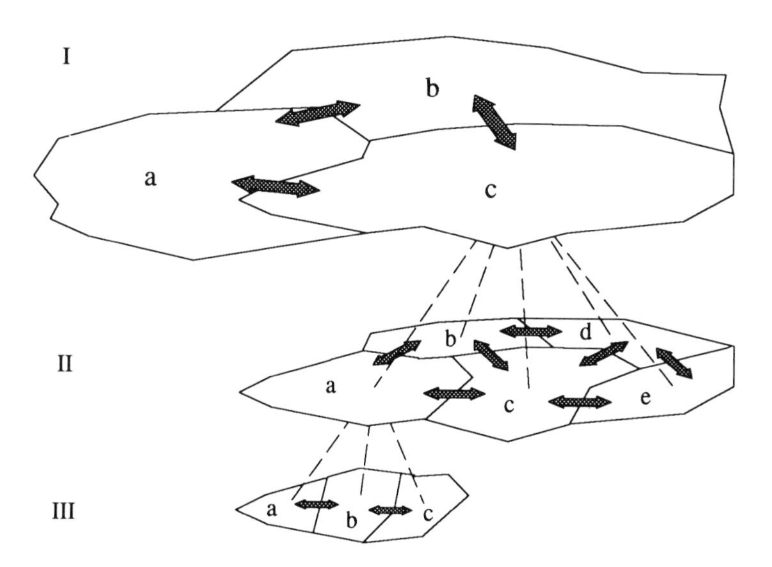

FIG. 9.3. Spatial contexts. The example region is divided into subregions on three levels of hierarchy. The vertical relationships between the subregions are shown with dashed lines and the horizontal relationships with double-ended arrows. For instance, the vertical context of the subregion a on level II is that it belongs in part to region c on level I. On the other hand, it divides on level III into subregions a, b and c. The horizontal context of the region a of level II is that it is a neighbour of regions b and c on the same level.

and the hierarchy of drainage basins of a river system. These kinds of divisions are relatively fixed, but the user could change them, to make a what if-type analysis.

Classification hierarchies are created by dividing the face of the earth into homogenous regions, often using some quantitative criteria. This way the world has been divided into zones on the basis of soil classifications, for example (for soil classification, see Strahler and Strahler 1987: 392–415). The main zones are divide into subzones which divide into subzones, and so on. A hierarchy of increasingly smaller and homogenous regions is formed.

Typical classification hierarchies are the climatic, soil and vegetation regions. Also cultural phenomena — languages as an example — have been classified in this way. These classificatory regions and hierarchies do not exist in reality. They are only conceptual generalizations of geographical reality. The North American prairie as such has nothing to do with the Central Asian plains, but on the basis of their shared characteristics they are classified to the same category: grasslands. In a visualization system, the user could change the classification criteria and produce a new hierarchy for browsing.

Horizontal contexts are relationships between features of equal importance. A county has vertical relationships to the state it belongs to and to its subordinate municipalities. Besides these, it also has horizontal relationships to other counties. Usually these are neighbourhood or connectivity relationships. Horizontal contexts form network-like structures. The "has common border" relationship is a typical horizontal context within a set of regions. Communications between cities is another. Various forms of interaction between regions, such as trade and migration flows, are a mixture of hierarchical and horizontal relationships.

A third type of context could be a problem or working context. The person doing the visualization has some problem to solve. A planner, for instance, could link to the map of his or her planning region the different incompatible land use requirements. A researcher performing a pollution study could link the polluted waters to the main factories nearby. In a problem context, the resulting structure is probably most often an associative network, but it could also be a hierarchy. A planner could build a version tree of alternative uses for a region, for example. Each user could define their own working environment or context.

The graphic narratives presented by Monmonier in Chapter 11 are one approach to hypermedia. A graphic script or an interaction point in it could form a node. On the other hand, the user could store a browsing or navigation sequence through the hypermedia network as an executable script. What Monmonier calls issue profiles are essentially the same as contexts here.

User Models

In addition to careful user interface design the usability of software can also be improved by incorporating knowledge of the user into the system. A user model commonly means a systematic description of the important characteristics of a user. At its simplest, it is only a set of user-definable default parameters in the system.

More advanced and potentially more useful are systems which can adapt themselves to different users at run time.

A user model is usually built by monitoring the user's input and inferring certain traits from it, such as the sophistication level of the user. Often the user is characterized with stereotypes. A stereotype consists of attribute-value pairs, called facets, which describe the different characteristics of the stereotype, such as experience and age (Rich 1983). The facets may be given weights to describe how strongly they belong to a given stereotype. An individual user may also be modelled with facets, the weights of which tell the system how much it relies on the value of the facet. Some easily observable facets work like triggers that activate the stereotype. All user input is fed into a kind of user processor, which infers from the input and its previous knowledge what the user might want to do and forwards this command to the application. If the user does not like the result, the model has to adapt itself. Brajnik *et al.* (1990) describe in detail how to build a user modelling system based on stereotypes.

User modelling has mostly been applied to text based applications such as editors (Zissos and Witten, 1985) and electronic news services (Allen, 1990). Holynski (1988) has determined a set of visual variables which can be used to express the users' preferences of display style: busyness, complexity, regularity, colour variety, shape variety, symmetry, grid size and balance. The variables are strongly interdependent and thus lend themselves well to the construction of

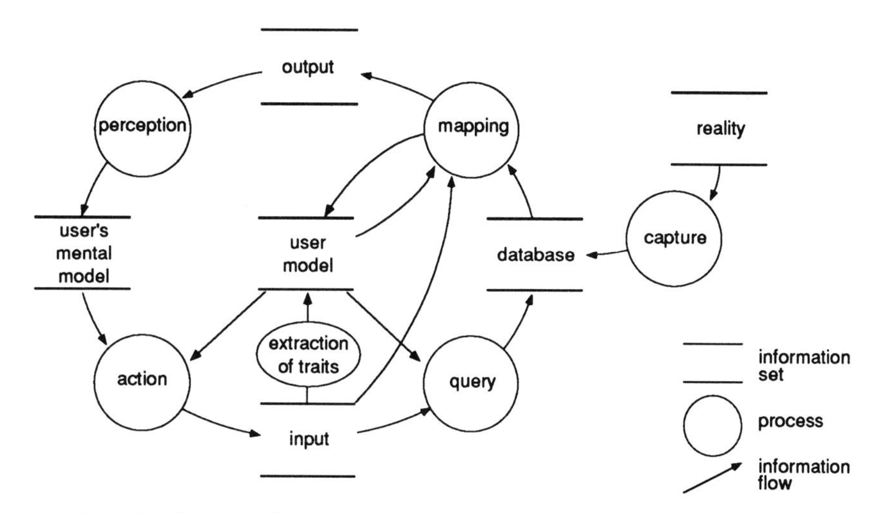

FIG. 9.4. Information flows in a visualization system with a user model. On the basis of user input, the user model makes some assumptions about his or her skills, information needs, habits, etc. The model then affects the way the subsequent user input is interpreted by the system and how the data manipulation and presentation are actually performed.

stereotypes in visual information systems, such as GIS. The cube of different map uses presented by MacEachren in the Introduction to this book could also be a basis for stereotype generation in a GIS with different stereotypes developed for different locations within the cube. We have discussed elsewhere (Lindholm and Sarjakoski 1992; Sarjakoski and Lindholm 1993) the implementation of user models in GIS by applying information theory.

Figure 9.4 presents the information flows in a visualization system with a user model. The model takes some of the responsibility of the system control from the user, but the user should be able to override the model at any time. The model may affect the translation of the user input, e.g. a long and tedious command sequence may be grouped into a single command. It may also formulate a query to the database, by modifying the user's input in light of the knowledge and assumptions it has of the user's preferences and needs. Finally, the user model may affect the presentation of the data. The same topographic data can be presented differently to a geologist and a landscape architect. The user model may be updated while creating the presentation to indicate that the user's current level of knowledge of the study area has been raised.

A practical example of user models is in Chapter 11, where Monmonier talks about user profiles which would adjust graphic scripts to the previous knowledge, preferences and interests of individual users. In a hypermedia system the user model could create working contexts automatically, linking related nodes to the main node of the work or create a path through the nodes which are most often used.

Conclusions

The birth of interactive multimedia cartography is at hand. Cartographers are becoming user interface designers. The presentation has to be adapted more to individual users and tasks, instead of creating static, generic maps. The methods described here provide a starting point for design, but more studies are needed in the formalization of communication and cartographic theories, among others. Such research is necessary for the creation of more intuitive input and output languages. More detailed task analysis and more precise user descriptions will aid in the development of more refined user models. Eventually this can lead to a personal cartography: each user can see the same data in his or her own terms.

Tufte (1983) points out in his book on statistical graphics, *Visual Display of Quantitative Information* that the reader should be led to think of only the data and forget all about the design. This is also the goal of visualization user interfaces. Hypermedia is one emerging method to make the form of the software reflect its contents more accurately. With recent developments in computing power and user interface methodology, the time has arrived for the presentation to be not the problem but the solution to communication between the user and the data.

References

Ackoff, R. (1958) "Towards a behavioral theory of communication", *Management Science,* Vol. 4, No. 3, pp. 218–234.

Alexander, J., C. Asmuth and N. Winarsky (1990) "The Sarnoff data analysis and visualization project", *Extracting Meaning from Complex Data: Processing, Display, Interaction,* SPIE, Vol. 1259, pp. 50–60.

Allen, R. B. (1990) "User models: theory method and practice", *International Journal of Man-Machine Studies,* Vol. 32, No. 5, pp. 511–543.

Bar-Hillel, Y. (1964) *Language and Information,* Addison Wesley and the Jerusalem Academic Press, Reading and Jerusalem.

Brajnik, G., G. Guida and C. Tasso (1990) "User modelling in expert man-machine interfaces: a case study in intelligent information retrieval", *IEEE Transactions on Systems, Man and Cybernetics,* Vol. 20, No. 1, pp. 166–185.

Camara, A. (1991) "Hypersnige — a navigation system for geographic information", *EGIS '91, Proceedings of the Second European Conference on Geographical Information Systems,* Brussels, pp. 175–179.

Carnap, R. and Y. Bar-Hillel (1952) "An outline of a theory of semantic information", *Technical Report No. 247 of the Research Laboratory of Electronics,* Massachusetts Institute of Technology. [Reprinted in Bar-Hillel (1964).]

Carroll J. M., R. L. Mack and W. A. Kellog (1988) "Interface metaphors and user interface design" in Helander, M. (ed.), *Handbook of Human–Computer Interaction,* Elsevier Science, Amsterdam.

Cherry, C. (1957) *On Human Communication,* MIT Press, Cambridge.

DeMarco, T. (1979) *Structured Analysis and System Specification,* Prentice-Hall, Englewood Cliffs.

Durrett, H. J. (ed.) (1987) *Color and the Computer,* Academic Press, Austin.

Foley, J. D., A. van Dam, S. K. Feiner and J. F. Hughes (1990) *Computer Graphics: Principles and Practice,* Addison-Wesley, Reading.

Gilmartin, P. P. (1988) "The design of choropleth shadings for maps on 2- and 4-bit color monitors", *Cartographica,* Vol. 25, No. 4, pp. 1–10.

Gooding, K. and D. Forrest (1990) "An examination of the difference between the interpretation of screen based and printed maps", *The Cartographic Journal,* Vol. 27, pp. 15–19.

Gould, M. D. and M. McGranaghan (1990) "Metaphor in geographic information systems", *Proceedings of the 4th International Symposium on Spatial Data Handling,* Zürich pp. 433–442.

Gould, P. and R. White (1974) *Mental Maps,* Pengùin Books, Baltimore.

Helander, M. (ed.) (1988) *Handbook of Human–Computer Interaction,* Elsevier Science, Amsterdam.

Holynski, M. (1988) "User adaptive computer graphics", *International Journal of Man-Machine Studies,* Vol. 29, No. 5, pp. 539–548.

Ichikawa, T., E. Jungert and R. R. Korfhage (eds.) (1990) *Visual Languages and Applications,* Plenum Press, New York.

Kuhn, W. (1991) "Are displays maps or views?", *Proceedings, AutoCarto 10,* Baltimore, pp. 261–274.

Laurini, R. and F. Milleret-Raffort (1990) "Principles of geomatic hypermaps", *Proceedings of the 4th International Symposium on Spatial Data Handling,* Zürich, pp. 642–651.

Lewis, S. (1991) "Hypermedia geographical information systems", *EGIS '91, Proceedings of the Second European Conference on Geographical Information Systems,* Brussels, pp. 637–645.

Lindholm, M. (1991) *Tietokonekartasto Hypermediaa Soveltaen.* (Applying hypermedia in a computer atlas, in Finnish), Geodeettinen laitos, tiedote 4/kartografia.

Lindholm, M. and T. Sarjakoski (1992) "User models and information theory in the design of a query interface to GIS", *GIS: From Space to Territory, Proceedings of an International Conference, Lecture Notes in Computer Science 639,* Springer, Heidelberg, pp. 328–347.

Lindholm, M. and T. Sarjakoski (1993) "User interface issues in a computer atlas", *Proceedings of the 16th International Cartographic Conference,* Cologne, pp. 613–627.

Maiguenaud, M. and M.-A. Portier (1990) "Cigales: a graphical query language for geographical information systems", *Proceedings of the 4th International Symposium on Spatial Data Handling*, Zürich, pp. 393–404.

Makkonen, K. and R. Sainio (1991) "Computer aided cartographic communication", *Mapping the Nations, Proceedings of the 15th Conference of ICA*, pp. 211–221.

Marcus, A. (1992) *Graphic Design for Electronic Documents and User Interfaces*, Addison-Wesley, New York.

Morris, C. W. (1938) "Foundations on the theory of signs", *International Encyclopedia of Unified Science Series*, Vol. I, No. 2, University of Chicago Press, Chicago.

Nauta, D. Jr (1972) *The Meaning of Information*, Mouton, The Hague.

Neisser, U. (1980) *Cognition and Reality*, Freeman, San Francisco.

Peterson, M. P. (1987) "The mental image in cartographic communication", *The Cartographic Journal*, Vol. 24, pp. 35–41.

Raper, J. (1991) "Spatial data exploration using hypertext techniques", *EGIS '91, Proceedings of the Second European Conference on Geographical Information Systems*, Brussels, pp. 920–928.

Rhind, D., P. Armstrong and S. Openshaw (1988) "The Domesday machine: a nationwide geographical information system", *The Geographical Journal*, Vol. 154, No. 1, pp. 56–68.

Rich, E. (1983) "Users are individuals: individualizing user models", *International Journal of Man-Machine Studies*, Vol. 18, No. 3, pp. 199–214.

Sarjakoski, T. and M. Lindholm (1993) "Modelling interactive cartographic communication with formal logic and Prolog", paper presented at the NCGIA Specialist Meeting for Initiative 8: *Formalizing Cartographic Knowledge*, Buffalo.

Shannon, C. and W. Weaver (1949) *The Mathematical Theory of Communication*, University of Illinois Press, Urbana.

Shneiderman, B. (1987) *Designing the User Interface*, Addison-Wesley, Reading.

Strahler, A. N. and A. H. Strahler (1987) *Modern Physical Geography*, 3rd edn, Wiley, New York.

Tufte, E. (1983) *The Visual Display of Quantitative Information*, Graphics Press, Cheshire.

Wallin, E. (1990) "The map as hypertex — on knowledge support systems for the territorial concern", *EGIS '90, Proceedings of the First European Conference on Geographical Information Systems*, Amsterdam, pp. 1125–1134.

Ward, P. T. and S. Mellor (1986) *Structured Development for Real-time Systems*, Vol. 1–3, Yourdon Press, New York.

Zissos and Witten (1985) "User modelling for a computer coach: a case study", *International Journal of Man-Machine Studies*, Vol. 23, No. 6, pp. 729-750.

CHAPTER 10

Expert/Novice Use of Visualization Tools

CAROL MCGUINNESS

School of Psychology

The Queen's University

Belfast, UK

Introduction

An important and defining feature of visualization, identified by MacEachren (this volume), is map use. He paints a three-dimensional picture of relationships within a human-map use space. The dimensions of the space refer to maps which are used as visualization tools for private thinking versus public communication, as tools which are high versus low in interaction, and whose goal is to reveal unknown information versus present already known information. The emphasis on the use of maps rather than on the characteristics of the visualization tools indicates the crucially psychological nature of visualization. As MacEachren *et al.* (1992) have previously argued, visualization is "an act of cognition, a human ability to develop mental representations that allow us to identify patterns and create or impose order".

Additionally, the emphasis on map use implies a map user — and users differ from one another. Although the traditional communication paradigm in cartographic research was user-oriented (Griffin 1983), users were often rather shadowy figures, not fully characterized and rarely allowed to be different from one another. In the emerging visualization research, it is important that user analyses hold their own (at least) in the rush for visualization tool design. And there are good reasons to believe that this will be the case. Historically, map communication research coincided with a stimulus-bound paradigm in psychological research which resulted in an over-emphasis on low level perceptual and behavioural measures of map use. In contrast, the visualization paradigm coincides with a more cognitive perspective in psychology in which comprehension and understanding is viewed as concept-driven as well as stimulus-driven. Within this perspective considerable variability can be expected between users doing the same task; these individual differences need to be conceptualized and their importance assessed. Currently, the dominant conceptual framework within cognitive psychology for user differences is in terms of the amount of domain-specific prior knowledge which is brought to a task. Comparisons are then made between those who have high prior knowledge in the domain with those who have little or no prior knowledge — between experts and novices.

The purpose of this chapter is to examine what the expertise paradigm has to offer visualization research in cartography. The following section analyses the paradigm and draws out some of its central assumptions and associated methods. This is followed by a brief and selective review of expert/novice studies which are relevant to visualization. Most of the studies report findings on static non-interactive displays. An example study from my own research on expertise and interactive visualization within a geographical information system (GIS) environment is then reported. The final section indicates a widening community of users which will require attention to user issues in the design of visualization tools.

Characteristics of the Expert/Novice Paradigm

Various methods are available to analyse and measure individual differences within psychology. Perhaps the best known method is the IQ test, which purports to measure differences in levels of general intellectual ability. In the context of map research, the assessment of specific abilities and aptitudes (for example, spatial ability, visual memory) has been important for understanding map reading and map learning (e.g. Thorndyke and Stasz 1980; Sholl and Egeth 1982). In contrast, the expertise approach does not focus on abilities or aptitudes which are inherent in the person (important though these might be). Rather, it examines how experience, prior knowledge and training in a specific domain affect perception, comprehension and problem solving in that domain. Its focus is on acquired knowledge effects and not on inherent traits and characteristics. Why is expertise important? It may seem an obvious prediction to make that experts are likely to be better than novices at a given task — they know more. But does this difference always affect performance? If not, when does it? When experts view or interact with a single display or a sequence of displays, do they extract the same information as novices do? Do they follow the same solution steps as novices or less experienced people? In terms of cognitive organization, we can ask whether the experts' mental representation of the task and their solution processes are similar to those of novices. Additional questions centre on the development and training of expertise. How is expertise acquired? Through what stages does it proceed? How does education and training impact on its development? Can support aids and tools affect how expertise is exercised? These are the kinds of questions which preoccupy expertise researchers irrespective of the domain under study.

Although most studies report comparisons between just two groups — experts and novices — implying two distinct categories, knowledge differences are more correctly considered as a continuum. Comparisons are then made by sampling different parts of that continuum; it is possible for a knowledge group to constitute the "experts" in one study and the "novices" in another, and care needs to be exercised when comparing results. In some studies experts include exceptional performers in a domain (e.g. grand masters in chess, professors of physics); in other studies the expertise being examined is at a more modest level and is defined as the number of years studying the domain (e.g. grade level), achieved educational status (e.g. student/teacher), self-assessed competence (e.g. frequency and competency in map use); and some studies sample the novice end of the continuum and compare those with some knowledge of the domain with those who have no knowledge at all. When the emphasis is on education and training, novice and intermediate stages of expertise are more likely to be included.

What methodologies are appropriate for expertise studies? Because a cognitive perspective is dominant in much of this research, the focus of measurement is usually the knowledge structures of the expert and novice users, together with their problem-solving strategies and tactics. A number of experimental techniques are commonly used.

Firstly, a mini-problem is selected which is very similar to the type of problem encountered as part of everyday practice in the domain (choosing and designing an experimental version of everyday practice which might yield expertise effects is not an easy task and is probably the most creative part of this type of research). Subjects in the study are then asked to exercise their skill in the normal way. Usually some quantitative measures of performance are collected — time taken to complete the task and/or number of solution attempts. But, because much of what is interesting about the cognitions of the subjects is going on inside their heads, the method of "thinking aloud" or verbal protocol analysis (Ericsson and Simon 1984) is often used to externalize these thought patterns. Using this method the subjects are asked to concurrently give a running commentary on their thoughts and decisions as they complete the task; this commentary is recorded verbatim and then analysed. (There are a number of variations on the basic methodology of concurrent verbalization, one of which is used in the study reported later in this chapter. There, the problem solving behaviour is videoed and, immediately on completion, the video is replayed and acts as a prompt for retrospective verbal protocol collection. The validity risks associated with the different methods are discussed by Ericsson and Simon 1984.) Verbal protocol analysis is clearly time consuming, but it does provide in-depth qualitative information on cognitive processes. It can yield data on expert/novice problem-solving strategies and tactics together with information on the reasons why certain solutions are generated.

Verbal protocols capture particularly the dynamic features of problem solving, but other methods are more suitable for identifying differences in cognitive structure — for example, memory tasks (subjects are asked to recall or reconstruct stimuli from a visual display which is specific to their domain) and sorting tasks (subjects are required to sort problems into similar categories and to explain the basis of the categorization). Although these may seem rather artificial in the sense that experts are not trained to do them, they have proved fruitful in examining the structure of expert knowledge and how that knowledge might sustain problem solving in a real-life task.

Most of these methods result in vast amounts of non-numerical data and additional analyses must be completed to extract the content and structure of knowledge and to identify and infer problem-solving strategies (e.g. problem behaviour graphs for analysing verbal protocols; cluster analysis for card sorting). For a comprehensive review and evaluation of the techniques which are currently available for representing expert knowledge, see Olson and Biolsi (1991).

Undertaking expertise research can be a hazardous business. Although real-life expertise may exist in a domain, not all experimental tasks will capture these differences, nor will differences be revealed in all measures of performance. In particular, quantitative measures (e.g. total time spent on task, number of solutions generated) seem less sensitive to expertise differences than qualitative measures (e.g. strategy, structure, allocation of attention, depth of understanding).

Notwithstanding the success of the expertise paradigm in studying the effects of prior knowledge in many domains (see next section), it does not readily yield the secrets of what it means to be an expert. Ericsson and Smith (1991) present a more detailed analysis of the prospects and limits of the empirical study of expertise.

Expert/Novice Differences and Visualization

One of the most striking and consistent findings in cognitive psychology over the past few years has been the importance of expertise in complex information processing, particularly in the interpretation of visual and graphical displays, including hardcopy maps. For the purposes of this review a small number of studies will be selected to demonstrate the range of domains which have been studied, the methods used and the nature of the findings. In terms of MacEachren's map use space, the visual displays tend to be single-view static displays (with the exception of the chess studies) and thus low on the interaction continuum. They also vary in a less systematic way on the other two continua; for example, chess playing is high on engagement in private thinking, whereas medical diagnosis of X-rays is clearly about revealing information not previously known.

The classic expert studies were first reported by de Groot (1965) in the mid-1960s when he compared master chess players with less experienced players. Chess is a good example of a dynamic visual display in the sense that the board positions are constantly changing and must be interpreted and re-interpreted as both players make their moves. (It differs from other types of display as control of the display is not in the hands of a single player — but that need not concern us here.) From thinking aloud protocols, de Groot reported surprisingly few differences between his master and less experienced players in terms of time and numbers of moves considered. What distinguished the experts was the "quality" of their moves, particularly the first move. This led de Groot to develop a technique to examine the knowledge structures of the players which he thought were determining the nature of the moves — memory for chess positions. He asked the players to briefly view (five seconds) chessboard positions in mid-game and then to reconstruct the board positions from memory. The master players' memory for the chess positions was far superior to the novice players; the superior memory could not be attributed to visual memory alone because, when random board positions were used, recall was equally poor for both masters and novices. Subsequent studies by Chase and Simon (1973) have confirmed that what characterizes expert chess playing is pattern recognition ability — the experts have the ability to "chunk" the pieces on the board into meaningful wholes. In other words, they do see a different board from the novices. This chunking sustains their memory performance in the memory task and, ultimately, determines the moves they make in the course of a chess game.

De Groot's findings about the superior pattern recognition abilities of the master chess players have now been replicated for many different types of experts. Egan and Schwartz (1979) reported that experienced electronic technicians can memorize and redraw symbolic drawings (electrical circuit diagrams) in ways which indicated that they were chunking the wires in the drawing into functional units or layers. For example, the skilled technicians knew that a power supply is likely to include a source, a rectifier, a filter, a regulator, etc. Knowing this conceptual category resulted in the experts grouping and chunking the units, which allowed them to search the drawings more systematically. Equally, when recalling building plans, experienced architects (Akin, 1980) produced a hierarchy of patterns — local patterns such as wall segments and doors were first recalled, then rooms and other spaces and finally clusters of spaces. In a thinking aloud study Lesgold (1984) reported that, compared with medical students, experienced radiologists showed different cognitive processes when making a diagnosis from an X-ray film. Radiologists "see" a patient when they look at a film, not just a complex visual stimulus. They zoom in on target features on the films, whereas the novices are preoccupied with properties of the X-ray itself.

Few cartographic studies have examined map users with different levels of geographical experience (e.g. Chang *et al.* 1985; Gilhooly *et al.* 1988; Williamson and McGuinness, 1990) and they all report differences between novice and expert map users. For example, Williamson and McGuinness (1990) simply asked their subjects to describe portions of ordnance survey maps. The less experienced geography students (and non-geographers) concentrated on the surface details of maps — on colours, names of places — and just listed the names of the map features without further elaboration. In contrast, the more experienced geographers evaluated the map features in terms of their quality, incidence and distribution, they located the map features within a spatial frame of reference, integrated and interrelated discrete features. Like the experienced radiologists, the expert geographers were able to "see" the reality behind the map. Both Chang *et al.* (1985) and Gilhooly *et al.* (1988) reported that previous map skill affected the interpretation and memory for topographic maps. In Chang *et al.*'s study, experts (geography, geology and aeronautical college students) had a better memory than novices for relative heights on contour maps; eye movement analyses collected when studying the maps showed that map interpretation was guided by familiar contours. Gilhooly *et al.*'s study showed that map skill affected memory for contour maps but not for planimetric maps and, from verbal protocols, it was clear that landform schemata (hills, rivers, valleys, interlocking spurs) sustained the recall. In a study of simulated wayfinding expertise (i.e. choosing routes on a map rather than traversing a physical environment), Crampton (1992) reported that expert orienteers differed from novices both in their mental representations of the terrain depicted by the map and in the strategies which they adopted to find routes between 10 locations. Through an intensive (and innovative) analysis of verbal protocols Crampton concluded that novices' mental representations of the terrain was sparse — like a string of landmarks — and that their wayfinding was characterized by efforts to analyse the locations and attributes of the destinations and to identify and overcome obstructions on the route. In contrast, experts had a richer and more detailed mental model of the geographical environment; their wayfinding strategies were "enabling", allowing them to analyse the surrounding landscape and identify good routes, to self-monitor progress and to position themselves well to approach the final destination. In effect, novices adopted a working backward strategy in contrast with the experts' working-forward approach, which is consistent with findings in other domains.

What these studies show is that, at least for single-view static visual displays, pattern recognition has a major impact on how a graphical display is interpreted and searched. Pattern recognition is not the whole story, however; Crampton's work, which analyses the more dynamic problem-solving processes involved in wayfinding, shows how expertise penetrates inferencing and problem-solving strategies even in a relatively simple everyday navigational task.

What additional factors might be important for multiple-view, multimedia computerization displays? Much has been published on user differences in human–computer interaction which is well beyond the scope of this chapter, but two phenomena are worth drawing attention to in the context of interactive visualization. Vicente and Williges (1988) pointed to individual differences between users "getting lost" in a network of visual displays while searching through a hierarchical file system. Low spatial ability users used commands such as "scroll up", "scroll down" and "zoom out" more often than high spatial ability users, indicating that the latter were better able to keep track of where they were in the system. The difference was interpreted in terms of visual momentum or the ease with which information is assimilated from one display to another. In that study spatial ability, not prior knowledge, was the important user variable; nevertheless, it does give some clues about the potential for individual differences when navigating through a multi-view system.

Another potential source of evidence for cartographic visualization tools comes from research on expertise effects in complex problem solving — in controlling computerized, dynamic, multivariate systems. These computerized complex problems are studied because they give an opportunity to examine the effects of domain-specific and factual knowledge as well as the development of strategic and tactical knowledge. An example is a computer scenario called MORO, which is a simulation of the government of a semi-nomadic tribe; the subject's task is to take care of the welfare of the tribe over a 20 year period. The tasks do not necessarily involve visual displays, but most of them do use maps, charts and graphs to communicate the state of the problem at any time and in response to requests for information. Like visualization tools, different views can be requested and the displays can be sequenced and juxtaposed. The results of an extensive series of experiments (mainly published in German) are reviewed by Dorner and Scholkopj (1991). Comparisons between experts (experienced executives in industrial and commercial sectors) and novices (students) showed a host of differences in terms of strategic and operational knowledge. For example, experts were capable of "doing the right thing at the right time"; they first attempted to gain an overall picture by asking questions and did not act until that was achieved; they co-ordinated information before making a decision. Experts consistently formed more hypotheses, they analyzing more intensely and did more planning thus gaining better developed knowledge of the interrelations in the system. Experts were not rule-bound but, through self-reflection, they adapted flexibly to new situations to which they did not have ready-made responses. It is likely that at least some of these differences will emerge as important for visualization tool use.

One point is clear — placing the study of visualization tool use within the mainstream of expertise research provides a rich source of hypotheses about the role of user variables in pattern recognition and inferencing, sequencing displays and getting lost, hypothesis testing, planning and so on.

Expert/Novice Differences and Interactive Visualization: an Example Study

This section reports briefly the methods and results of a study which is part of an ongoing research programme on expert/novice responses to map displays within the context of a GIS (for a fuller report of the study see McGuinness *et al.* 1992). The study concentrates solely on the users' interactions with the map displays for the purpose of exploring data. As such it serves as a good example of one type of visualization tool use — map use which is characterized by high interaction to reveal previously unknown information for the purposes of private thinking and understanding. Additionally, it demonstrates some of the characteristics of the expertise paradigm which were outlined earlier.

The emphasis of this exploratory study was on the cognitive processes of users as they interacted with the map; we were particularly interested in potential expertise effects on pattern recognition, integration of information across patterns, sequencing and planning.

Eighteen users (nine experts and nine novices) were recruited as expert or

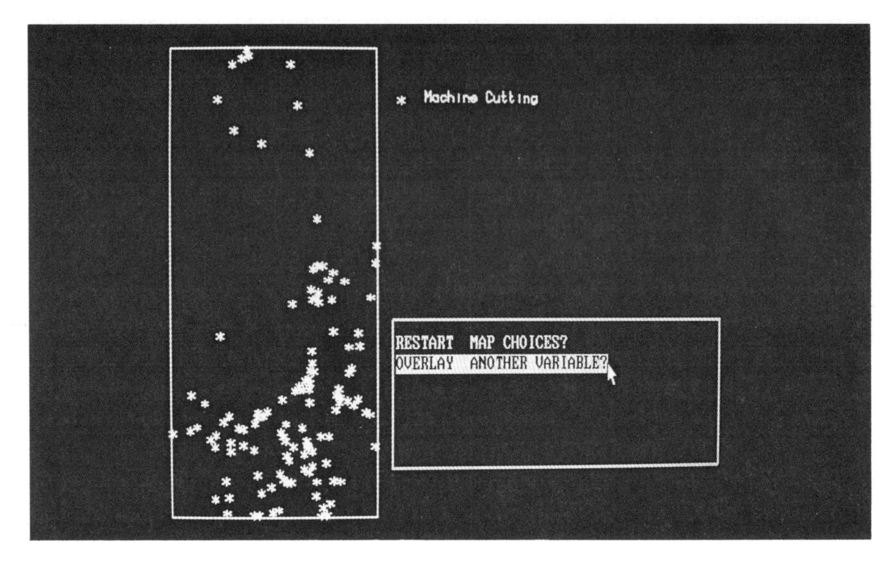

FIG. 10.1a. Peat scenario: user-generated map display of machine peat-cutting.

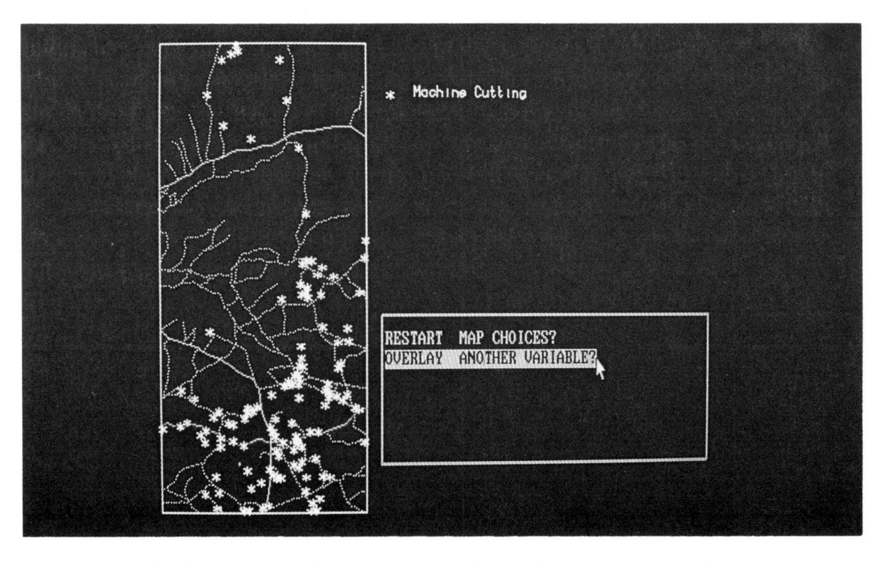

FIG. 10.1b. Peat scenario: user-generated map display of two variables — machine peat-cutting and road pattern.

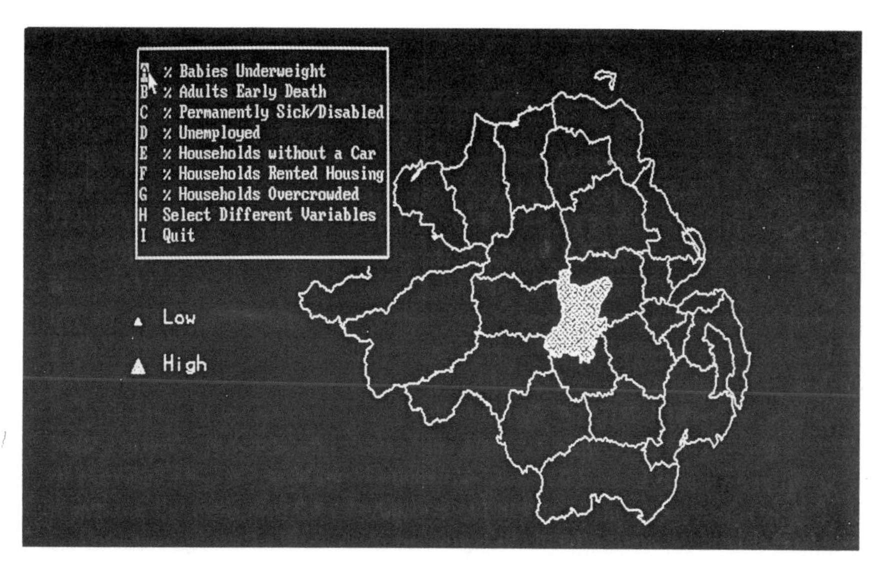

FIG. 10.1c. Health scenario: initial map display and menu choices.

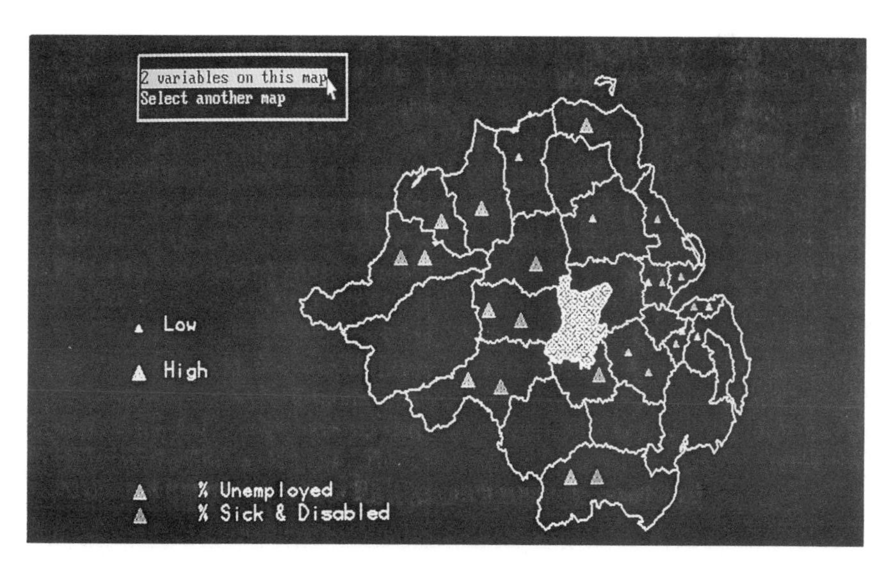

FIG. 10.1d. Health scenario: user-generated map display of two variables — percentage unemployment and percentage sick and disabled.

novice depending on the amount of previous GIS experience and training. The experts' GIS experience was not extensive; seven were geography undergraduates who had completed a 10 week hands-on course on GIS and two were research fellows who had self-taught GIS skills. The other nine users, the novices, were psychology undergraduates/postgraduates with no knowledge of GIS. All users were computer literate. The experts had significantly more formal education in geography than the novices; they used maps more frequently and they rated themselves as marginally more competent than the novices. Both groups said that they enjoyed using maps.

There was no significant difference between the user groups on two psychometric tests of spatial ability (map memory and hidden figures from the kit of factor-referenced tests, Ekstrom *et al.* 1976), although the experts did score higher on the hidden figures test (this test measured the ability to find simple forms in more complex forms, which we thought might be related to pattern recognition and overlaying).

Users were tested individually and completed two interactive problem scenarios (called the peat scenario and the health scenario) in a single two to three hour testing session. Both tasks involved data exploration, plotting coverages or variables on maps. Nine variables formed the databases for both scenarios, which were part of ongoing projects in the Northern Ireland Regional Research Laboratory. They ran on ArcInfo (display module ArcPlot) on an IBM PC.

The purpose for which the peat data was originally collected was to examine how environmental and social variables were related to the incidence of machine peat cutting, with particular reference to conflicts which might exist between economic and conservation pressures. This general storyline formed the background to the scenario and the experimental tasks mimicked as closely as possible what real-life map users might do when they first viewed the data — examine the distribution of variables (incidence of machine cutting, type of peat, contour, roads, rivers, rainline, rainstations, areas of natural beauty/scientific interest, density of wading birds), explore the relationships and begin to test for patterns. Users were invited to prepare a brief for a talk to the local conservation society on the impact of machine peat cutting in the area. They were presented with a blank screen together with the menu of variables to be plotted. Essentially the users had to chose what variables to plot, how many variables (out of nine) to overlay on a single map, what combinations to view and the order in which to view the displays. For example, Fig. 10.1a shows a screen display for a user who has just plotted the distribution of machine peat cutting in the area; Fig. 10.1b shows a screen display for a user who is checking the relationship between the incidence of machine peat cutting and the road pattern. The menu choices show that the user can overlay another variable or begin again.

The second task drew on a statistical database which consisted of health and deprivation variables collected at the level of district council in Northern Ireland (Fermanagh district council is excluded because of incomplete data). Nine variables could be plotted: rate of premature mortality, percentage permanently sick and disabled, percentage low birth weight (these were the health indices): percentage unemployed, percentage households without a car, percentage rented accommodation and percentage overcrowding (the deprivation variables); additional variables were social class and religion. To simplify the data, each variable was dichotomized high/low and a large or small triangle signified the status of variables for each area. Users were asked to act as research assistants for their local member of the European Parliament and to prepare a brief for a debate on the relationships between health and deprivation indices in regional areas of Europe (i.e. Northern Ireland). Figure 10.1c shows how the screen is presented to a user at the beginning of a session — a blank map of the region and a menu of variables to be plotted. Only two of the statistical variables could be plotted at any one time. Users had to choose what variable to plot, how many variables (one or two) to plot simultaneously, what combinations to view and what order to view the displays. Figure 10.1d shows a screen display for a user who has just examined the relationship between unemployment and percentage permanently sick and disabled.

For the two scenarios we were interested in how the users proceeded, what variables they plotted, the reasons they gave for their choices and the conclusions they reached; these constituted "the interaction" with the map display which was the focus for the study.

Three sources of data were available for analysis: the on-screen performance, which included the physical interaction with the database — the time spent on the task, the type, number and order of the variables which were plotted; the summary notes, which were the conclusions the user reached on the basis of the exploration (these notes were rated by two independent judges on quality criteria to do with meeting the task requirements, references to spatial relationships, level of detail and interpretation, degree of planning, etc.); the verbal protocol performance — subjects were videoed while doing the tasks and, under experimental conditions, they immediately watched their own performance, and provided a running commentary which was dubbed onto the video. These verbal protocols on the video were subsequently intensively analysed both at the level of individual statements and at a more global level for "search strategy".

Although the differences between the groups in terms of prior knowledge specific to the visualization tool was not great, a consistent profile of expert and novice interactions with the map displays did emerge. Experts did not spend much more time on the tasks than novices, nor did they plot a substantially greater number of maps. They did, however, explore the database more extensively than the novices; they consistently plotted all nine variables at least once for both scenarios; the novices plotted fewer variables (four to six). Additionally, the experts plotted and replotted the same combinations of variables more frequently than novices; 30 repetitions compared with nine repetitions in the peat scenario, 20 repetitions compared with 11 in the health scenario. In the peat scenario where it was possible to overlay large numbers of variables, the experts were conservative about overlaying. In contrast with the novices, experts were intent on keeping the map display simple and on plotting and replotting variables in sequential displays rather than cluttering a single map display with too many variables. The novice performance was more variable; three novices overlaid very few variables, whereas three others overlaid many variables — one novice plotted all nine variables simultaneously and produced a very cluttered map!

From the database the experts gained more knowledge and were judged by the independent raters to have given better quality answers to the general problem posed by both tasks. Compared with the novices, the expert answers (from summary notes) met the task requirements more adequately, made more frequent references to spatial relationships and were more planned and interpreted.

From the verbal protocols, experts were more likely to make systematic searches through the databases and to explore in a planned and careful way (plotted more variables); they consistently tested hypotheses, checked their interpretations (high number of repetitions) and reached comprehensive conclusions (quality of answers). In contrast, the novices often randomly searched, missed out on variables and engaged only in local and superficial hypothesis testing. With regard to viewing variables some novices believed that it was possible to successfully "read" many overlays and were frustrated by the restriction to two variables in the health problem. In contrast, experts commented that two variables were adequate for the type of information being presented: they appeared to know the limitations on their own information processing capacities. Experts engaged in what appeared to be an "optimizing" style of interaction; by finding some optimal balance between the level of their own understanding and the visual complexity of the screen display. The more inconsistent performance of the novices implied that this balance evaded them.

Although these profiles are merely suggestive they do give some indication of how expert and novice use of visualization tools can differ, not only in terms of extracting meaning from single patterns, but in composing patterns, sequencing them, repeating them, hypothesis testing and planning. Such differences need to be further documented and their consequences for visualization tool design and education need to be explored.

Future Directions

Defining visualization in terms of map use carries an implication about the map user: that map users have characteristics which ultimately define whether a tool is a visualization tool or not. The purpose of this chapter was to draw out the consequences of that position with regard to the importance of expertise. Previous research with static single-view displays reported that prior knowledge in the domain being visualized has considerable impact on the quality of visual thinking and communication achieved. What little we know about user differences and interactive cartographic visualization points to similar conclusions, although this research territory is relatively uncharted.

Many authors in this volume and elsewhere (Buttenfield and Mackaness 1991; Slocum and Egbert 1991; Taylor 1991; MacEachern and Monmonier 1992) are excited about the general "visualization revolution" in science and especially in cartography. They point to the opportunities which computerization affords to transform visual thinking and communication; to move away from the single optimal map to multiple views, three-dimensional images, user interactions, time and movement animation, multimedia, and so on. In common with first-generation users of other computer technologies, many of the new visualization tools are designed for, and used exclusively by, experts. Although their application might be currently restricted to an expert group, second generation users are likely to include a wider community who may not be as adept at spatial reasoning as geographers, environmental scientists, surveyors and cartographers. Increasingly the new tools will be used in education and to communicate with, and persuade, colleagues, policy-makers, planners and the public. Hence taking users into account in terms of previous knowledge of the domain (and computer skill) becomes important for designing the tools and for education and training in their use.

For example, with a time series animation of climatic change, DiBiase *et al.* (1992) noted that colleagues at conferences and students found it confusing, whereas the originators of the database were able to identify previously unknown relationships through the animation. Although DiBiase and his coworkers concluded that for the design of displays for exploratory analyses "we can assume expert highly motivated viewers" (p. 213), this is likely to be the case only at the very frontiers of scientific exploratory data analyses. In contrast, Monmonier (1992) assumed that the navigational option (blank screen, all options available) may be daunting for some users and, through a narrative metaphor, constructed a guided tour through a sequence of maps and statistical charts. He recommended customizing the tour to a user profile which might include preferences and knowledge about statistical techniques, focus on certain variables or geographical areas, as well as contain updates on previous history with the system. Similarly, Ormeling (1992), commenting on structure in multimedia electronic atlases, referred to the need for users to navigate through a specific route or to operate in a browse mode. He speculated on how interactive journeys through electronic atlases could be turned into adventure games and become powerful educational tools for children (and adults). In this regard, the role of expert systems to guide the novice user is relatively unexplored, although some interesting examples are being developed in relation to the *Electronic Atlas of Canada* (Taylor *et al.* 1991; Yang *et al.* 1993).

However, for such recommendations and developments to have a widespread impact, the knowledge base on user interactions with visualization tools needs to be considerably extended. The expertise paradigm provides a productive theoretical framework where such research could be positioned.

Acknowledgements

This study is part of a research programme supported by the United Kingdom's

Economic and Social Research Council Grant No. R-000-23-2337. I thank Dr Roy Tomlinson and Dr Margaret Cruikshank, School of Geosciences, The Queen's University, Belfast, for permission to use their database on machine peat cutting for the Peat Scenario; Ms Paula Devine, Northern Ireland Regional Research Laboratory, for technical assistance in developing the problem scenarios on ArcInfo; and Ms Vilinda Ross who joined the research team during the second half of the project.

References

Akin, O. (1980) *Models of Architectural Models*, Pion, London.

Buttenfield, B. J. and W. A. Mackaness (1991) "Visualization" in Maguire, D., M. F. Goodchild and D. W. Rhind (eds.), *Geographical Information Systems, Principles and Applications*, Vol. 1, Longman, London, pp. 427–456.

Chang, K., J. Antes and T. Lenzen (1985) "The effect of experience on reading topographic relief information: analysis of performance and eye movements", *The Cartographic Journal*, Vol. 22, pp. 88–94.

Chase, W. and H. Simon (1973) "Perception in chess", *Cognitive Psychology*, Vol. 4, pp. 55–81.

Crampton, J. (1992) "A cognitive analysis of wayfinding expertise", *Cartographica*, Vol. 29, pp. 46–55.

DiBiase, D., A. MacEachren, J. B. Krygier and C. Reeves (1992) "Animation and the role of map design in scientific visualization", *Cartographic and Geographical Information Systems*, Vol. 19, pp. 201–214, 265–266.

de Groot, A. (1965) *Thought and Choice in Chess*, Morton, The Hague.

Dorner, D. and J. Scholkopj (1991) "Controlling complex systems: or, expertise as 'grandmother's know-how'" in Ericsson K. A. and J. A. Smith (eds.), *Toward a General Theory of Expertise*, Cambridge University Press, Cambridge, pp. 218–239.

Egan, D. and B. Schwartz (1979) "Chunking in recall of symbolic drawings", *Memory and Cognition*, Vol. 7, pp. 149–158.

Ekstrom, R. B., W. French and H. Harmon (1976) *Kit of Factor-Referenced Cognitive Tests*, Educational Testing Service, Princeton.

Ericsson, K. A. and H. A. Simon (1984) *Protocol Analysis: Verbal Reports as Data*, MIT Press, Cambridge.

Ericsson, K. A. and J. A. Smith (1991) "Prospects and limits of the empirical study of expertise: an introduction" in Ericsson K. A. and J. A. Smith (eds.), *Toward a General Theory of Expertise*, Cambridge University Press, Cambridge, pp. 1–38.

Gilhooly, K. J., M. Wood, P. R. Kinnear and C. Green (1988) "Skill in map reading and memory for maps", *Quarterly Journal of Experimental Psychology*, Vol. 40, pp. 87–107.

Griffin, T. L. C. (1983) "Problem-solving on maps — the importance of user strategies", *The Cartographic Journal*, Vol. 20, pp. 101–109.

Lesgold, A. (1984) "Acquiring expertise" in Anderson, J. R. and S. M. Kosslyn (eds.), *Tutorials in Learning and Memory*, Freeman, San Francisco, pp. 31–60.

MacEachren, A. and M. Monmonier (1992) "Introduction", *Cartographic and Geographical Information Systems*, Vol. 19, pp. 197–200.

MacEachren, A., B. P. Buttenfield, J. C. Campbell and M. S. Monmonier (1992) "Visualization" in Abler, R., M. Marcus and J. Olson (eds.), *Geography's Inner Worlds: Pervasive Themes in Contemporary American Geography*, Rutgers University Press, Rutgers, pp. 99–137.

McGuinness, C., A. van Wersch and P. Stringer (1992) "User differences in a GIS environment: a protocol study", *Proceedings of the 16th International Cartographic Conference*, Cologne, Vol. 1, pp. 478–485.

Monmonier, M. (1992) "Authoring graphic scripts: experiences and principles", *Cartographic and Geographical Information Systems*, Vol. 19, pp. 247–260, 272.

Olson, J. R. and K. J. Biolsi (1991) "Techniques for representing expert knowledge" in Ericsson K. A. and J. A. Smith (eds.), *Toward a General Theory of Expertise*, Cambridge University Press, Cambridge, pp. 240–285.

Ormeling, F. (1992) "Ariadne's thread — structure in multimedia atlases", *Proceedings of the 16th International Cartographic Conference*, Cologne, Vol. 2, pp. 1093–1100.

Sholl, A. J. and H. E. Egeth (1982) "Cognitive correlates of map-reading ability", *Intelligence*, Vol. 6, pp. 215–230.

Slocum, T. A. and S. L. Egbert (1991) "Cartographic data display" in Taylor, D. R. F. (ed.), *Geographic Information Systems: the Microcomputer and Modern Cartography*, Pergamon Press, Oxford, pp. 167–199.

Taylor, D. R. F. (1991) "Geographic information systems: the microcomputer and modern cartography" in Taylor, D. R. F. (ed.), *Geographic Information Systems: the Microcomputer and Modern Cartography*, Pergamon Press, Oxford, pp. 1–20.

Taylor, D. R. F., A. Zhang and E. M. Siekierska (1991) EASSA: an expert advisory system for statistical analysis for the electronic atlas of canada, *The Canadian Conference on GIS: Proceedings, Canadian Institute of Surveying and Mapping*, pp. 981–993.

Thorndyke, P. W. and C. Stasz (1980) "Individual differences in procedures for knowledge acquisition from maps", *Cognitive Psychology*, Vol. 12, pp. 137–175.

Vincente, K. J. and R. C. Williges (1988) "Accommodating individual differences in searching a hierarchical file system", *International Journal of Man-Machine Studies*, Vol. 29, pp. 647–668.

Williamson, J. and C. McGuinness (1990) "The role of schemata in the comprehension of maps" in Gilhooly, K. J., M. T. G. Keane, R. H. Logie and G. Erdos (eds.), *Lines of Thinking*, Vol. 2, Wiley, London.

Yang, J., E. M. Siekierska and D. R. F. Taylor (1993) "Concept, design and knowledge acquisition for thematic map design advisory system (TMDAS)", *Proceedings of the Canadian Conference on GIS, Canadian Institute of Surveying and Mapping*, pp 474–483.

Linking the Tool to the Use: Prototypes and Applications

CHAPTER 11

Graphic Narratives for Analyzing Environmental Risks

MARK MONMONIER*

Department of Geography
Syracuse University, Syracuse
NY 13244, USA

Introduction

Two metaphors guiding the development of dynamic cartography are navigation and narration. Although they represent different strategies for geographic visualization, these metaphors are complementary rather than competitive. Navigation, the dominant metaphor, is readily apparent in graphic interfaces and software tools for helping users locate images and other facts about places, regions, geographic distributions and spatial relationships. Although what I call navigation might well be labeled exploration, I prefer the connotation of a user at the helm or steering wheel guiding the analysis toward a destination, however ill-defined at the outset. A typical navigational strategy is geographic brushing (Monmonier 1989a), whereby an analyst exploring data with an interactive graphics system uses a rectangular or circular brush to select clusters of places on a map. The brush provides a dynamic link between the map and a statistical graph, such as a scatterplot or vertical bar chart, on which highlighted symbols represent places concurrently within the perimeter of the movable brush. The user learns about regions and geographic relationships by moving the brush around the map and observing instantaneous changes in highlighted symbols on the statistical graph. In contrast, a graphic narrative can introduce the user to a new data set by moving the brush automatically from south to north, say, or from east to west — with an

*email: mon2ier@syr.edu

appropriately orchestrated variation in the brush's shape and orientation. The presentation is narrative rather than navigational because the user is now a comparatively passive viewer, who watches while the system drives the brush around the screen and controls the succession of scenes. In a further effort to help the analyst detect meaningful patterns, the system might canvass a sequence of irregularly shaped yet potentially meaningful regions such as Appalachia, the Midwest and the Pacific Rim states. Navigational and narrative approaches are complementary insofar as an analyst introduced to data by a narrative "guided tour" might then explore whatever patterns or parts of the map seem especially interesting. Moreover, if a navigational examination reveals significant relationships worth recording and sharing, the analyst could compose and edit an interpreted graphic narrative, with a voice-over, for presentation to a group of scientific peers or public officials.

This chapter consists of two sections. The first part examines basic concepts for the development and efficient use of graphic narratives. Especially important in the authoring of graphic narratives is the need to integrate cartographic displays with text, numerical data, photographic images and explanatory diagrams without overwhelming the viewer — key concerns for effective design include coherent sequencing and an interface that employs sound as well as vision, promotes customized content and images, and allows the user to control the direction and pace of the presentation. The second section explores the role of narration in multimedia cartographic presentations designed to analyse and promote an understanding of environmental risk.

Scripts, Graphic Scripts, and Graphic Narratives

Fundamental to an understanding of graphic narratives is the graphic script, a concept derived from the notion in artificial intelligence (AI) of a script. As defined by AI pioneers Schank and Ableson (1977:41), a script is "a structure that describes appropriate sequences of events in a particular context". Schank's favorite example is the dialog between a waiter and a customer ordering a meal in a restaurant: both actors' lines reflect a traditional exchange of civilities, questions and answers (Schank 1991: 82–91). Like other conversational scripts, the restaurant script works efficiently because the players not only know their parts but also anticipate each other's concerns. Although people learn scripts through experience, these sequences of mutually anticipated queries and responses have a natural coherence that facilitates communication — it is logical, after all, to order a cocktail before selecting the entrée and to order a salad before choosing a salad dressing. The menu's organization and layout typically reinforces the restaurant script's canonical sequence of messages.

In a similar way, a graphic script promotes efficient communication by exploiting logical sequences and viewer experience and by anticipating information needs. Defined as a coherent sequence of maps, diagrams, text and other images that address a particular task or goal, a graphic script necessarily presents information to the viewer gradually, so that each new display reinforces or builds on the previous display (Monmonier 1989b). For example, a script addressing the geographic and statistical correlation of two variables might begin by examining each variable individually, move on to a treatment of their covariation, and conclude by exploring a map of residuals. This sequence reflects an increased complexity found in textbooks and lectures on spatial analysis as well as numerous articles in geographical journals. And when the bivariate correlation involves a cause–effect relationship, the graphic script also anticipates a typical series of underlying questions: What is the dependent variable (the effect, or response variable) like? What is the independent variable (the cause, or stimulus variable) like? How similar are the two variables statistically? Geographically? What is the geographic pattern of the dependent variable if the influence of the independent variable is removed?

With appropriate text and labels the script can explain the analysis to an intelligent, interested viewer unfamiliar with correlation and regression.

Graphic Phrases

Graphic scripts and shorter, more focused sequences called graphic phrases can be especially useful in avoiding the need to display just one or two representations of a geographic distribution. In situations requiring a choropleth map, for instance, tradition or space limitations commonly restrict the map author to a single view with fixed category breaks. This single-map mentality can have unfortunate consequences because an author who fails to experiment with the visual effects of varying the class intervals might overlook a potentially meaningful geographic pattern, and thus present a jumbled, incoherent pattern of gray tones that ignores a causal link with coastal areas, Appalachia, dry summers or some other important influence. Another ill-effect of the single-map strategy arises when a biased map author deliberately manipulates the classification to favor a particular ideological position. In both examples, viewers might benefit from the opportunity to examine a dynamic series of views portraying the mapped pattern's stability or coherence — a series of views easily generated by a graphic phrase designed to explore alternative representations.

And that's not all. Other graphic phrases might informatively integrate a dynamic choropleth map with histograms or univariate scatterplots to show the relative homogeneity of the map's categories as well as to explore the visual effects of such inherently meaningful breaks as the statewide mean, the national mean or an estimated worldwide mean. A graphic phrase linking a pair of maps, one based on rates and the other based on counts, can suggest places where abnormally high rates merely reflect chance occurrences in small populations. And other graphic phrases might provide meaningful disaggregations of the data — for example, by treating infant death rates separately for a region's white and African–American populations. Moreover, when monthly or yearly data are available, graphic phrases can address the temporal stability of patterns mapped by years or decades. In a similar sense, if data are available for smaller areal units, a graphic phrase might explore the effects of areal aggregation or spatial autocorrelation. And for census data, still other graphic phrases might probe the effects of a change in definition or method of measurement as well as the effects of random rounding, allocation, substitution and other strategies for safeguarding privacy or compensating for missing data.

Experiments with Graphic Scripts

To experience the challenges of designing a graphic script, I composed and programmed two prototype scripts, both based on state-level US data (Monmonier 1992). A 10 minute graphic script (titled "Women in Politics") examined the bivariate relationship between women's share of elected officials in local government and the statewide labor force participation rate for women. A 19 minute graphic script (titled "Daily Newspapers in the US: 1900–1990") described and summarized change over a 90 year period in the number of firms publishing daily newspapers. I divided each script into three acts, each with multiple scenes. The correlation script's three acts addressed, in turn, the variables' individual distributions, their similarities and dissimilarities and a map of residuals based on least-squares linear regression, whereas the historical script examined the spatial–temporal distribution of number of firms in the first act, rates of change in the second act and dynamic centrographic maps for newspaper firms and population in the third act. (People sending an e-mail request to mon2ier@syr.edu will receive via Internet a compressed file containing Macintosh versions of the prototype scripts.)

I extended the stage and play metaphor by treating each raw or re-expressed variable as an actor identified by a brief title and a unique signature hue. In the correlation script, for instance, red always signified the dependent variable, whereas blue represented the independent variable and magenta identified the residuals. To minimize visual noise, I adopted standard locations on the screens for maps, statistical diagrams and text blocks. Following guidelines developed by communication theorist Williams (1990) for composing coherent, reader-friendly sentences and paragraphs, I moved from less complex to more complex representations, used text to announce the point, or purpose, of each act or scene, and added new information one element at a time by building, where appropriate on old information introduced in the previous scene or screen. Because of structural similarities between written and graphic narratives, Williams's strategies for helping writers anticipate the needs of readers are also useful in helping map authors accommodate both the cognitive limitations and the information expectations of people viewing graphic scripts.

Group interviews with people experienced in either electronic information systems or geographic analysis indicated that my graphic narratives, although engaging, were at times frustratingly rapid. Although they generally understood and appreciated the scripts' ability to integrate maps and statistical graphs, viewers wanted additional time — pauses in some cases, repeated scenes in others — to comprehend unfamiliar representations, ponder intriguing patterns, or examine specific places more closely. Equally frustrating to many of these "focus-group" participants was the need during some scenes to read text while simultaneously watching a rapidly changing screen. Written or oral text was important to understanding what they saw, and most participants would have been even more perplexed without the running commentary provided by the focus-group facilitator. In short, the focus groups identified the need to supplement vision with audible text and to allow viewers to control the direction and pace of the script.

"Openness" and User Interaction

Although the concerns and complaints of focus group participants might suggest a fundamental flaw in the notion of narrative graphics, the prototype graphic scripts are "closed" scripts representing but one end of a much broader spectrum of "openness," that is, the user's freedom to control the graphic sequence. To make a graphic script markedly more open, a script author might insert pause points or interaction points before each scene or graphic phrase. When the script halts at a pause point, the viewer can examine a text block, map or graph for as long as he or she chooses, and then proceed with the narrative by pressing the upward-arrow key. The script author might provide further openness by allowing the viewer to press the downward-arrow key and repeat the previous graphic phrase. An interaction point makes the script even more open: in addition to the choices provided by the pause point, the viewer can use the mouse or track ball to highlight symbols for particular places on a map or graph, move a brush across a map or scatterplot, move the entire display forward or backward in time, vary the length of the time period shown or in some other way explore the data or alter the analysis.

Further openness requires embedding the script in a system that is primarily navigational. One strategy is a set of narratively organized pull-down menus, in which the leftmost menu represents the graphic script's first act the menu immediately to its right represents the second act and so on. Within each pull-down menu, operation phrases representing individual scenes or graphic phrases are also arranged in narrative order, from top to bottom. Given these narratively ordered operation phrases, a user can advance through the entire script, operation by operation, beginning at the top of the leftmost pull-down menu and concluding at the bottom of the rightmost pull-down menu. Of course, the viewer who wants a strict canonical sequence of graphic phrases would probably prefer to run the script fully automatically in player–piano mode, which obviates the need to pull down a menu and chose the next operation at every pause point. Player–piano mode would be also useful at other times if a STOP key allows the viewer to halt the graphic narrative at will and then use the menus to jump to another part of the sequence. Control keys especially useful in a hybrid narrative–navigational system include STOP, SHOW MENU (to automatically lower the appropriate menu and identify the current operation), PREV OP (to repeat the immediately preceding operation), NEXT OP (to advance to the next graphic phrase) and PLAY MORE (to resume the narrative sequence in player–piano mode).

A single list of operation phrases displayed in a long, thin vertical command window is a potentially helpful variation of this narrative–navigational strategy. The system I envision would provide control keys that allow the user to show or hide the command window at will, and a vertical scroll bar for moving forward or backward in the list of operations. Showing the command window while running the script in player–piano mode can be useful for introducing new users to the vocabulary of the system's operation phrases. When the command window is visible, the system would highlight the current operation phrase as well as scroll down the list automatically, whenever necessary.

Hypermedia tools can also promote openness by providing links among related graphic phrases and graphic scripts. For users already familiar with a particular hypermedia interface, this approach could facilitate the integration of graphic narratives with other relevant text and images.

Graphic Scripts in a Visualization Support System

In addition to presenting a sequence of maps, text, and other images addressing such specific analytical concepts as bivariate correlation or spatial–temporal trends, graphic scripts might serve a wide variety of analytical, expository and pedagogic needs. For example, narrative graphics might be especially useful for helping viewers understand unfamiliar analytical techniques or use new software. Other important applications include an overview of an unfamiliar data set or geographic database, alternative graphic scripts proposed by experienced users with different approaches to exploratory data analysis and briefings to inform elected officials, the media and the general public about a locally important issue, such as a flood hazard, an evacuation plan a radical change in the region's master plan or a proposed facility for the storage, treatment or disposal of hazardous waste. Also significant are graphic scripts that use pattern analysis or template matching methods to screen data and select for presentation only images a viewer is likely to find interesting or meaningful. The variety and complexity of this broad range of possible graphic scripts suggests the need for a visualization support system able not only to generate the required displays but also to tailor images, information content, analytical techniques and terminology to the interests, experience and understanding of the viewer.

A key concept is the user profile, with which a visualization support system can adjust a graphic script to the needs of a particular viewer (Monmonier 1990: 28, 39). Defined as a collection of relevant facts about the user's experience, graphic preferences, regional interests and disciplinary biases, the user profile directs the automatic generation of customized graphic scripts. As an example, a tailor-made script could introduce a new database by drawing upon the viewer's area of residence in central New York State whenever detailed examples are needed; highlighting other areas the viewer knows well through research or residence in

Pennsylvania and coastal New England; favoring the jargon of cartographers and geographers over that of geologists or landscape architects; accommodating an explicit preference for trend-surface maps, centrographic maps and other highly generalized summary graphics; and calling attention to univariate and bivariate relationships with regional patterns in which Appalachia or New England figure prominently. (See Chapter 9 section on "User models" for related ideas.)

An efficient visualization support system with a competent user profile uses the viewer's time efficiently by providing displays he or she would most certainly want to see and avoiding redundant displays of what is already known. Although this strategy might appear to sacrifice the chance of serendipity for a higher level of vigilance, the system need not present only what a viewer wants to see. The cautiously catholic viewer could, after all, borrow user profiles and view alternative graphic narratives customized for other users with radically or marginally different goals and preferences. Moreover, the system might even be directed to identify anomalous spatial patterns in the data that challenge or contradict beliefs represented implicitly or explicitly in the viewer's user profile.

Issue profiles and metadata are also useful in tailoring a graphic script to a specific context. An issue profile might exclude irrelevant variables or coverages, focus on specific places, attributes or relationships, and call for graphic phrases that address analytical or conceptual concerns found to be important in earlier analyses of similar situations (Armstrong *et al.* 1992). Metadata, which describe the origin and limitations of the database, are especially useful with graphic phrases designed to examine data quality and assess the effects of interrelating two or more slightly flawed, low resolution indicators. A system module that points out deficiencies in data and likely misinterpretations is as important to computer-assisted geographic visualization as a module that automatically screens data for interesting patterns and relationships.

Graphic Scripts and Environmental Hazards

Visualization support systems can be especially valuable in helping users comprehend and assess risks associated with environmental hazards (that is, specific threats, such as earthquakes). As this section of the chapter explains, graphic narratives can help viewers understand a hazard's underlying processes and thereby develop insights about relative risk, methods for assessing risk and the difficulty of forecasting natural and technological hazards. The section also examines cartographic issues in portraying environmental risk and warning viewers about hazards, and suggests potential roles for graphic scripts in risk communication.

Risk and Environmental Hazards

What is risk? Simply defined, risk is the probability of the occurrence of an uncertain event with unpleasant consequences. Success in estimating risk varies with the regularity or inevitability of the event, the availability of reliable data covering a representative period of time, the effectiveness of preventive measures and the size of the area, population or period of time for which a forecast is desired. Consider, for example, the risk of death. Because death is inevitable and health statisticians can rely upon decades of mortality data, the probability of dying can be estimated with considerable accuracy, even for narrowly specific age–race–sex cohorts. Given mortality tables and an individual's demographic characteristics and general medical history, an actuary can estimate the likelihood he or she will die of natural causes within, say, five years; these estimates are sufficiently accurate for insurance firms that hedge their bets by selling policies to large numbers of people in each category. Insurance companies are cautious, of course, to exclude suicide, known pre-existing fatal illnesses and unpredictable, widely catastrophic events such as war. Where epidemiologists provide relevant data, actuaries can also estimate the risk of dying from such specific causes of death as lung cancer, leukemia, lightning or car–pedestrian motor vehicle accidents. But for rare catastrophic events, such as a core meltdown at a nuclear power plant, the lack of good data prevents an actuarially competent estimate of risk.

Perceived risk can differ radically from epidemiological estimates of risk. It is natural, after all, to fear diseases and types of accident that have killed relatives, close friends or beloved celebrities. And it is equally common as well to underestimate the significance of causes that have never affected anyone we know. It is understandable too that people overestimate risk when the media have a heightened awareness of a particular hazard and that cynics might be unduly wary of calming announcements based on official statistics. Such seemingly irrational fears might be quite justified at times, particularly when a hazard seems more threatening than usual. Parents who have observed crumbling asbestos insulation around pipes suspended from the ceiling of a basement classroom ought not be mollified by otherwise reassuring data showing a negligible "average risk".

Risks associated with natural hazards are seldom measured in mortality rates and reduced life spans. Seismic research, for example, is far more concerned with forecasting earthquakes than with estimating the direct impact of tectonic movement on human life. The seismologist's immediate goal is an accurate prediction of the magnitude, location and time of occurrence of the next earthquake. If studies of seismic risk indicate that a large earthquake is more or less inevitable within, say, 30 years, surficial geologists can then point out unstable slopes, deposits of unconsolidated materials and other areas subject to landslides and severe shaking. Emergency management officials can, in turn, translate this information into warnings for local residents and public officials. Best case and worst case estimates of fatalities are by-products of seismic forecasting, especially useful rhetorically in arguing for land use restrictions or stricter, more rigorously enforced building codes.

For some environmental hazards, such as air pollution, statistical models can portray risk as a function of distance from a known or suspected source (Committee on Advances in Assessing Human Exposure to Airborne Pollutants 1991; Zannetti

1990). Atmospheric dispersion models can predict the general pattern of exposure to atmospheric contaminants emitted at a given concentration from a stack of known height: the basic model's concentric rings typically reflect a doughnut-shaped pattern in which the risk of exposure increases with distance within a comparatively narrow zone around the source and then declines steadily outward. Because environmental impact statements for incinerators, power plants and other industrial facilities with stack emissions use dispersion models to estimate the geographical extent of effects on air quality, graphic narratives could be especially useful in helping local officials and the public understand the model's assumptions and limitations as well as its adaptation to an area with irregular terrain and a dominant wind direction. (See Chapter 13 for more on visualization of air.)

More complex models that simulate the dynamic behavior of groundwater can predict or reconstruct the migration of contaminants through an aquifer (Committee on Ground Water Modeling Assessment 1990). These models supplement a digital map of subsurface geology with data from a network of wells at which officials sample water quality as well as monitoring the pumping rate and the height of the water table. If a new contaminant is detected at one of the wells, the model might pinpoint the likely source or at least suggest where test drilling should prove most informative. Groundwater models can also forecast the spread of a contaminant under various patterns of pumping and identify wells that should be shut down, at least temporarily. Because high withdrawal can lower the water table around a well and even reverse the direction of flow past a nearby well, officials might divert contaminated water from a municipal water supply well by encouraging increased pumping at carefully selected locations. A customized graphic script describing the area's groundwater model can be useful for describing the monitoring program, justifying a denser sampling network, evaluating permit applications for new wells and exploring the possible impact of a proposed landfill. A competent visualization support system can increase the value of a groundwater model by making it intelligible to more than a handful of geologists and planners.

Cartographic Issues

Guidelines for designing static environmental maps (Monmonier and Johnson 1991) are useful in developing dynamic graphic narratives addressing environmental risk. Among the more relevant guidelines is the functional distinction between the locator map linking a hazardous site to a broader geographic frame of reference and the risk map describing threats to life or property in the vicinity of the hazard. The relative location and environmental threat are distinctly different cartographic themes, requiring treatments different in content, level of detail and scale. A customized map generated by a visualization support system can use graphic space efficiently if the system places the hazardous site and the viewer's place of work or residence on opposite sides of the screen. Zoom animation can then link the locator map with a more detailed representation of the site and the area at risk. For a hazard with an exceptionally broad impact, a dynamic zoom-in/zoom-out sequence might begin with a large scale locator map showing the viewer's residence, move to a smaller scale map describing the hazard's wider scope and regional impact and conclude with a more detailed, larger scale cartographic treatment that warns of danger in the vicinity of the site. In this three-stage portrayal, a smaller scale regional map is needed to explain the hazard's cause or development, whereas a more detailed view with a smaller geographic scope is required to describe local impact.

Another relevant design guideline is that maps communicating environmental risk must be interpreted for local officials, news editors and other viewers unfamiliar with scientific jargon and technical details. An important strategy, especially in print communications, is juxtaposition, which affords a ready link among one or two maps, an interpretative profile diagram and a block of explanatory text. However, electronic formats can provide more effective interpretations because a skillful customized animation can link a much larger number of revealing views while a voice-over describes relevant physical processes in language suited to the viewer's background and experience. Animated presentations can also be useful in describing the likely sequence of events and in relating a general explanation to the unique character of the locality. Multiple cartographic views are especially useful for volcanic hazards, which involve several types of risk, including rock fragments scattered during an explosive eruption, molten rock ejected from the vent, overland flows of viscous lava mudflows carried tens of miles beyond areas inundated by lava and ash falls covering hundreds of thousands of square miles downwind from the eruption. At least one summary map with a clear title and key should identify areas that will definitely be affected by the hazard, areas that might be affected and areas not likely to be affected.

Script authors must be wary of misinterpretation in communicating risk and uncertainty. If a map shows varying degrees of risk, for instance, printed text or a voice-over should warn viewers that "low risk" is not "no risk". And where boundaries are uncertain, the map should point out this uncertainty with dashed boundary lines, interdigitated fill patterns, question marks or other readily interpreted symbols. Dynamic effects can be useful too, as MacEachren (1992) has illustrated. On a flood hazard map, for example, an animated symbol oscillating between maximum and minimum flood lines provides a dramatic readily understood representation of uncertainty — once the system explains the symbol to uninitiated viewers. Other iconic strategies for communicating risk include using red to represent danger, a black "X" or a well known red circle with a diagonal line to indicate prohibition, an equally familiar red "+" to show a hospital, circles with increasing radii and decreasing thickness to portray impact as a declining function of distance, animation to describe the advance of floodwaters or mudflows and moving arrows or pictorial cars to show direction of evacuation.

Event-specific cartography can be especially important when a hazard requires an evacuation or other restrictions. Because the appropriate response can vary substantially from neighborhood to neighborhood, house to house or person to person, *ad hoc* customization of a map's title, geographic scope, features and symbols is especially valuable in generating maps showing where people may not drink well water or build homes, or where evacuation is necessary. If contaminated groundwater is discovered in one part of an aquifer, for instance, only residents served by the affected wells need to switch to bottled drinking water. For a flood emergency, customized maps showing where homes might have six feet of water in the living room and where houses are likely to sustain only mild to moderate wind damage can assure an orderly evacuation that doesn't needlessly overwhelm regional shelters. Customized maps distributed to police and fire officials might prevent tragic deaths by identifying elderly or disabled persons living alone and other vulnerable households. Because script-generated dynamic graphics are readily customized for a timely response to a specific threat to a specific area, scripting promises significant improvements in risk communication and emergency management.

During a disaster a graphic narrative might serve as a checklist to remind emergency management officials about needs and priorities. Developing and evaluating such a script could be a useful strategy for assessing the adequacy of a community's disaster planning. Moreover, graphic scripts that promote awareness of hazards can be rhetorically useful in convincing local officials and property owners to restrict land use on steep slopes or to invest in a municipal water system and sewage treatment plant.

Scripts for Environmental Analysis

Graphic scripts can address a variety of themes in environmental analysis. For local officials concerned with the impact of either a proposed waste disposal facility or a newly discovered hazardous waste dump, a script based on appropriate environmental models and credible assumptions might clarify important planning issues by presenting best case and worst case scenarios. To help officials explore the range of plausible outcomes, a graphic script might contrast results obtained with different models, different assumptions and different remediation strategies. Where uncertainty is primarily a reflection of incomplete or otherwise suspect data, graphic narratives might help scientists and engineers better understand the implications of a broad range of low probability conditions and occurrences.

Trained analysts are not the only potential users. By making environmental models accessible to non-technical people as well as to competent scientists unfamiliar with arcane geographic software, a visualization support system can extract greater returns from costly efforts to collect data and build databases. Moreover, graphic scripts designed to introduce inexperienced viewers to environmental models can ensure fuller citizen participation in local planning decisions. Scripts might even promote a dialog between state or federal officials responsible for siting a locally objectionable facility and local officials willing to negotiate safety standards and an appropriate benefits package. As Shiffer (1992) observes, electronic collaborative planning systems based on associative information structures and graphical interfaces can overcome barriers to group decision-making. An interactive system serving multiple users could encourage various parties to a controversy not only to state clearly their own objectives, but to understand the concerns and objections of others. A graphic script that illustrates a threat to life or property might encourage the developer to either modify his or her plans or propose a conciliatory benefits package for local residents. Similarly, a script demonstrating clear benefits with minimal risk might encourage opponents to develop counterproposals that ensure safety and aesthetics. By welcoming and empowering citizens groups and individuals, such systems might establish a mutually beneficial dialog that replaces mistrust and hysteria with understanding and rational skepticism.

Not everyone would see it that way, however. In commenting on an earlier draft of this chapter, one reviewer noted:

> The role of graphic scripts in 'establishing a mutually beneficial dialog...' is hard to buy. Are not graphic scripts an idea for those in power to control public opinion by creating the pretense of showing multiple perspectives while at the same time inhibiting access to those perspectives that are furthest from their own? The whole idea of user profiles could be turned against the citizen advocate if the government power structure is the one with control over both the data and the provision of graphic scripts.

Although my knee-jerk reaction was glibly to dismiss this rather unsound objection as creeping Harleyism, the possibility exists that a government, a corporation or some special interest group might use graphic scripts arrogantly and unfairly to highlight its own views and stifle opposition. It's not only possible but quite likely. Yet the evidence, however anecdotal, suggests that in a modern (or post-modern) democracy such heavy-handed efforts to manipulate public opinion ultimately backfire. Moreover, the multimedia technology required seems not so sophisticated that only the rich and powerful will have access to it. If cartographic and geographic educators are effective in promoting informed skepticism and encouraging users to question the map-maker's authority, the benefits will clearly outweigh the threats.

Concluding Remarks

Although visualization support technology is still largely speculative, its philosophical approach, feasibility and potential benefits seem quite clear. Graphic narratives can be a powerful tool for generating informative maps tailored to the expertise and concerns of individual users as well as for addressing important questions that call for more than a handful of maps. Graphic scripts cannot only integrate maps with text and other displays, but also organize information coherently for efficient understanding. There are limits, of course, to an individual's endurance as well as to his or her will and capacity to grasp complex issues. Customized graphic narratives are, however, promising tools for exploiting and extending the viewer's comprehension.

Visualization support systems seem likely to evolve gradually, building on advances in cartographic animation, interface design and spatial cognition. The inevitable implementation of these systems suggests several productive avenues for basic cartographic research, including cartographic complementarity and geographic informativeness. By cartographic complementarity, I mean efficiently integrated multiple graphics. Some of these designs might portray relationships jointly or serially in the geographic space of the map and the attribute space of the statistical diagram, whereas others might provide a parsimonious and coherent treatment of spatial–temporal data multivariate data or multiple representations at various scales. By geographic informativeness, I refer to strategies for making the content, spatial pattern and visual design of a map or statistical diagram interesting and meaningful. Visualization support systems must address effectively the moderately vague queries of the user who begins an exploratory session by asking, "What environmental hazards exist in my neighborhood?" or "What's wrong with this environmental impact statement?" Hardly unreasonable requests, are they?

References

Armstrong, M. P., P. J. Densham, P. Lolonis and G. Rushton (1992) "Cartographic displays to support locational decision making", *Cartography and Geographic Information Systems*, 19, pp. 154–164.

Committee on Advances in Assessing Human Exposure to Airborne Pollutants, National Research Council (1991) *Human Exposure Assessment for Airborne Pollutants: Advances and Opportunities*, National Academy Press, Washington.

Committee on Ground Water Modeling Assessment, National Research Council (1990) *Ground Water Models: Scientific and Regulatory Applications*, National Academy Press, Washington.

MacEachren, A. M. (1992) "Visualizing uncertain information", *Cartographic Perspectives*, No. 13, pp. 10–19.

Monmonier, M. (1989a) "Geographic brushing: enhancing exploratory analysis of the scatterplot matrix", *Geographical Analysis*, Vol. 21, pp. 81–84.

Monmonier, M. (1989b) "Graphic scripts for the sequenced visualization of geographic data", *Proceedings of GIS/LIS '89*, Orlando, pp. 381–389.

Monmonier, M. (1990) *Atlas Touring: Concepts and Development Strategies for a Geographic Visualization Support System*, New York State Center for Advanced Technology in Computer Applications and Software Engineering, Syracuse.

Monmonier, M. (1992) "Authoring graphic scripts: experiences and principles", *Cartography and Geographic Information Systems*, Vol. 19, pp. 247–260, 272.

Monmonier, M. and B. B. Johnson (1991) "Using qualitative data gathering techniques to improve the design of environmental maps", *Proceedings of the 15th International Cartographic Conference*, Bournemouth, pp. 364–373.

Schank, R. C. (1991) *The Connoisseur's Guide to the Mind*, Summit Books. New York.

Schank, R. C. and R. Ableson. (1977) *Scripts, Plans, Goals, and Understanding: An Inquiry into Human Knowledge Structures*, Lawrence Erlbaum Associates, Hillsdale.

Shiffer, M. J. (1992) "Towards a collaborative planning system", *Environment and Planning B*, Vol. 19, pp. 709–722.

Williams, J. M. (1990) *Style: Toward Clarity and Grace*, University of Chicago Press, Chicago.

Zannetti, P. (1990) *Air Pollution Modeling: Theories, Computational Models, and Available Software*, Van Nostrand Reinhold, New York.

CHAPTER 12

Designing Interactive Maps for Planning and Education

HARTMUT ASCHE

Department of Cartography
Map Design Laboratory
Berlin Polytechnic University
Fachhochschule, Germany

CHRISTIAN M. HERRMANN

Department of Cartography
Karlsruhe Polytechnic University
Fachhochschule, Germany

Introduction

Against the background of a worldwide spread of modern information and communication technologies, the earth sciences, during the last two decades, have witnessed a momentous increase in all kinds of geographic data. To structure and exploit this largely alpanumeric information, transformation into graphic form is a particularly effective way of spatial data handling (Bertin 1981; Tufte 1990). However, the rapid propagation of information graphics has clearly multiplied the demand for effective computer-assisted methods of graphic data presentation and analysis collectively known as visualization. Apart from its technological element, visualization is not a completely new concept in the earth sciences, where maps are well established tools to visualize, evaluate and communicate geographic information. With the use of digital information technology becoming increasingly widespread in cartography, a general trend can be observed to broaden the concept of the conventional static map graphic towards interactive electronic applications.

Changing Framework of Cartographic Modelling

In cartographic information processing, the adequate visualization of geographic information is of fundamental importance for the clarity and comprehensibility of the resulting map graphic. This decisive process of transforming spatial data into two- and three-dimensional graphic models is known as cartographic modelling. Professional map modelling is essential for the creation of meaningful maps that facilitate effective visual communication and intuitive cognition of geographic data structures. During the era of conventional map production technologies, cartographic data processing has largely been restricted to modelling static two-dimensional map graphics. In the age of automated cartography, modern computing and information technologies are removing this restriction by presenting a wealth of options for data selection, manipulation and display, including multimedia integration. Combining high level processing with a great potential for the cartographic visualization of geodata, standard computer graphics systems support complex interactive modelling tasks, including multidimensional map graphics, dynamic presentations of environmental change or spatial mobility, and spatio-temporal simulations.

Conventional to Digital Map Design

In conventional map modelling the transformation of spatial data into a map graphic involves a number of design tasks, including the preparation of a compilation manuscript which is manually rendered into a sketch map as the graphic reference for map production. Because of the manual techniques applied, alterations or complete redesign are problematic in terms of the time and costs involved. This rather inflexible map design process clearly limits changes in the graphic structure of conventionally designed maps.

The implementation of digital mapping has substantially increased the creative potential of map design. Desktop mapping environments provide the cartographer with a variety of flexible and intuitive drawing and design functions (Whitehead and Hershey 1990; Asche and Herrmann 1991). By directly interacting with the digital mapping system, he/she is able to manipulate and experiment with the map data and their symbolization on the screen. However, flexible, digital map design holds additional, although limited, benefits of spatial data use for the earth scientist. Owing to the overlay concept of most desktop graphics packages, maps are designed by placing related graphic elements in separate layers. By turning specific sequences of layers on and off these map elements can be selected, displayed and hardcopied in any combination, enabling the analyst to derive a number of different maps from only one set of geographical data.

Both conventional and digital map modelling ultimately aim at producing thematic maps for analog output (Fig. 12.1). Like their conventional counterparts, digital paper maps allow for all forms of analog map use from general information and data exploration to in-depth analysis. However, all of these tasks exclude direct interaction with the information presented in the map on the user's part.

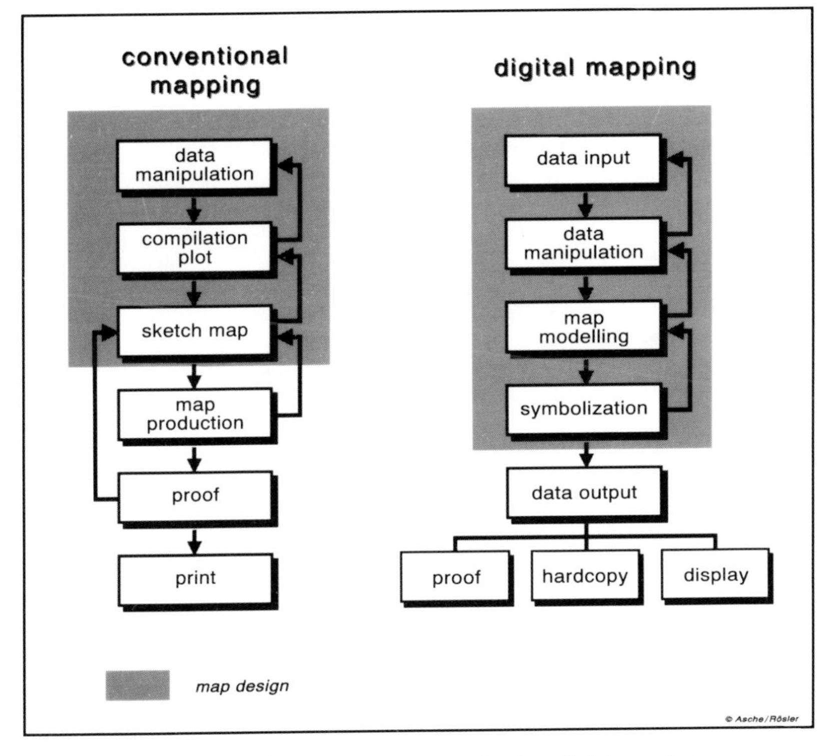

FIG. 12.1. Map design process: conventional to digital mapping.

Interactive Mapping

Once digital technology is not only utilized for map design and production but also for publication and use, digital map data allow for subsequent extension into a map-based multimedia product termed the "electronic" or "interactive" map. Unlike conventional paper maps, these maps are primariliy designed for softcopy display on high quality monitors and not for hardcopy output, i.e. for use on graphics workstations or electronic mapping systems. Fully conforming to the principles of map modelling, the visualization of spatial data is central to interactive mapping, and the map graphic is the essential part of an electronic mapping environment. Interactive mapping supports the integration of time-based data, of animation and sound as well as system guidance by virtual "agents" or guides (cf. Ormeling 1993). In addition, the data structures of interactive mapping systems can be manipulated through interactive controls, annotated with graphics, images and text supplements. By providing user-defined access to complete sets of geographic information, interactive mapping enables the user as well as the cartographer to experiment with the cartographic representation of these data.

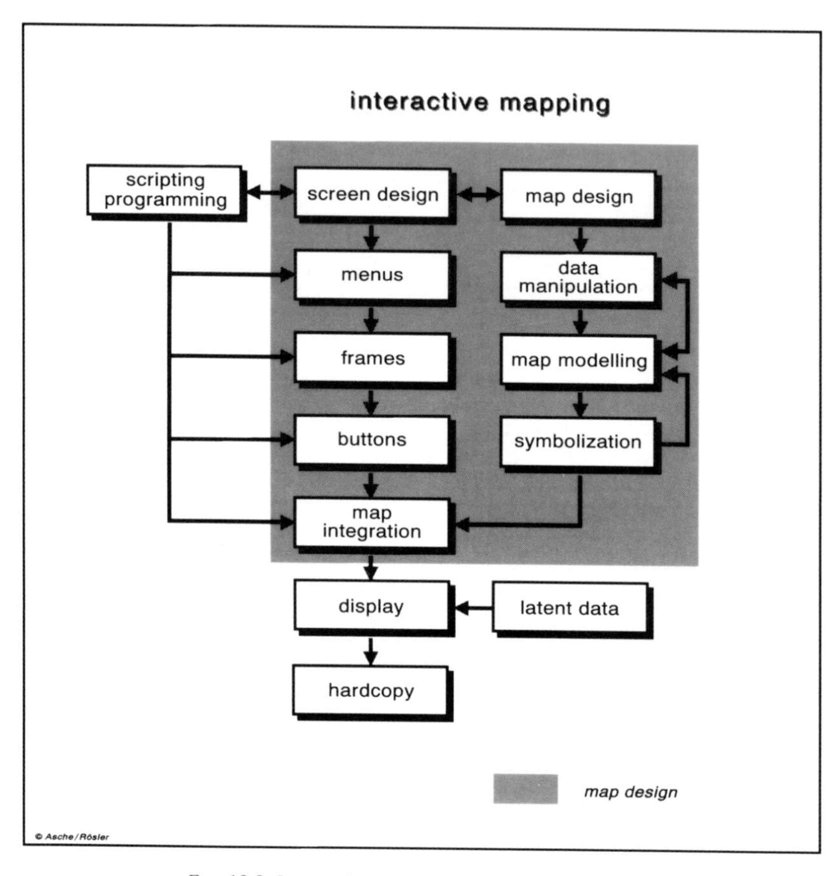

FIG. 12.2. Interactive mapping: map design process.

Interactive Map Design: Basic Issues

Like static maps, interactive multimedia maps are graphic instruments to analyse and communicate geographic data. Designing interactive maps therefore involves the tackling of issues typical of any cartographic communication process including information content, visualization and data extraction. At the same time, relevant principles of multimedia such as interactivity, navigation and transparency will have to be observed (Hoffos 1992). Coupled with the need to develop an application-specific user interface, the challenge of interactive map design is to merge both demands into intuitive and effectively visualized cartographic applications (Fig. 12.2). Some of the basic design strategies employed in creating interactive mapping applications include application consistency, system interactivity, data navigation and application-specific screen map design.

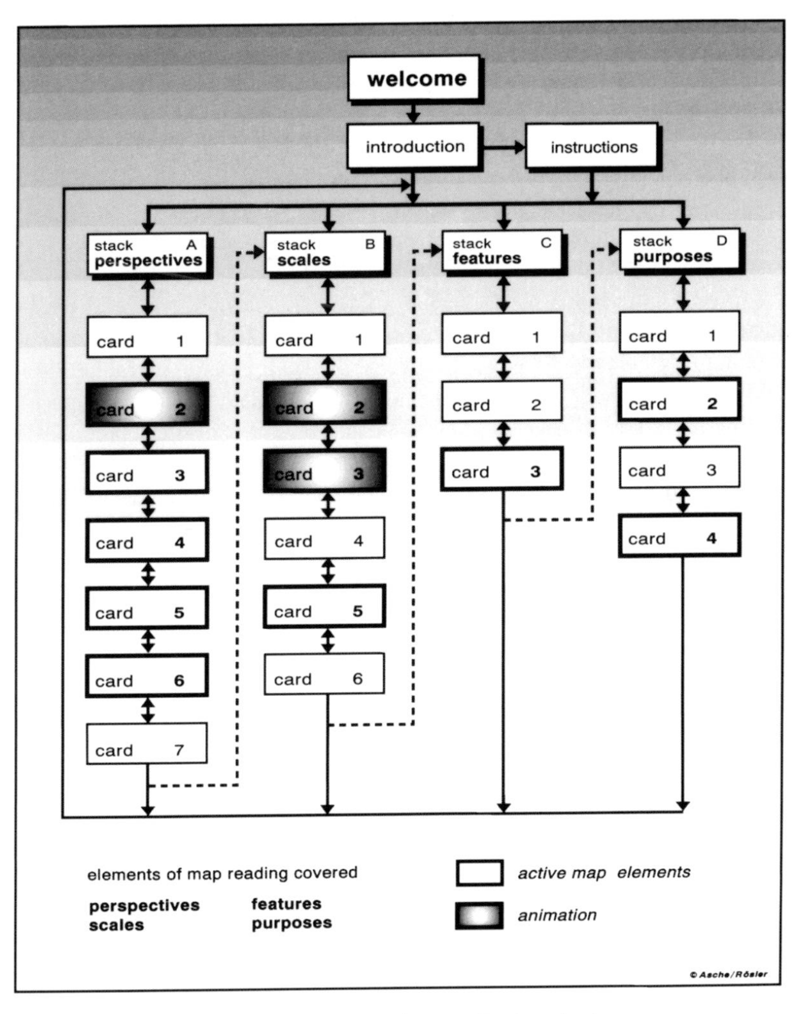

FIG. 12.3. Interactive mapping: application structure.

Consistency

At a conceptual level, interactive map design requires the development of a set of conventions for each and every part of the application, specifically for (1) the visual presentation of the information or screen design and (2) ways to access and explore the information or database navigation. It is the cartographer's task to transform these conventions into an interactive map design that effectively conveys the options of data access and navigation to the user. Thus simple and consistent screen designs are a prerequisite to facilitate easy user operation without further thoughts about the basic system functions of menus and buttons. A well established way to achieve consistency of an application is to adopt a strictly hierarchical structure of map content and its visualization regardless of its possible non-linear presentation to the end user (Fig. 12.3).

Interactivity

In the context of computer technology, interactivity is concerned with the flow of information between people and machines. One of the major challenges facing the cartographer in interactive map design is to achieve the right type and degree of interactivity for many different potential users (cf. Lindholm and Sarjakoski 1993, this volume). Interactive systems allow the individual user to determine the level and flow of information he/she wants to access, setting his/her own pace and branching to different data as they interest him/her. For that purpose, truly interactive mapping systems should present the user with a number of choices, options and decisions including (1) on-screen choices, (2) resources to make this choice and (3) appropriate program response. To encourage and challenge the user to explore the creative potential of an interactive mapping application, interactive features should be as transparent as possible and provide constant feedback.

Navigation

To access the information contained in interactive maps, the user is presented with multiple ways to jump between associated data. This concept of variable data access and navigation around the database is fundamental to interactive systems

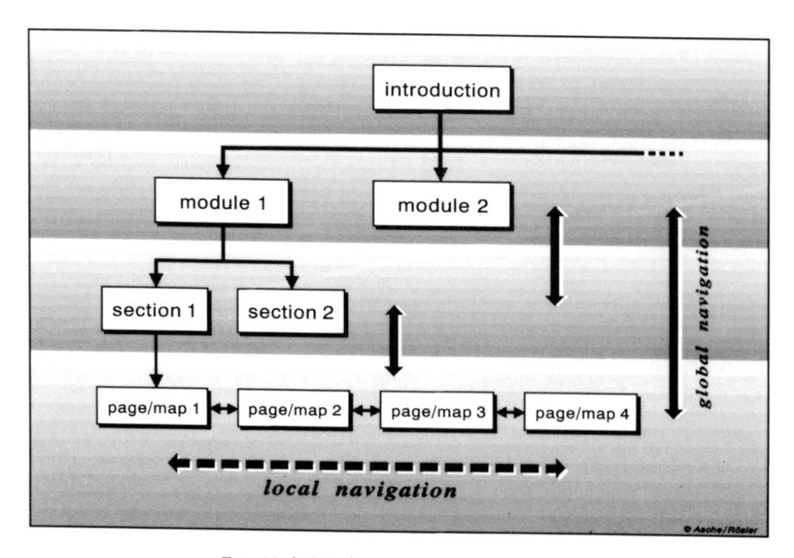

Fig. 12.4. Database navigation concepts.

and the key to interactive mapping. However abundant navigation possibilities may seem in an interactive map, the options are ultimately limited by the data stored in the system and the navigation paths offered. Interactive mapping systems should support at least two different kinds of database navigation: (1) "global" navigation which enables the user to access different types of information and (2) "local" navigation which allows the user to move within the same type of data (Fig. 12.4).

To move through the database, interactive maps provide a number of navigational aids which, when activated by the user, display windows containing additional information on a specific topic, switch to other screens, start animations or play video sequences. Navigational tools used in interactive map design belong to the groups of (1) text-based aids, of which pull-down menus are the most important and (2) buttons or icons in the form of navigation buttons or graphic buttons as the main means of interaction (cf. Lindholm, Sarjakoski, this volume). To ensure wide acceptance from many different users including computer novices expecting system guidance and professionals familiar with database navigation, the cartographer will have to design different ways of interactive control and navigation.

Screen Design

Interactive map modelling is based on the concept of links between the data visualized on the map and additional spatial information in the database which otherwise could not have been integrated into the map by conventional methods because of modelling problems. Screen map design, as well as hardcopy design, is concerned with the adequate visualization of point, line and area objects. However, interactive map graphics are primarily designed for on-screen use, necessitating the dimensions of point symbols, line widths and map lettering to be adapted to the respective screen resolution to achieve clarity and legibility of these softcopy maps.

Screen design, like hardcopy design, serves the primary purpose of communicating geographical information effectively. Whereas paper map modelling concentrates on the single optimal graphic presentation of spatial data, interactive map design requires the full multidimensional, multi-perspective visualization of geographical databases. Therefore interactive maps consist of a variable number of graphic screen displays or "screens". Active elements or "hot" spots on each screen facilitate user-controlled access to this latent information (which can be displayed on-screen in addition to the actual screen map) and navigational tools provide choices of which data to move to next (Fig. 12.5). Effective screen design helps the user comprehend the respective application structure enabling him/her to utilize the creative potential of the application while within an intuitive framework.

Although interactive mapping presents the cartographer with a multitude of new and creative possibilities of map design, he/she is faced with the risk of overloading the user with information and allowing the presentation to become more important

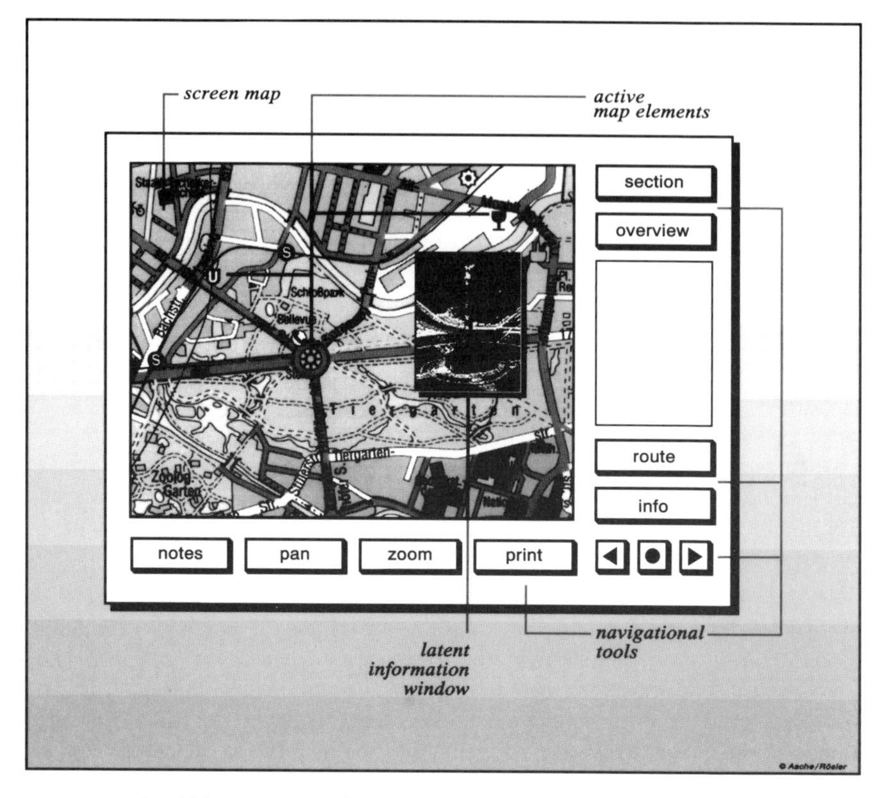

FIG. 12.5. Screen map design: map elements and interactive controls.

than the content of the map. Crowding the screen with detail frequently indicates an information overload on one screen that might confuse or even discourage the non-professional user. Breaking the information into multiple screens, or overlaying portions of the actual screen display instead, seems an appropriate way to reduce visual confusion and increase the efficiency of communication. Ideally, screen design should attract, encourage and keep a diverse range of users engaged while operating an interactive application.

Interactive Mapping Applications

Within the scope of an application-oriented multimedia mapping project, some key issues about the development, design and use of interactive maps are being investigated to assess the cartographic potential of interactive multimedia. Based on a range of entry-level to full-featured multimedia packages, several interactive mapping prototypes have been developed in cooperation with potential system users and field tested to cover strategic geo-related applications in the fields of environmental planning, citizen-based information and education (Rauner 1992; Trummer 1992; Cebulla 1993; Mann 1993; Rachfall 1993; Wurm 1993). In keeping with potential system users' requirements of performance, ease of operation and cost effectiveness, the interactive mapping prototypes discussed below have all been developed on Apple Macintosh hardware and have made use of commercial authoring software components (cf. Murie 1993).

During the development of all prototypes, constant feedback of potential users' has been secured through the evalution and discussion of concepts and development versions of the respective applications. The goal of obtaining this feedback was to achieve interface and map designs that facilitate broad user acceptance, application and use of the respective prototypes.

Environmental Planning Prototypes

In collaboration with the environmental agency of the German federal state of Brandenburg, one feasibility study investigated the application potential of interactive mapping for complex regional and environmental planning applications (Rauner 1992). To fit user requirements of integrating data stored in geographical information systems (GIS) into an interactive map-based information system, the Swiss–German cart/o/info software company's spatial cart/o/info modelling package was chosen to develop the interactive map prototype. Aimed at regional planners who are familiar with conventional map use and, partly, with computing, the mapping application attempts to combine cartographic visualization and GIS functionality with user-controlled system interactivity.

Program Environment

Like comparable GIS software, the cart/o/info package consists of modules or layers providing the user with a common graphical interface for uncomplicated application development (Fig. 12.6).

1. The "database" comprises a relational databank storing all geographically referenced information of the application. Data can be entered either by tablet or keyboard or imported from external databases via relevant exchange formats (e.g. ArcInfo, Sicad formats).
2. The "background" layer contains graphic data of all kinds (topographic and thematic maps, aerial photos, satellite imagery) to be used as base maps for the display of object data. Like their object counterparts, all background data are referenced to a uniform geodetic grid (e.g. Gauss–Krüger, UTM).
3. The "display" layer facilitates the visualization of data selected from layers (1) and (2) for screen display. Transparent graphic buttons can be assigned to selected map symbols, providing direct access to the attributed information (map, graphic, chart, spreadsheet, image, text) when activated.

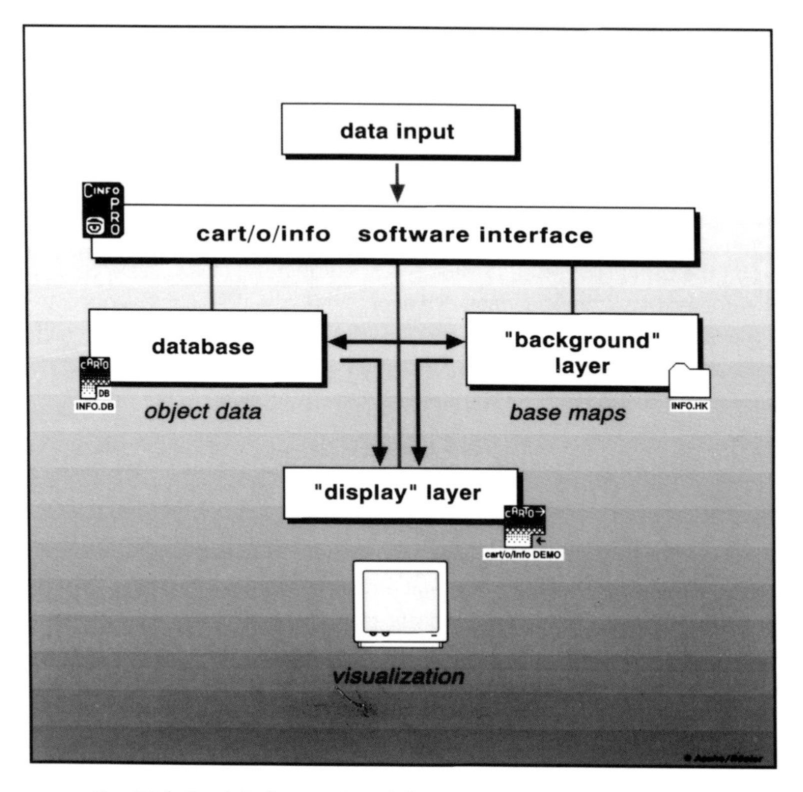

FIG. 12.6. Cart/o/info spatial modelling package: program structure.

Among the range of application development tools included in the program is the powerful cart/o/graphix map construction module, which provides flexible and intuitive interactive map design functions in line with the quality demands of professional mapping.

Prototype Design

Owing to the fact that a full-featured version of the program was not available for the development of the Brandenburg protoype, the functionality of this application differs slightly from systems subsequently developed with the complete package. All application-specific interactive controls — the most important of which are menus and graphic buttons — are designed on the prototype's programming layer. At present, controls are mouse-operated, but can easily be adapted to touch-screen operation if required. Alphanumeric and graphic data on the environmental status (groundwater, surface water, air pollution) of the Kleinmachnow test area are stored and manipulated interactively in the prototype's database layer. Here each object is

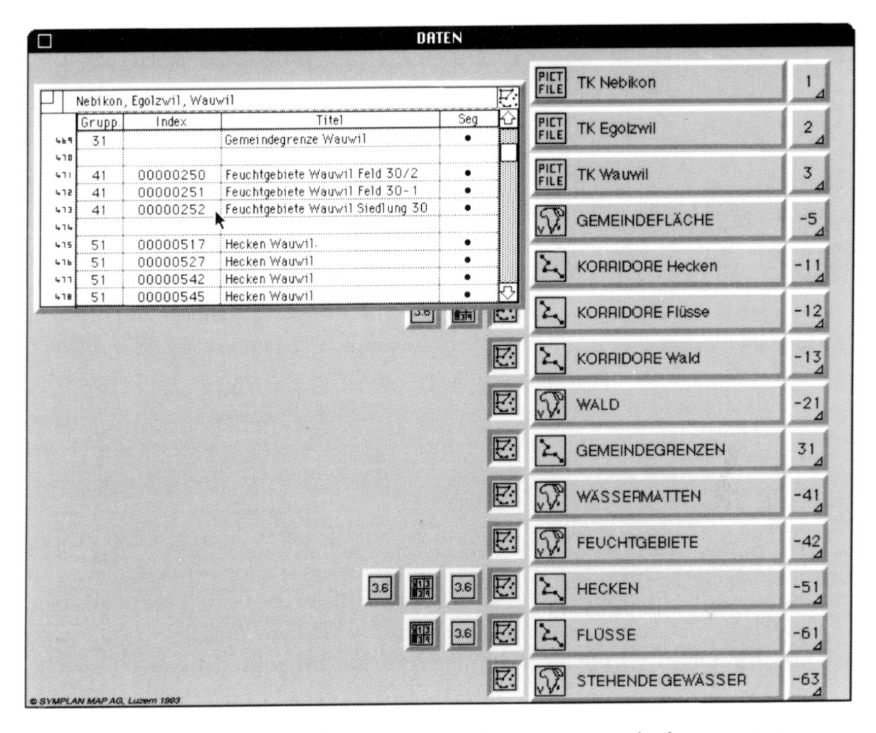

FIG. 12.7. Environmental change prototype Lucerne canton: database content.

identified by its respective Gauss–Krüger coordinates, to which object number, attribute information and symbolization are assigned (Fig. 12.7). As required, the majority of the GIS data for the test area can be imported into the prototype from existing ArcInfo files through a specially programmed format exchange interface. Large and medium scale topographic maps (1 : 10,000, 1 : 50,000 scales) and a reference map (1 : 500,000 scale) to mark the areas covered by the application are included in the prototype's background layer. Map data have been integrated into the prototype by scanning the originals at double screen resolution (144 dpi).

Saved as 24-bit colour TIF files, all data were converted to 8-bit PICT files to compress the file size before being imported into the application via the cart/o/graphix module. Additional thematic screen maps were created with the cart/o/graphix map construction module. A run-time version of the cart/o/info programme is integrated into the interactive planning protoype, enabling the user to run the application without a program license.

Parallel with the Brandenburg prototype a similar case study has investigated the integration of GIS and multimedia data into an electronic atlas of environmental change (Trummer 1992). Developed in cooperation with a Swiss environmental planning enterprise, the application visualizes environmental change in the Swiss canton of Lucerne by communities from 1900 to 1990, allowing the user to select different states of change and access complete data sets for individual evaluation and visualization (Fig. 12.8).

Prototype Operation

To facilitate intuitive and effective interactive map use, the prototype's user interface has been customized to the specific requirements of the application. Compiled by a group of five regional planners and geographers of the environmental agency with various degrees of computer literacy, requirements included menu-operated interactive controls and graphic buttons or hot symbols for database access. That is why user interaction with the application is through active map symbols and pull-down menus, the latter enabling the user to select map scales, map types, map content and map use tools (Fig. 12.9). Regional map coverage of the scale selected is indicated in the reference map from which areas of specific interest can be selected interactively. For visual comparisons or regional setting, the current area can be shown in other base maps displayed in additional windows on-screen (Fig. 12.10). Buttons integrated into these windows allow the user to zoom the scale of the window map to the screen map scale and, more importantly, to overlay or merge window maps with the screen map. When selected via the menu bar, environmental topics can be symbolized on the respective background map. In any of these maps, hot symbols can be activated to pop up a menu-like field next to the symbol displaying predefined object data (name, location, characteristics) and additional information available from the database (Fig. 12.11). Controls included in this field allow the user to add or update data linked to the activated symbol.

Providing regional planners with a full-featured map-based information system, the Brandenburg prototype allows for in-depth interactive data evaluation and database manipulation in a variety of ways. Thus the environmental planning application presents the user with an intuitive, highly interactive framework for the compilation and production of professional planning maps individually tailored to project-specific demands. As a large proportion of regional and environmental planning tasks still require printed map documents, often at short notice and in a limited number of copies, the prototype supports high quality hardcopy output on a variety of devices and formats from cost effective inkjet plotters to colour separated offset films.

Education Prototype

To study the cartographic potential of entry level multimedia programmes, two case studies have been carried out (Rachfall 1993; Wurm 1993), of which the one discussed below investigated their application to interactive map reading. Intended for primary school pupils, a low interaction introduction to map reading has been developed with the Supercard multimedia package (Rachfall 1993). Based on an

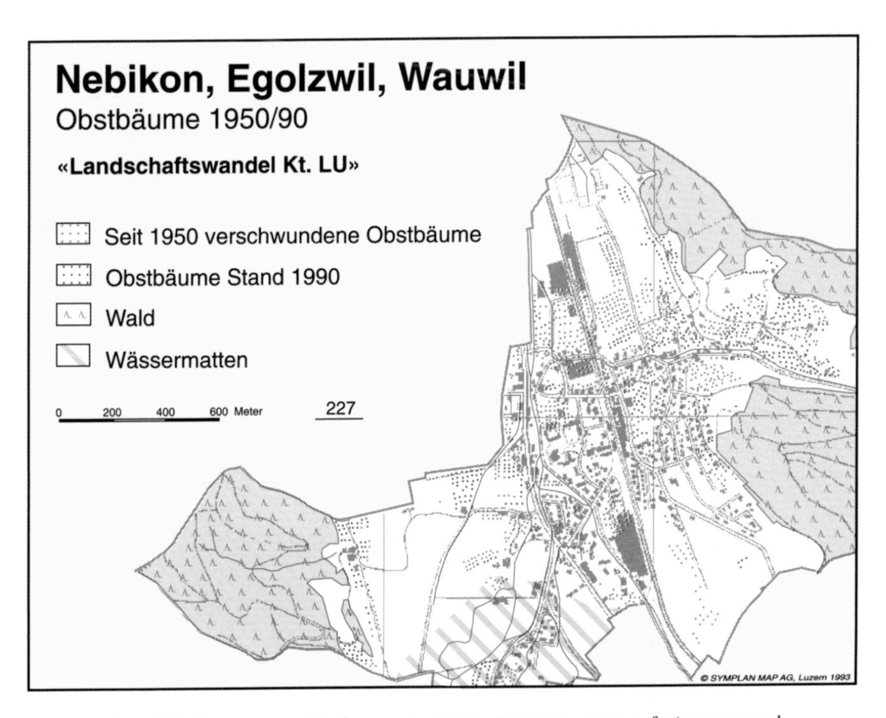

FIG. 12.8. Environmental change prototype Lucerne canton: fruit tree stand 1950–1990, Nebikon area.

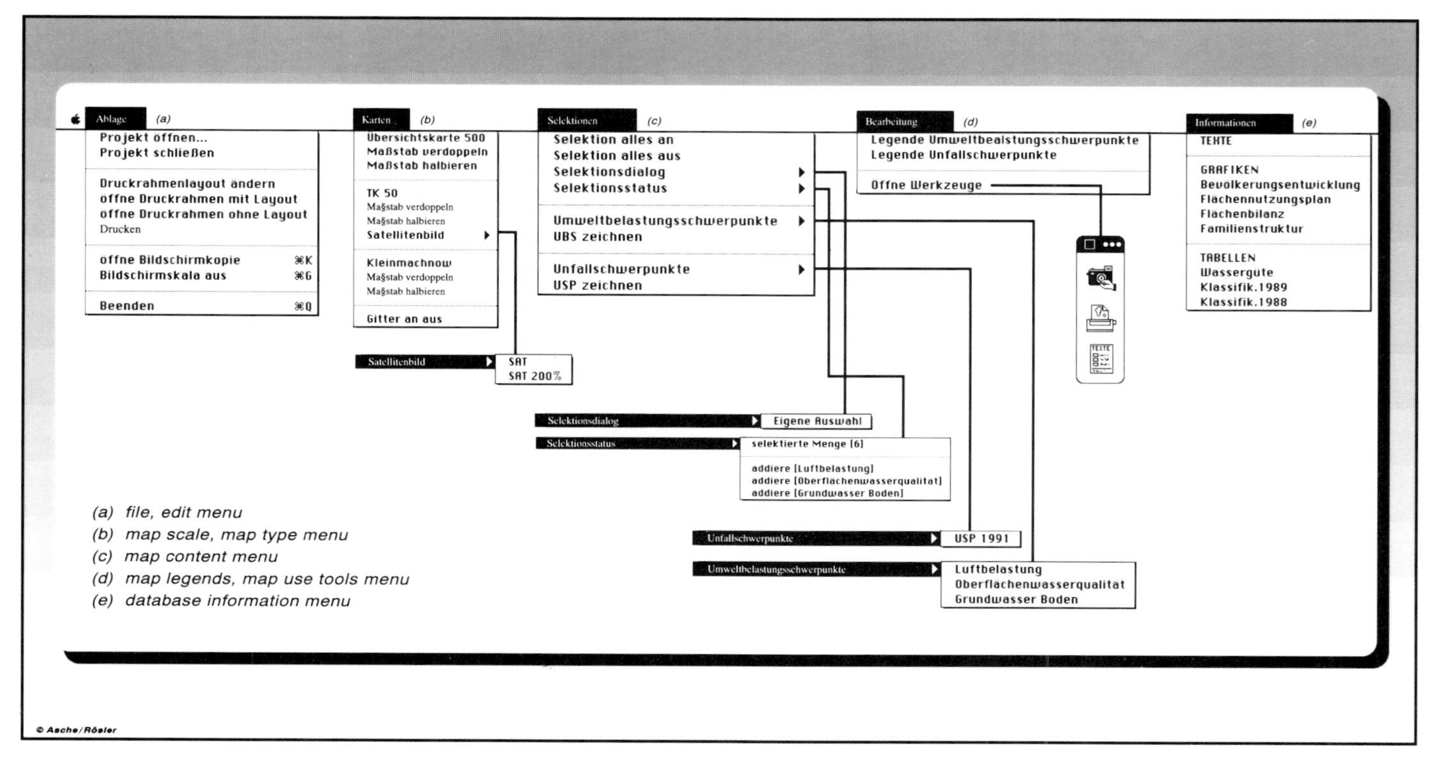

FIG. 12.9. Environmental planning prototype Brandenburg: application structure and menu controls.

existing conventional introduction for a Berlin primary school atlas produced in co-operation with a German map publishing house (Töns 1990), this interactive prototype aims to communicate the principles of cartographic data processing, symbolization and map reading to young map users in a multidimensional, yet easy to understand way.

Program Environment

The Supercard package is a stack-oriented programming environment for interactive multimedia applications, consisting of (1) the main program to run and control an application, (2) the authoring part and (3) a scripting language based on the Hypertalk standard to control the program's interactive features. Interactive control devices such as menus, navigation buttons and hot spots can easily be created, as an icons library is included in the program. In addition, the program supports elementary graphic design functions. All data in the application are stored on cards, each of which holds pieces of information and buttons for programme navigation. Collections of cards form stacks which contain information about

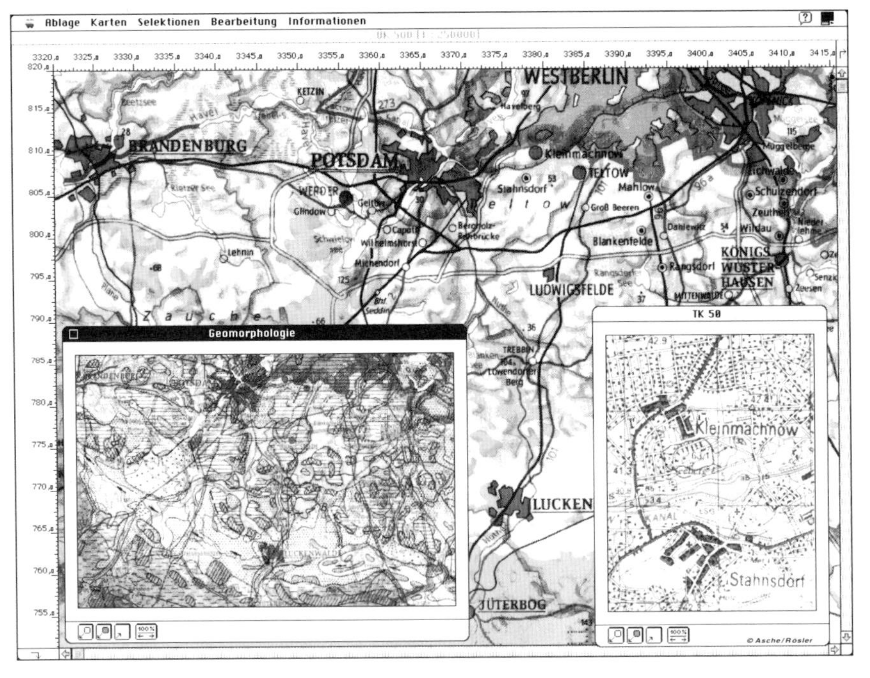

FIG. 12.10. Environmental planning prototype Brandenburg: screen map (physical, 1 : 250,000 scale) and window maps (geomorphology, 1 : 500,000 scale; topographic, 1 : 50,000 scale).

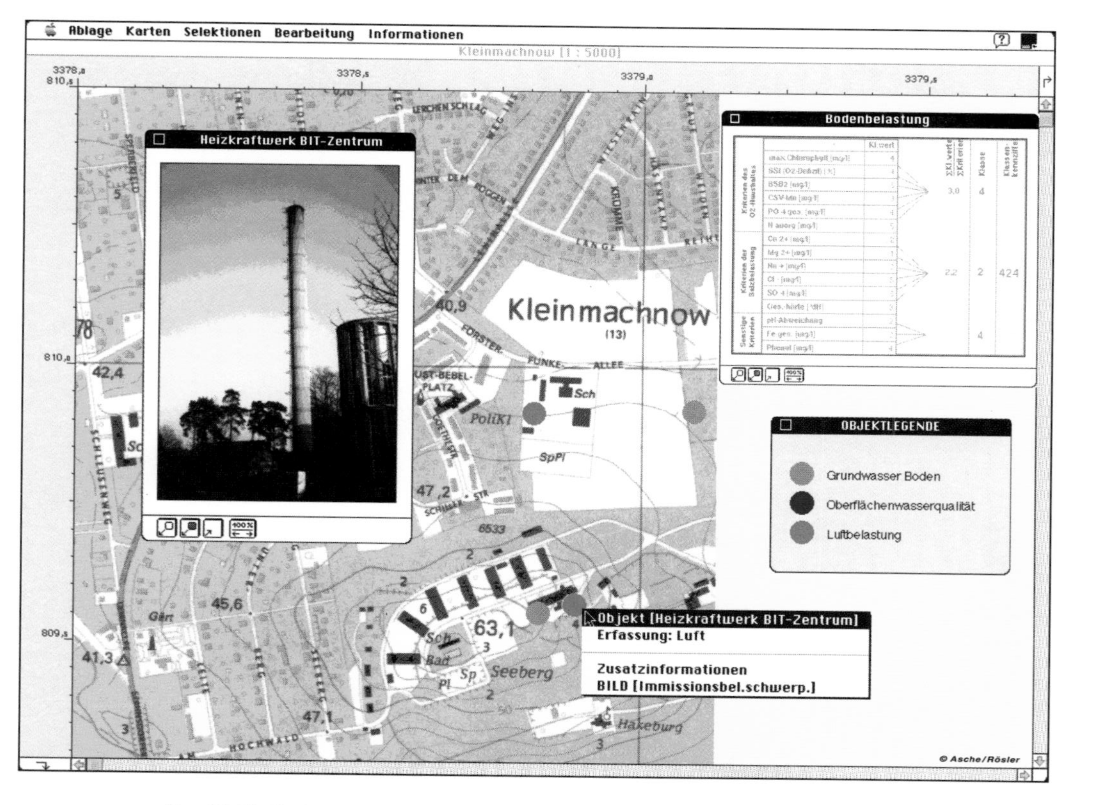

FIG. 12.11. Environmental planning prototype Brandenburg: user-defined map, soil and air pollution status, Kleinmachnow area.

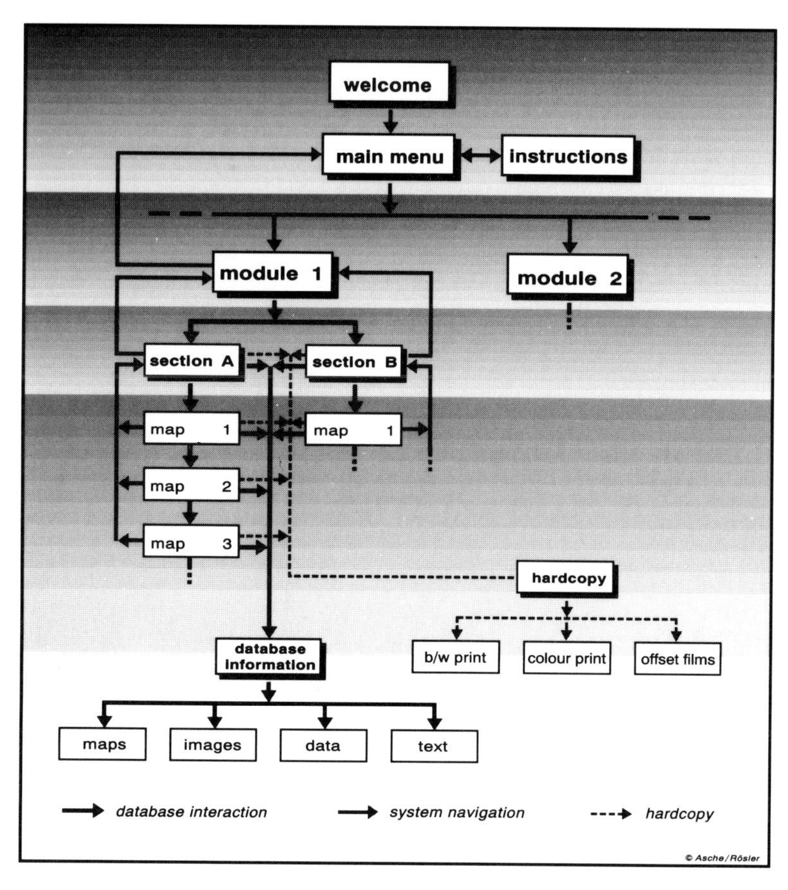

FIG. 12.12. Map reading prototype: application structure.

different aspects of map reading. A player module, integrated into the program, allows for the creation of standalone applications that can be run on any colour Macintosh without the original program.

Prototype Design

For ease of operation and maximum transparency, the education prototype follows a strictly hierarchical design. Operated from a start stack including the welcome screen, user instructions and program information, the application consists of four stacks, each of which covers a different aspect of map reading: perspectives, scales, features and purposes (Fig. 12.12). Topics presented in each stack are preceded by a text card explaining the topic and the accompanying map reading exercises. Each card of each stack contains two navigation buttons to access the preceding and following cards, which are placed on a common graphic background. Depending on the information presented, up to three buttons have been added on an individual card to facilitate context-specific operations.

In the development process, Supercard's graphic functions have proved inadequate for cartographic modelling tasks, necessitating all spatial information including objects and screen maps to be designed outside the program. Owing to its widespread use in cartography and its intuitive and flexible drawing, design and editing features, the Freehand desktop design package has been selected as the appropriate software for designing the screen maps required for the prototype. Integrating them into Supercard, however, turned out to be problematic as the only exchange format shared by Supercard and Freehand was PICT. Nonetheless, even the import of Freehand PICT files into Supercard resulted in significant colour changes and accidental distortions in the graphic structure of the screen maps, preventing direct integration into the application. In contrast, scan data imported as PICT files did not present similar integration problems. To facilitate integration into the Supercard application without major loss of quality, all Freehand screen maps had to be hardcopied as high quality colour proofs, scanned and saved as PICT files, which could be imported into the application along with the associated colour look-up tables to avoid colour flashing.

FIG. 12.13. Map reading prototype: virtual tour guide "KartoMax".

FIG. 12.14. Map reading prototype: animated sequence on perspectives.

Prototype Operation

To attract young users to interactive map reading, to keep them engaged and allow for their identification with the application, the prototype provides a virtual tour guide called "KartoMax" for user guidance throughout the application. Shaped as a pupil, the guide explains the use and purpose of the application, encouraging the user to explore the stacks, access latent information, work on map reading tasks or start animated sequences (Fig. 12.13).

In compliance with the target audience, screen design of the prototype has been kept comparatively simple to enable young map users to concentrate on the information presented rather than on the medium. Consequently, with the exception of the welcome stack, only three different types of screens are presented to the user for foolproof spatial data exploration. The amount of graphic data required for animation severely impedes the interactivity of the application. That is why only one sequence, the change of perspective from terrestrial to bird's eye view, is animated. This sequence can be played by clicking on a balloon manned by the virtual guide (Fig. 12.14). Activating the guide's camera will produce a view of the identical object from the perspective selected.

Discussion of Prototypes

To evaluate the above interactive map prototypes, their design, functionality and interactive map use potential, no formal testing of users under controlled conditions has been carried out to date. Both prototypes were, however, informally tested and discussed with potential users in addition to a critical analysis by all involved in their development.

For assessment of the Brandenburg prototype the application has been operated by more than a dozen computer literate regional planners and five computer novices of the environmental agency. With all of them the application has afterwards been discussed in informal interviews on several occasions in 1992–93. In these discussions, planners emphasized the prototype's visual presentation and access to spatial data, intuitive and flexible operation, ability to produce user-defined, project-specific interactive maps and capability of instant hardcopy output, particularly of good enough quality hardcopy for field tasks.

A number of regional planners suggested the implementation of reduced versions of the application on colour notebook computers for field use to complement and eventually substitute, for paper maps.

To evaluate the education prototype, a field test was carried out in 1993 with 28 eight to ten year old primary school pupils hitherto unacquainted with educational computer applications. The test revealed no major problems in operating the prototype which, despite its preconceived low interaction profile, greatly attracted the users to master the information presented in the prototype. A brief evaluation of the pupils' on-screen performance and informal discussion in small groups of four pupils each gave the following results.

Independent Learning

Pupils could individually control the pace of learning, repeat a card or stack or activate additional explanations. The semi-private learning situation of the application significantly helped to reduce pupils' fear to openly admit they did not comprehend the course topic the first time.

Hands-on Learning

Pupils were actively involved in the process of learning. In particular, animated sequences on map scales with the virtual guide gave pupils the impression of actually controlling or experiencing the dynamic process of changing map scales. As topics such as this are difficult to present in static maps or text, the interactive application allows pupils to repeat the process as many times as it takes them to grasp the underlying concept.

Course Differentiation

From a teacher's perspective, the individual learning environment of the application allowed for better course differentiation. Quick learners were offered the opportunity to explore the topic beyond the basics whereas slower learners could further their understanding of the topic by working through additional explanations contained in the application. In a conventional learning environment, this need to provide more detailed explanations to a few pupils would often slow down the rest of the class.

Interactive Map Use: Spatial Data Exploration

Whatever their specific application, interactive maps share some basic functional characteristics essential for interactive map use. Any interactive mapping requires an electronic mapping system, the heart of which is a high performance graphics computer, optionally complemented by optical data storage facilities (Burger 1993; Cartwright, this volume). By integrating the abilities of the computer, the user is provided with a variety of intuitive tools to access and interact with the multidimensional information presented in interactive maps in many different ways.

The scope and scale of extensions added to the conventional forms of analog map use by the extra dimension of interactivity depend essentially on the range of options for operation and navigation implemented in the application and the presumed abilities of the potential users. Based on the applications developed and user experience with other interactive systems (among them the Swiss infoplan regional planning system, the Grisons information system of Switzerland, city information systems including those of Berlin, Flensburg, Potsdam and Vienna, the Planetscapes educational system, PC atlases including those of Arkansas, Catalonia and Sweden), a range of applications can be identified. The scope of spatial data exploration may vary from screen map interaction by a diverse crowd of non-professional users typical of point of information systems and computer-based training applications to extensive database manipulation of complex spatial information systems by professional users such earth scientists.

From development experience of the prototypes discussed and informal analysis of prototype users, a set of spatial data exploration features can be identified as essential for effective interactive map use. To provide for a broad range of user interaction (cf. Ormeling 1993), the following options, ranging from viewing, browsing and exploration to manipulation and visualization functions, should be included in interactive mapping systems. Depending on the specific application, system features can either be restricted to view-only use or include the full range required for complex high interaction data manipulation tasks.

1. Interactive maps of the *view-only type* allow for the user-defined selection of map area and map scale. For that purpose, the user is able to scroll or move through the complete area modelled for screen display, selecting different regions for screen display as they interest him/her and zoom for detail or improved legibility (cf. Dorling 1992). Interactive controls provided to the user should include standard devices for interactivity, among them menus to make choices for zooming on any region of the screen map, branching to different map scales or topics available in the application, accessing indexes or printing documents. Various buttons allow for local navigation to following or preceding map graphics. For comparative map data evaluation it should be possible to view two or more maps simultaneously in separate windows, allowing analysis without the necessity to jump between different screens. At this elementary level, users are provided with limited means to exploit the interactive power of multimedia mapping. Consequently, this electronic book-like interactive map use is characterized by a low degree of user interaction.
2. More complex interactive mapping applications of the *database interaction type* present the user with additional, more flexible interactive controls in the form of buttons and hot areas in screen maps. These elements allow the user to interact with the content and graphic structure of the application. While moving around the database, the user is able to access different layers of information contained in the application (cf. McGuinness *et al.* 1993; McGuinness this volume). All data can be displayed on the screen in a user-controlled fashion, facilitating the composition of fine-tuned maps geared to serve individual spatial data use requirements. Depending on the individual user, interactive maps of one mapping system can differ in the amount of detail displayed on the screen. In addition, different user privileges to access all or restricted amounts of data contained in a system enable regional planners, for example, to handle more detail than citizens asked to review a specific plan, or scientists to access more data than students. Enabling the user to exploit the

associative powers of the medium, database interaction stimulates his/her motivation to explore the evaluation potential of interactive maps to acquire increasingly specific information. Additional functions allow the user to summarize his/her single- or multidimensional evaluation of the data, including annotations or summaries in user-defined map form (cf. Monmonier 1992b). If required, all interactive mapping data displayed on-screen can be printed on a variety of hardcopy devices.

3. Combining the visualization of geodata with efficient functions of data management, manipulation and analysis familiar from GIS applications, advanced interactive mapping systems of the *analytical type* provide the user with sophisticated functions for the parallel use of all graphic and alphanumeric information contained in an application. Access to the complete sets of geographical and attribute data allows for detailed evaluation, extraction and analysis of all kinds of spatial information in the database. Just like the cartographer, the user is able to manipulate, supplement and link selected data in the application with external information for user-controlled visualization of spatial data, including the redesign of existing screen maps. Characterized by a high degree of user interactivity, the concept of "intelligent" interactive spatial data use (cf. MacEachren 1987) enables the user to become actively involved not only in the evaluation of spatial data, but also in the visualization of geographic information.

Conclusions

A first assessment of the interactive mapping prototypes discussed in this paper and of other prototype applications (Cebulla 1993; Mann 1993; Wurm 1993) shows that professional applications can be developed with commercial multimedia packages, from entry level to state of the art interactive mapping systems. Althouth map-based spatial information systems provide the range of drawing and design functions indispensible for quality mapping, general purpose multimedia platforms, for the time being, offer limited graphic design tools not complying with the elementary requirements of thematic cartography. For adequate screen design, the cartographer will therefore have to resort to flexible desktop mapping packages. Depending on the application and the multimedia program, advanced map construction packages (of the cart/o/graphix type) or graphics software (of the Freehand type) can be employed for spatial data visualization. For integration into multimedia applications these digital maps have to be imported into the multimedia platform where they are furnished with all interactive controls planned for the application. Regrettably, most multimedia programs offer limited possibilities of importing graphic formats relevant in thematic cartography. Therefore the development and design of application-oriented interactive mapping requires a good deal of experimentation with both graphic design options and screen map integration into the multimedia environments and, if necessary, the development of workarounds.

In the context of rapidly increasing access and use of digital technology in cartography, desktop concepts and interactive multimedia are already exerting a substantial influence on the development of the discipline. Effective geographic visualization is central to this development, as it helps to bring the visualization of spatial data closer to the actual perception habits of map users who are surrounded by worlds of multidimensional, multimedia data presentations. Interactive map design, in particular, is adding novel forms of spatial data presentation to the map graphic. Promoting a multi-perspective treatment of geographic information, interactive mapping provides both the cartographer and map user with a world of intuitive options for the multidimensional visualization, manipulation and use of geographic data. Although there is sufficient evidence of the cartographer's benefits from interactive mapping, little research has been published on the map users' response to screen map graphics and on-screen user interaction. However, evidence from prototype tests and other information indicate the users' readiness for experimentation with the presentation and evaluation of spatially related information, as long as ease of use, simplicity of map graphics and ease of understanding are maintained.

Acknowledgements

The research work described here is the result of a concerted effort involving a number of collaborators from Berlin and Karlsruhe. Of particular note is the work of Marina Rösler, Map Design Laboratory, TFH Berlin. Thanks are also due to Alan MacEachren, Terry Slocum, Fraser Taylor and an anonymous referee who read earlier versions of this paper and provided valuable comments.

References

Asche, H. and C. M. Herrmann, (1991) "Professional map design in desktop mapping environments", *Proceedings of the 15th International Cartographic Conference*, Bournemouth, Vol. 2, pp. 440–445.

Asche, H. and C. M. Herrmann, (1993) "Electronic mapping systems – a multimedia approach to spatial data use", *Proceedings of the 16th International Cartographic Conference*, Cologne, Vol. 2, pp. 1101–1108.

Bertin, J. (1981) *Graphics and Graphic Information Processing*, Walter de Gruytes, Berlin-New York.

Burger, J. (1993) *The Desktop Multimedia Bible*, Addison-Wesley, Reading.

Cebulla, V. (1993) "Stadtinformationssystem Maskat. Konzeption und Programmierung eines interaktiven Stadtplanprototyps", *Unpublished Diploma Thesis*, TFH Berlin.

DiBiase, D., A. M. MacEachren, J. B. Krygier and C. Reeves (1992) "Animation and the role of map design in scientific visualization", *Cartography and Geographic Information Systems*, Vol. 19, No. 4, pp. 201–214, 265–266.

Dorling, D. (1992) "Stretching space and splicing time: from cartographic animation to interactive visualization", *Cartography and Geographic Information Systems*, Vol. 19, No. 4, pp. 215–227.

Forer, P. (1993) "Envisioning environments: map metaphors and multimedia databases in education and consumer mapping", *Proceedings of the 16th International Cartographic Conference*, Cologne, Vol. 2, No. 2, pp. 959–982.

Hoffos, S. (1992) *CD-I Designer's Guide*, McGraw-Hills, Maidenhead.
Huffmann, N. H. (1993) "Hyperchina: adventures in hypermapping", *Proceedings of the 16th International Cartographic Conference*, Cologne, Vol. 2, No. 2, pp. 959–982.
Lindholm, M. and T. Sarjaskoski (1993) "User interfce issues in a computer atlas", *Proceedings of the 16th International Cartographic Conference*, Cologene, Vol. 2, pp. 613–627.
MacDougall, E. B. (1992) "Exploratory analysis, dynamic statistical visualization and geographic information systems", *Cartography and Geographic Information Systems*, Vol. 19, No. 4, pp. 237–246.
MacEachren, A. M. (1987) "The evolution of computer mapping and its implication for geography", *Journal of Geography*, Vol. 86, No. 3, pp. 100–108.
MacEachren, A. M. and J. H. Ganter (1990) "A pattern identification approach to cartographic visualization", *Cartographica*, Vol. 27, No. 2, pp. 64–81.
MacEachren, A. M. and M. Monmonier, M. (1992) "Introduction. Geographic visualization", *Cartography and Geographic Information Systems*, Vol. 19, No. 4, pp. 197–200.
Makkonen, K. and R. Sainio, (1991) "Computer aided cartographic communication", *Proceedings of the 15th International Cartographic Conference*, Bournemouth, Vol. 1, No. 1, pp. 211–222.
Mann, C. (1993) "Stadtinformationssystem Celle: ein kartengestütztes Auskunftssystems für Städtetouristen", *Unpublished Diploma Thesis*, TFH Berlin.
McGuinness, C., A. van Wersch and P. Stringer (1993) "User differences in a GIS environment: a protocol study", *Proceedings of the 6th Interational Cartographic Conference*, Cologne, Vol. 1, pp. 478–485.
Monmonier, M. (1992a) "Summary graphics for integrated visualization in dynamic cartography", *Cartography and Geographic Information Systems*, Vol. 19, No. 1. pp. 237–246.
Monmonier, M. (1992b) "Authoring graphic scripts: experiences and principles", *Cartography and Geographic Information Systems*, Vol. 19, Vol. 1, pp. 247–260, 272.
Monmonier, M. (1993) "'Navigation' and 'narration' strategies in dynamic bivariate mapping", *Proceedings of the 16th International Cartographic Conference*, Cologne, Vol. 1, pp. 645–655.
Murie, M. D. (1993) *Macintosh Multimedia Workshop*, Hayden, Carmel.
Ormeling, F. (1993) "Ariadne's thread — structure in multimedia atlases", *Proceedings of the 16th International Cartographic Conference*, Cologne, Vol. 2, pp. 1093–1100.
Rachfall, S. (1993) "Konzeption und Programmierung einer elektronischen Karteneinführung für die Grundschule", *Unpublished Diploma Thesis*, TFH Berlin.
Rauner, A. (1992) "Untersuchungen zur Erarbeitung einer Konzeption für die Kartographische Umsetzung von Umweltdaten", *Unpublished Diploma Thesis*, TU Dresden.
Steinbeck, V. (1991) "Konzept für einen Elektronischen Atlas am Beispiel eines thematischen Atlasses für das Sultanat Oman", *Unpublished Diploma Thesis*, TFH Berlin.
Taylor, D. R. F. (1991) "A conceptual basis for cartography/New directions for the information era", *Cartographica*, Vol. 28, No. 4, pp. 1–8.
Töns, G. (1990) "Entwicklung einer Karteneinführung für einen Grundschulatlas", *Unpublished Diploma Thesis*, TFH Berlin.
Trummer, M. (1992) "Cart/o/info-Atlas der Gemeinden des Kantons Luzern", *Unpublished Diploma Thesis*, FH Karlsruhe.
Tufte, E. R. (1990) *Envisioning Information*, Grahpics Press, Cheshire, CT.
Whitehead, D. C. and R. R. Hershey (1990) "Desktop mapping on the Apple Macintosh", *Cartographic Journal*, Vol. 27, No. 2, pp. 113–118.
Wurm, C. (1993) "Konzeption und Prototyp für einen elektronischen Fahrradatlas Berlin", *Unpublished Diploma Thesis*, TFH Berlin.
Zimmermann, H. (1990) "Kanton in Konstruktion", *Macup*, Vol. 6, pp. 150–155.

CHAPTER 13

Spatial–Temporal Analysis of Urban Air Pollution

ALEXANDRA KOUSSOULAKOU

Department of Cadastre

Photogrammetry and Cartography

University of Thessaloniki, Greece

Introduction

The subject of this chapter is the analysis and cartographic visualization of the elements involved in an environmental problem, namely air pollution. The specific subject of urban air pollution is analysed and visualized as an example; the methodology followed, however, can be applied for any other scale of the air pollution problem (i.e. local, regional, continental or global).

Visualization in Environmental Applications

Earth science applications in general and environmental applications in particular generate large amounts of data describing the development of physical phenomena (both natural and man-made) in space and time. The multidimensional nature of these large data sets (observed and/or simulated) makes necessary, apart from their spatial representation, the development of innovative computer visualizations for exploration and interpretation.

The environmental system is a dynamic system: spatial models can simulate environmental impacts and interactions, predict future situations and compute changes as a function of time; the need for the analysis, interpretation and transfer of information about such spatiotemporal phenomena offers cartographic visualization an interesting and expanding field for research and the creation of new cartographic products.

The creation of maps which display the status of elements and their relations within the environmental system is of interest in environmental cartography. Growing interest in this field has been witnessed during the last 15 years, due mainly to increasing concern about the environment — factors related to it — and the advanced capabilities of computers. Environmental cartographic visualizations can be divided in three major groups (although in practice combinations are common): (1) maps/visualizations of the status of physical elements (e.g. water quality/temperature, soil erosion, slope stability); (2) maps/visualizations of natural environmental processes (earthquakes, volcanic eruptions, meteorological processes); and (3) maps/visualizations of environmental processes induced by humans and the resulting environmental degradation (air, water, soil and noise pollution, acid rain, ozone layer depletion).

Some of the earlier environmental cartographic visualizations can be found in Pape (1977), where analytical and composite maps of urban air pollution are proposed, together with temporal (although static) displays of changes. An example of more specific demands for environmental (computerized) visualizations is given by Hayes (1979). Since the ICA cartographic conference in Tokyo in 1980, which was dedicated to environmental cartography, environmental mapping and visualization work has increased (ICA 1980; Vent-Schmidt 1980; Pape 1980; Grotjan 1982; Kadmon 1983; Lavin and Cerveny 1987; EuroCarto 7 1988). By the late 1980s environmental cartography was also concerned with environmental data integration, the spatial characteristics of the data to be mapped and environmental databases (Ormeling 1989). Computer capabilities have, in the meantime, promoted the development of systems for efficient environmental monitoring and impact studies and more realistic visualizations (IEEE 1984, Papathomas *et al.* 1988; Wolfe and Liu 1988; CGW 1990; DiBiase *et al.* 1991; Abel *et al.* 1992; Emmett 1992; Gantz 1992; Lang 1992; Lang and Speed 1992; Kruse *et al.* 1993; Leipnik 1993; Stenbert 1993).

Air Pollution Visualization

Air pollution is a multidimensional spatial phenomenon. It involves the three dimensions of geometric space, the temporal dimension and the value of (at least) one more quantitative variable. This multidimensional nature is a characteristic of earth science and particularly environmental phenomena. The temporal component, which has attracted increasing attention in the cartography and geographical and information system (GIS) literature, is of particular importance to air pollution because of the higher rate of change of the elements involved compared with other environmental phenomena (such as soil or water pollution). Apart from the visualization need created by this temporal component, the various thematic variables related to the air pollution phenomenon introduce more possibilities and demands for data analysis and mapping. Cartographic visualization and — as defined and re-emphasized in this book — offers the optimum means for analysing and displaying (and consequently understanding and explaining) air pollution and all the elements contributing to its development and presence.

Although cartographic visualization can be "visualized" as the area complementary or opposite to communication within the broader context of a conceptual map use space (see Chapter 1), it can be argued that the fundamental question raised for the purposes of cartographic communication: "HOW do I say WHAT to WHOM" (Koeman 1971) can also be utilized, from a new perspective, for the purposes of cartographic visualization. As MacEachren points out in the introductory chapter, the keyword for cartographic visualization is "map use". The use (or the different uses) of a cartographic product will determine its content and form. It is therefore the word "WHOM" — in the fundamental question given above — that has to be answered first to establish a basis for analysis and visualization of the themes that have to be mapped (and which offer the answer to the word "WHAT"). Finally, the word "HOW" poses interesting challenges to the cartographer for designing the appearance of the maps and/or diagrams (in terms of a variety of parameters such as the utilization of the graphic variables, views, sequencing, interaction, animation and sound). The answer to this word ("HOW") dictates the way in which cartographic visualization is implemented so that the optimum use of the created maps can be made.

Users and Uses of Urban Air Pollution Maps (Visualization for WHOM)

The need to visualize air pollution (and the elements related to it) covers the whole continuum of map use as conceptualized by MacEachren's cube (Chapter 1) from private map use for exploration purposes which requires interaction possibilities, to public display of the air pollution status in the city for informing the general public.

A variety of professional categories with numerous tasks and functions is required in air pollution monitoring and control agencies (Stern *et al.* 1984). These categories can be divided in two main groups: the air pollution specialists (e.g. engineers, chemists, computer scientists, inspectors, meteorologists, physical scientists) and the urban planners/administrators (concerned with the effects of the problem on the urban structure and life).

Consequently, in an urban context, three main groups of map users exist: the air pollution specialists, the urban planners/administrators and the general public, who are often affected by air pollution and its consequences for urban life (high concentrations, traffic control).

The tasks and/or visualization needs of the above groups of users of the urban air pollution maps are given in Fig. 13.1.

Elements to be Visualized (Visualization of WHAT)

The status of air quality in an area is determined by the existence and interaction of a number of elements (both natural and man-made). These elements are either directly involved in the production and transport of air pollutants, or they do not generate pollution directly but influence the course and development of the air pollution phenomenon. These elements are grouped into the sources of pollution

POLLUTION OFFICE	URBAN PLANNERS	PUBLIC
- overview of distribution	- overview of distribution and sources	- overview of distribution
- overview of sources and their contribution	- assess effects of urban development plans	- public display e.g. at shops in CBD; on TV (such as weather report)
- locate monitoring network stations	- assess effects of modifying city structure	- inform about 'hot spots'
- monitor local influence of sources	- assess local effects from projected developments	
- traffic control: instruct drivers along arteries or in a ring of roads	- assess impact of modifying neighborhood structure	
- take control actions	- monitor significant spots	
- inform about 'hot spots'	- inform about effects on population	
- inform about effects on population		

FIG. 13.1. Tasks and/or needs of the map user groups.

and their emissions in the area, the atmospheric (meteorological) conditions and the topography of the area. All these element groups determine the status of air quality (pollutant concentrations) in the area. The urban activities of the area also influence the air quality status.

The emissions generated by various sources (industries, traffic, heating) are influenced by atmospheric transport and diffusion mechanisms (caused by the meteorological conditions in the area), by the topography and by the rest of the urban activities. All these elements produce spatial variations in pollutant concentrations and visibility (about which information is obtained from sources of pollution information, including monitoring stations and reports from the public). Information about pollutant concentrations can trigger reactions to pollution events (advice about alarm conditions, short-term traffic and industrial controls).

The effects of air pollution (on humans, animals and buildings) determine the control actions to be taken by the air pollution office of the urban area. These control actions are either tactical (short-term, e.g. a scenario for pollution episodes) or strategic (basically concerned with establishing and maintaining air quality and emission standards) (Stern *et al.* 1984). Figure 13.2 shows how the groups of air pollution generating elements relate to each other and to air pollution control.

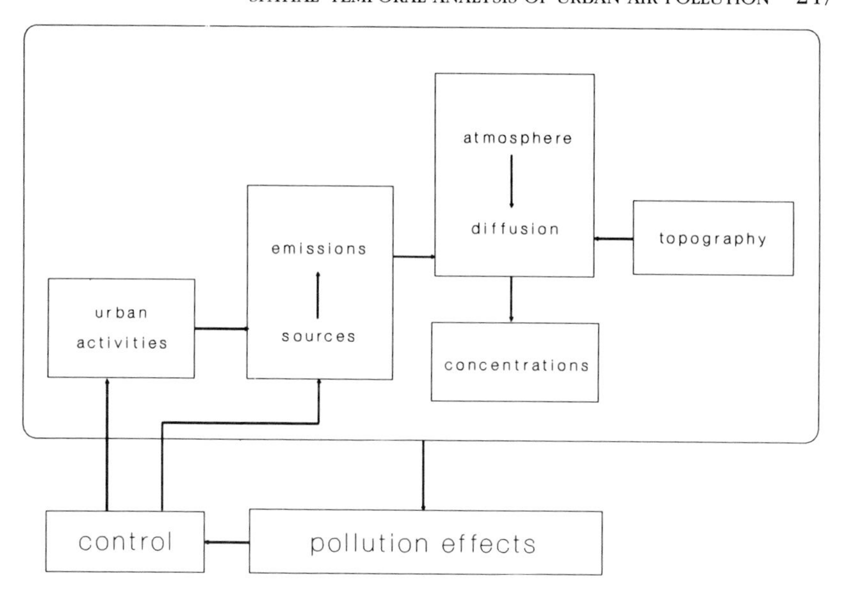

FIG. 13.2. Urban air pollution: generating elements and control.

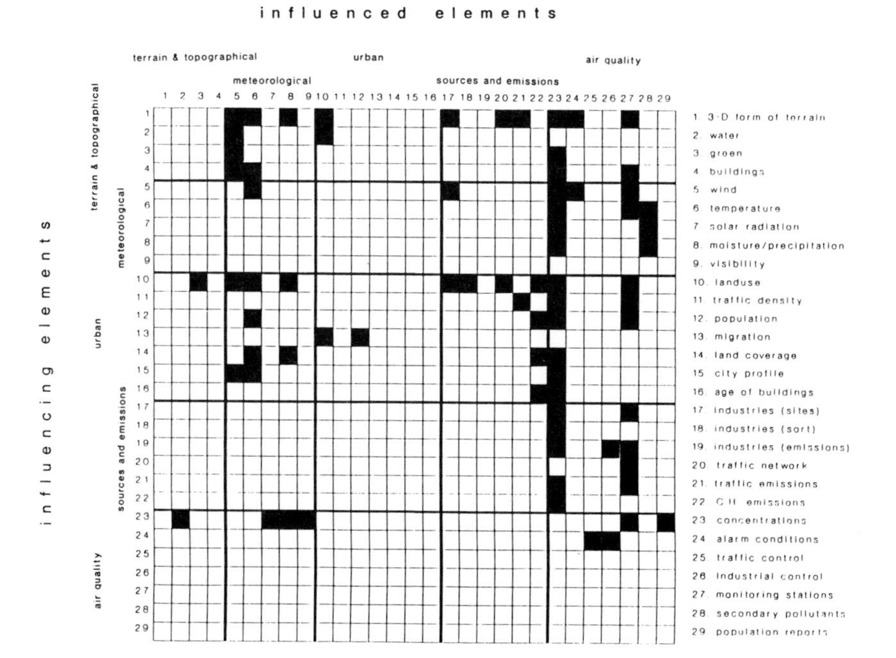

FIG. 13.3. Elements involved in urban air pollution and their interrelations (for detailed explanation, see Koussoulakou 1990).

These five basic urban air pollution generating groups consist of the following elements:

- Sources and emissions group
  Industries (site, sort, emissions)
  Traffic (network, emissions)
  Central heating emissions

- Meteorology group
  Wind
  Temperature
  Solar radiation
  Moisture/precipitation
  Visibility

- Air quality group
  Pollutant concentrations (various, primary and secondary air pollutants)
  Alarm conditions
  Traffic control
  Industrial control
  Monitoring stations
  Population reports

- Topography group
  Three-dimensional form of terrain
  Water
  Green (open spaces within a city: parks, gardens)
  Buildings

- Urban conditions group
  Land use
  Traffic density
  Population
  Migration
  Land coverage
  City profile
  Age of buildings

All these elements are in practice closely linked by influences and interrelations. These influences and interrelations (together with references) are given in detail in Koussoulakou (1990). A first impression of these interactions is given by the matrix of Fig. 13.3.

Such a matrix is a first visual indication of the complexity of the phenomenon to be mapped and the many variables involved, but also a guide for selecting variables which can be correlated visually (through map overlays and/or other visualization techniques).

Implementation of Visualization (Visualization HOW)

The subjects to be analysed and visualized for the purposes of urban air pollution monitoring and control are determined by the two major factors given in the previous paragraphs: the users of the maps and the elements involved in the physical phenomena that these maps depict. The cartographic visualization of the phenomenon is, therefore, the outcome of the combined consideration of these two factors, as shown in Fig. 13.4.

Themes of the Maps

The combination of the physical elements and the map user needs, as described above, determines the need for the visualization of a phenomenon and the way to implement the visualization. For the purpose of urban air pollution this combination is summarized as follows (Koussoulakou 1990).

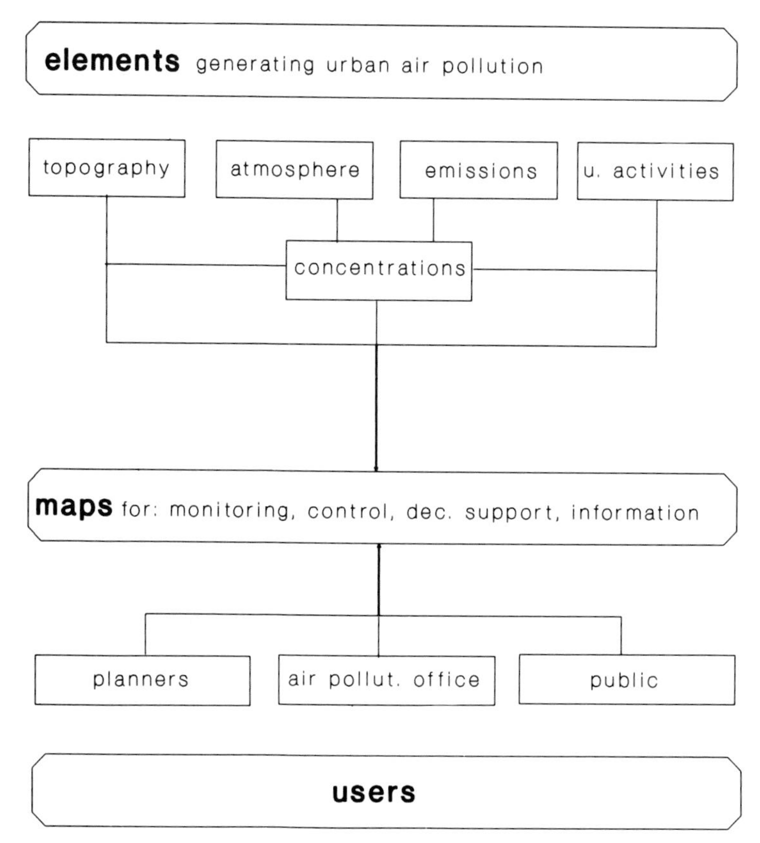

FIG. 13.4. The map users and the elements to be mapped determine the products of cartographic visualization.

The influence of the physical elements on each other can be observed in both the whole urban area and locally. For instance, the form of the terrain influences the overall appearance of the wind field and other meteorological factors, which, in turn, influence air pollution events: a city located inside a basin is more likely to witness critical air quality conditions than a city on flat terrain. Local features, such as hills, change the wind field locally and buildings contribute to the development of turbulent eddies, which influence the pollutant concentrations in their neighbourhood. By its nature, therefore, the urban air pollution problem implies the concept of scale, as inside an urban area it is influenced by factors of varying spatial significance and, therefore, it develops over areas of varying spatial extent. Furthermore, man-made elements such as sources have an overall and a local influence on the pattern of emissions and concentrations. Consequently, the visualization of the air quality status can involve the whole area, but also certain municipalities or neighbourhoods (industrial areas, areas with busy traffic, the CBD) and specific points of interest (archaeological sites, monuments, street intersections). Even scales outside the urban range (e.g. regional scales) might be useful for visualizing other factors related to the urban air pollution problem (such as, for instance, migration from the countryside to the urban centre).

Because of the combined influence of the considered physical elements on the air quality status and, in general, on each other it is obvious that apart from displaying their status separately they should also be visually combined on more complex maps, allowing for visual correlations. In general, the maps which are useful for monitoring and controlling urban air pollution and generally enhancing the decision support process are series of analytical maps, i.e. maps presenting only one characteristic (Pape 1980). From these series of maps inventory (or composite) maps can be derived (inventory maps present more characteristics by exhaustive superposition or by juxtaposition of the original analytical maps). Another useful sort of map, especially for planning purposes, is the so-called synthesis (or communication) map that results from the simplification (i.e. structural/conceptual generalization) of exhaustive analytical maps and their consequent simplified superposition. A general discussion of the three types of maps mentioned above (inventory, analytical, synthesis) can be found in Rouleau (1984). A more specific discussion of the use of synthesis and composite maps/visualizations in the earth sciences is given by DiBiase (1990). The generation of the latter type of map (synthesis), however, was outside the scope of the present study.

For urban air pollution purposes, analytical maps showing elements such as the concentration of air pollutants, the industries emitting the pollutants and the inhabited areas affected by them are of interest to planners (Pape 1973). Composite maps combining the above with more information (particularly on urban settlements and population, e.g. the residential density, the professional structure of the inhabitants, the types and age of the urban settlements and the kind of industry or location of pollution sources that might be well upwind of the urban area), would also be valuable for planning purposes as there are connections between these factors. Planners actually prefer to combine all these elements, and even more, on a single map: this is a common "planners' trend", which, however, is not acceptable from the cartographic point of view (on overloaded maps only questions at the elementary reading level are possible, which is not sufficient for a general overview such as a planner actually requires). Instead, either selective (i.e. relatively simple) inventory maps or synthesis maps have to be produced, although the latter are highly generalized, resulting in a considerable loss of information. Alternatives to synthesis/communication maps have to be explored by experimenting with cartographic visualization (e.g. unconventional maps with a simultaneous display of multiple variables, user interaction and queries).

The interests of pollution specialists comprise the visualization of all elements involved in the urban air pollution problem, as the pollution office is the main authority concerned with them. Series of exhaustive analytical maps showing the concentrations of a number of air pollutants over the area are of primary importance. Emission maps are also required. Composite maps combining air quality and emission distributions with meteorological, topographic and urban factors are also useful. Synthesis maps are not really necessary for the tasks of the air pollution specialists, mainly because of the maps' high degree of generalization (as already mentioned, they are more suitable for communicating gross information).

Finally, a number of maps created mainly for pollution specialists and urban planners can be selected to supply this information to the general public. These are mostly analytical maps displaying the distribution of the most important air pollutants, or simple inventory maps showing additional information such as wind or traffic densities.

Another interesting point with respect to the cartographic representation of the air pollution elements is their three- and four-dimensional (i.e. temporal) aspects. The three-dimensional space of the urban area is involved in the generation of air pollution. It is useful, therefore, to display the elements involved in three-dimensional space, either on ordinary two-dimensional maps (e.g. slices at different heights or at different vertical planes) or, for more comprehensive visualization, to display aspects of the area (e.g. terrain, urban profile or their combination) in three-dimensions. Another possibility is the combination (superposition) of three-dimensional map layers with two-dimensional layers (e.g. layers viewed in three-dimensional space but containing only two-dimensional thematic information).

The visualization of air pollution dynamics is of interest to all user groups. Most important for all users are the changes in the air quality status; changes in the meteorological conditions and traffic emissions are also useful, especially for pollution specialists. Such changes can be displayed either on series of static maps or on dynamic (i.e. animated) maps. They could reflect either the present situation (which might be described by direct measurements of the pollutants' concentrations and/or simulated on the basis of available meteorological and emission data), or they could portray possible developments in air quality status resulting from different control or planning scenarios ("what if"). The combination of dynamic maps displayed onto three-dimensional map layers allows for even more realistic and comprehensive visualizations.

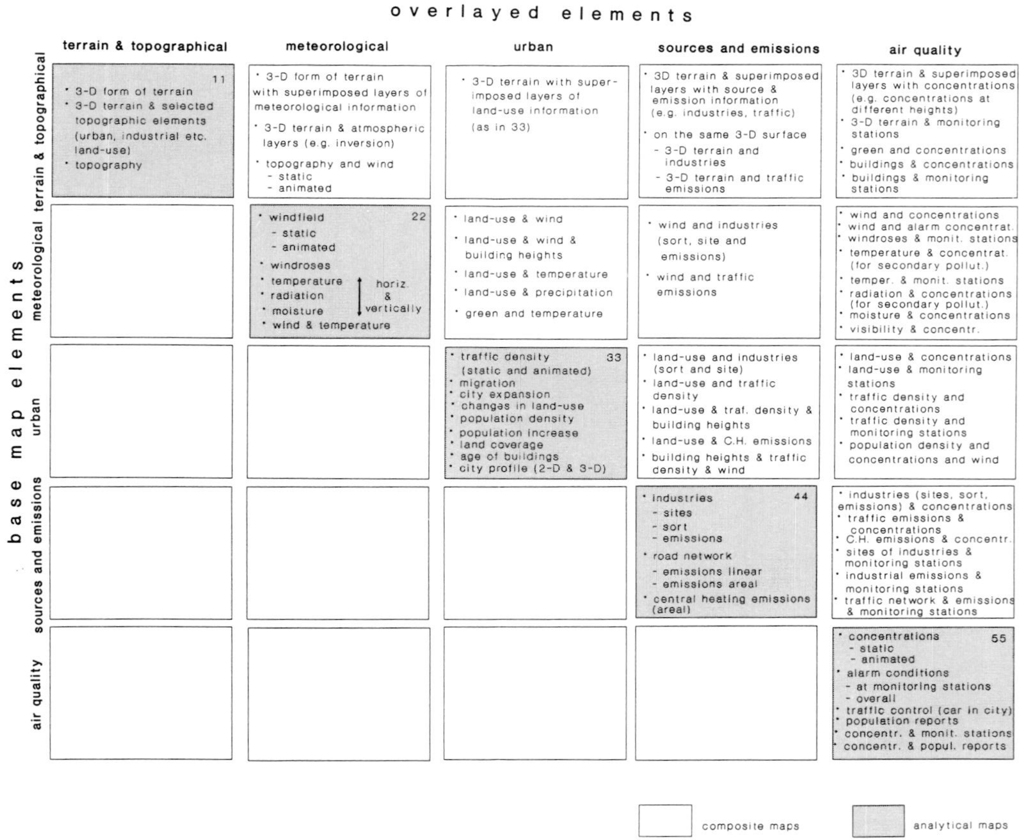

overlayed elements (columns) / base map elements (rows)

base map elements	terrain & topographical	meteorological	urban	sources and emissions	air quality
terrain & topographical	11 • 3-D form of terrain • 3-D terrain & selected topographic elements (urban, industrial etc. land-use) • topography	• 3-D form of terrain with superimposed layers of meteorological information • 3-D terrain & atmospheric layers (e.g. inversion) • topography and wind - static - animated	• 3-D terrain with superimposed layers of land-use information (as in 33)	• 3D terrain & superimposed layers with source & emission information (e.g. industries, traffic) • on the same 3-D surface - 3-D terrain and industries - 3-D terrain and traffic emissions	• 3D terrain & superimposed layers with concentrations (e.g. concentrations at different heights) • 3-D terrain & monitoring stations • green and concentrations • buildings & concentrations • buildings & monitoring stations
meteorological		• windfield 22 - static - animated • windroses • temperature • radiation • moisture (temperature, radiation, moisture: horiz. & vertically) • wind & temperature	• land-use & wind • land-use & wind & building heights • land-use & temperature • land-use & precipitation • green and temperature	• wind and industries (sort, site and emissions) • wind and traffic emissions	• wind and concentrations • wind and alarm concentrat. • windroses & monit. stations • temperature & concentrat. (for secondary pollut.) • temper. & monit. stations • radiation & concentrations (for secondary pollut.) • moisture & concentrations • visibility & concentr.
urban			• traffic density 33 (static and animated) • migration • city expansion • changes in land-use • population density • population increase • land coverage • age of buildings • city profile (2-D & 3-D)	• land-use and industries (sort and site) • land-use and traffic density • land-use & traf. density & building heights • land-use & C.H. emissions • building heights & traffic density & wind	• land-use & concentrations • land-use & monitoring stations • traffic density and concentrations • traffic density and monitoring stations • population density and concentrations and wind
sources and emissions				• industries 44 - sites - sort - emissions • road network - emissions linear - emissions areal • central heating emissions (areal)	• industries (sites, sort, emissions) & concentrations • traffic emissions & concentrations • C.H. emissions & concentr. • sites of industries & monitoring stations • industrial emissions & monitoring stations • traffic network & emissions & monitoring stations
air quality					• concentrations 55 - static - animated • alarm conditions - at monitoring stations - overall • traffic control (car in city) • population reports • concentr. & monit. stations • concentr. & popul. reports

Legend: composite maps (white boxes); analytical maps (shaded boxes: 11, 22, 33, 44, 55)

FIG. 13.5. Themes of the proposed maps.

On the basis of the characteristics discussed (i.e. the elements involved in air pollution and their interrelations, the needs of the users for mostly analytical and composite maps and the spatiotemporal characteristics of the involved elements), the most important themes proposed to be mapped for urban air pollution monitoring and control are summarized in Fig. 13.5.

The Implementation

The software modules developed for implementing the visualization of urban air pollution are presented in the following in terms of their functionality. The data that are visualized cover the area of Athens, Greece. The area faces a serious air pollution problem, which has developed and accumulated during the last two decades.

As urban air pollution is a relatively old environmental problem, considerable work has been carried out, not only on a research level but on the implementation of monitoring and control systems by local and/or regional authorities. Nevertheless, even in cases where well developed sophisticated systems are used for urban air pollution data collection and processing the utilization of maps is restricted — computer maps in particular are rarely used. The purpose of the present work has been to indicate the importance and usefulness of visualizing elements already present in such monitoring and control systems.

The existing work on mapping urban air pollution, or elements related to it, was considered [e.g. Giddings 1983; Grotjan 1982; Kadmon 1983; Papthomas *et al.* 1988; Pape 1973; 1977; 1980; Tufte 1983; yearly reports from Dutch and German local authorities, such as the RIVM (Dutch National Institute for Health and the Environment), the DCMR (Central Environmnetal Service of Rijnmond, the area around the port of Rotterdam) and the Environmental Service of the German State of Hessen].

Based on this existing mapping experience, a system was developed to include all the important aspects of air pollution, to propose further possibilities for their visualization and to combine all the involved elements so that a global view of the problem could be gained through its detailed visualization.

The cartographic visualization software developed for the case study, was part of a broader project for simulating and mapping air pollution in the area for monitoring and control purposes. Apart from the visualization functions that are of interest in the cartographic visualization of the phenomenon, software for processing data on air pollution and simulating the air pollutants' concentrations within the urban area was used. These processing/simulating modules were developed to test model runs for the area and to compare the simulated results with those measured at stations of a monitoring network established by the pollution office of the city. The visualization modules make use of the data generated by the processing/simulation modules.

The software was developed on an IBM PS/2 (Model 60) using Fortran with the graphics subroutine packages GKS and HALO and on an IBM 9370 (Model 80) mainframe computer using Fortran and graPHIGS (IBM version of PHIGS: Programmer's Hierarchical Interactive Graphical Standard).

The visualization functions are presented in terms of their subject and functionality (mainly user interaction with the data, the presence of animation possibilities and two-dimensinal or three-dimensional display).

For all two-dimensional maps a basic introductory function for scale selection is used at the beginning of each visualization. The base map of the whole GAA (Greater Athens Area) is drawn and the user is asked to select, with a cursor, the window (i.e. the subarea) to be viewed in the following display.

As described in the section above, the maps can be divided into analytical and composite maps. The maps in each group can be static or animated, two-dimensional or three-dimensional. In practice, composite maps are rather confusing when animated. This is especially true for the two-dimensional maps, where all the layers are viewed on one plane. For the three-dimensional composite maps, on the other hand, the layers can be viewed separately, but because of the complex and overloaded images animations are slow, unless very powerful computers are available.

The visualization functions developed for the urban air pollution visualization prototype presented in this chapter are briefly described in the following.

Visualization Functions for Analytical Maps

Static Maps

Emissions From Point Sources (Industrial Sources)

Proportional or range-graded point symbols (circles, pies) are used for the emissions of each industry. Initially all the industrial sources of the GAA are displayed, classified into four types (basis, heavy, middle, light industries). The user is asked to select interactively, with the help of a menu, one or more type(s) of source(s) for viewing their emissions and also to select the emitted pollutant(s) to be viewed. The display option of proportional or range-graded point symbols is also offered in the menu. The selection of proportional symbols produces an impressive pattern of emissions: it is the industries in the western part of the GAA that release the largest amount of industrial pollutants. Although this was empirically known, the quantitative visualization of the fact reveals the large variation in industrial emissions in the area.

After the selected combination (i.e. of industry type(s), pollutant(s) and symbol) has been displayed, the user can go back to the main menu to make another selection or can exit (Fig. 13.6).

Emissions From Line Sources

Emissions from linear sources (e.g. roads) are displayed with the help of symbols of varying thickness. Real data about emissions from linear sources, however, were

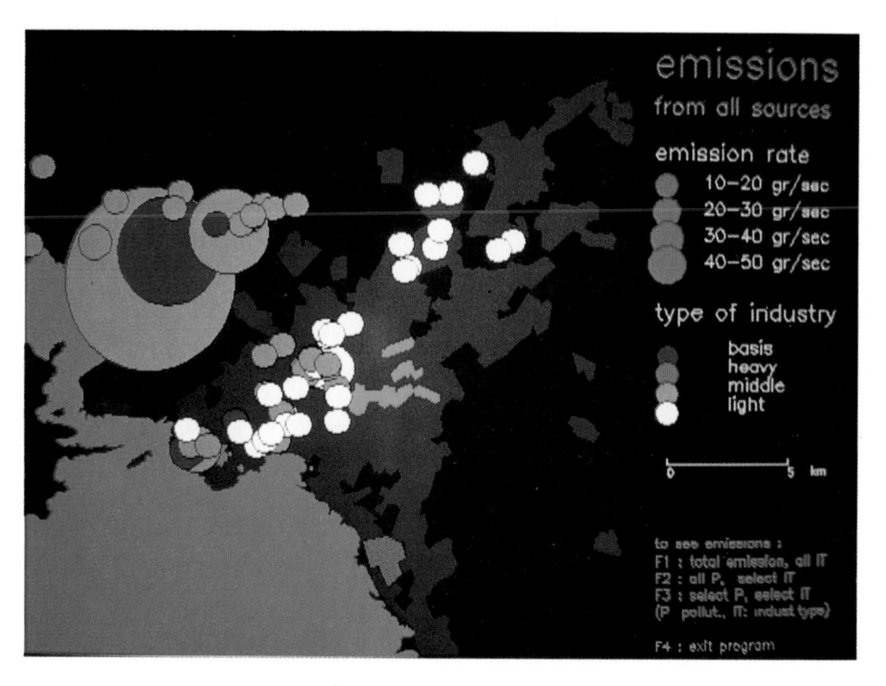

FIG. 13.6. Emissions from point sources.

not available for the area. The emissions from the urban road network had been converted and made available in the form of a matrix of area sources (cells) covering the whole area (this is a common practice in urban air pollution studies). The function was developed as a test, for a limited part of the network.

Emissions From Area Sources

Area sources (a matrix of square cells covering the whole of the GAA) are displayed with the help of this function. Colour values are used to show the various emission levels. As both urban roads and central heating are available as area sources, the user can select between displaying each or both of them. The emissions are displayed with the help of a transparent overlay (XOR drawing) over the base map of the GAA (in shades of grey). When only street emissions are selected, the pattern of colour generated on the cell matrix fits with the underlying road structure in a way that indicates that it is the roads in the centre of the urban area that produce the most emissions (the visual overlay can be used as a check for the original linear emissions which are converted to area cells).

Wind Field

Point symbols (arrows) of varying magnitude and direction are used to display the wind field above the area. It is the static version of a dynamic function described later in this paragraph. The user selects the hour of the day for which the wind field will be displayed in the form of a matrix of arrows above the area. The dimensions of the point symbols (arrows) can be manipulated by the user (i.e. the thickness and length of the arrow); more specifically, the user can manipulate the "unit size" against which all arrows are scaled rather than manipulate the size of each arrow. This was considered useful for the display of parameters of varying magnitude (i.e. for the different hours of the day), as the adoption of a fixed ratio for the magnitude of the arrow symbol could occasionally produce very small or very large arrows.

Long-term Meteorological Conditions (Wind, Pollution and Stability)

Point symbols in the form of wind-roses, pollution roses and stability-roses are used for the display of long-term meteorological conditions and air quality levels. The user is asked to select the type of rose (combinations of wind + pollution or wind + stability are also possible), the season (winter/summer) and the part of the day (day/night). Yearly and/or daily averages of the wind field and the stability over the area can be displayed at selected points.

Pollutant Concentrations at Points

This is a display of pollutant(s) concentrations with point symbols. The selection of range-graded or proportional symbols is possible, as well as the selection of one or more pollutants and of the hour of the day to be displayed.

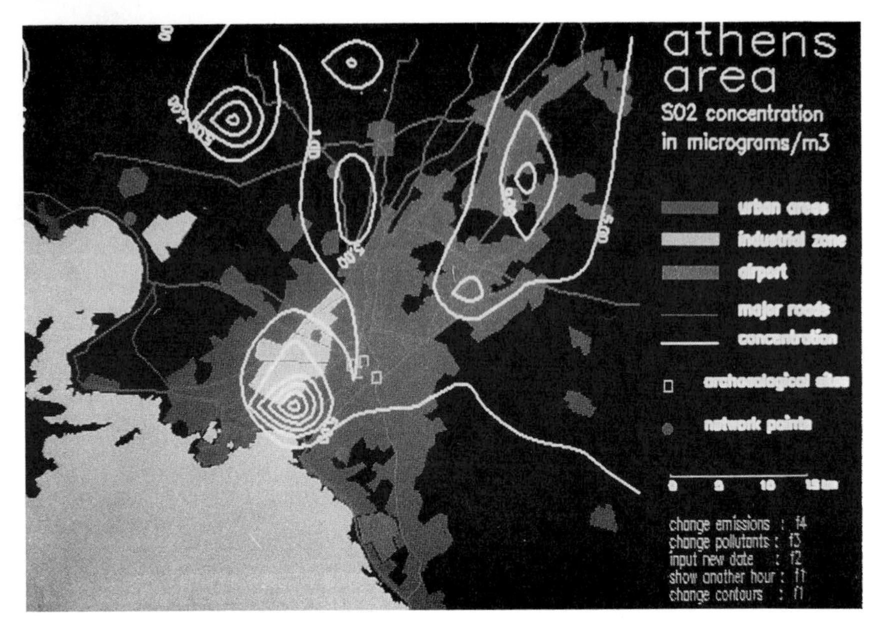

FIG. 13.7. Pollutant concentrations over the GAA.

Population Reports

This is a visualization of the complaints of the population to the pollution office. Point symbols indicate the number of complaints per 1000 inhabitants for each municipality for a date and hour selected by the user.

Pollutant Concentrations Over Area

This is a display of pollutant concentrations (pollutant selected by the user) with the help of isoconcentration lines, or the respective layer tints for a clearer visualization of the distribution. The user can select the date and hour of the day to be displayed and the interval of the isolines to be drawn (Fig. 13.7).

Animated Maps

Wind Field

This is the same type of display as generated by a previously described function (display of wind field with arrows); this version, however, is an animated visualization of the wind field. The magnitude and direction of the arrows change with time, indicated by an analogue clock. Daily changes are displayed. Interaction is possible by going backward in time, freezing the display and going forward. A static version with small multiples (frames of animation) is also generated.

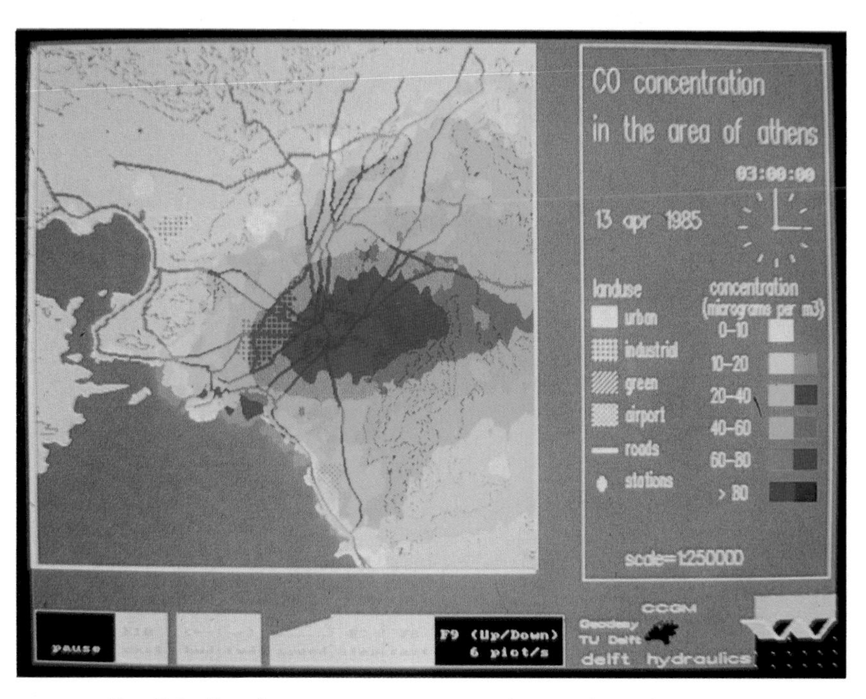

FIG. 13.8. Air pollutant concentrations over the GAA (static frame of an animated sequence).

Traffic

This is a display of the variations in traffic emissions or traffic load of the network during one day. This display is similar to that of linear emissions described previously, but is animated. The displayed theme (load or emissions) is depicted by line segments of thickness or colour value varying through time, indicated by a digital clock. The user can interactively select groups of traffic loads/emissions to be displayed as well as the classes of the colour values (five or 10 classes can be displayed and animated). It is possible to stop the animation at any time and then continue.

Movement of Vehicle on Road Network

This shows the movement of a vehicle on the road network of Athens (car navigation-like display). This type of visualization could be utilized to indicate alternative routes within the city when traffic measures are imposed (e.g. during critical pollution situations). Selection between two visualization options is possible: (a) the car (depicted by a point symbol, e.g. a red circle) moves along the roads of the (static) base map; or (b) the point symbol is actually stable in the middle of the display while the base map moves. This creates a more "dynamic" sense of movement; the viewer gets the impression of "following" the street.

Air Quality Status at Points

Blinking point symbols (circles) indicate the status of air quality at certain locations (here the monitoring stations). The user can select the date, hour and pollutant; after the base map and the locations have been displayed, the user is asked to select from a menu to view either the status of air quality or the concentrations at the above locations. If "status" is selected the points begin to blink in various colours according to the classification of the existing air quality status in "alert" (yellow), "warning" (orange) or "emergency" (red), as defined by international standards (these criteria state how the air pollution effects are associated with the various air pollutant concentration levels; various standards have been issued, both national and international). Points where no limits are exceeded remain static (in green). The user can go back and ask to view the concentrations; two options (proportional or graduated point symbols) are available for the display.

Air Quality Status Over Area

A display similar to the previous function is produced, but concentrations are visualized with isoconcentration lines. A line will blink (in one of the colours described above) when the concentration it depicts exceeds the respective limit.

Air Pollutant Concentration Over Area

This is a visualization of the changes in air pollution concentrations above the city in the course of a day, as simulated by an air diffusion model. The concentration levels are displayed with the help of transparent layer tints (XOR drawing) which evolve with time (indicated by a digital and analogue clock). The user can control the animation to move fast-forward, fast-backward, backward, forward, freeze it, and move frame by frame (Fig. 13.8). A static version with small multiples (frames of the animation) is also generated (Fig. 13.9).

The dynamic display is an example of a non-real-time animation (i.e. the individual frames have to be precalculated before the display of the animation starts). For a real-time animation with interaction with the diffusion model's parameters a more powerful computer (than the PC used here) is required. Such a real-time interaction with the model's parameters and consequently animation of the results of the model's run was attempted in an earlier version of the program, implemented on an IBM 9370 Model 80 mainframe computer, where changes in air pollutant concentrations were shown by circles of changing size through time. These changes were caused by interaction with and modification of the model's parameters by the user, the consequent real-time run of the model was followed by real-time animation of the results.

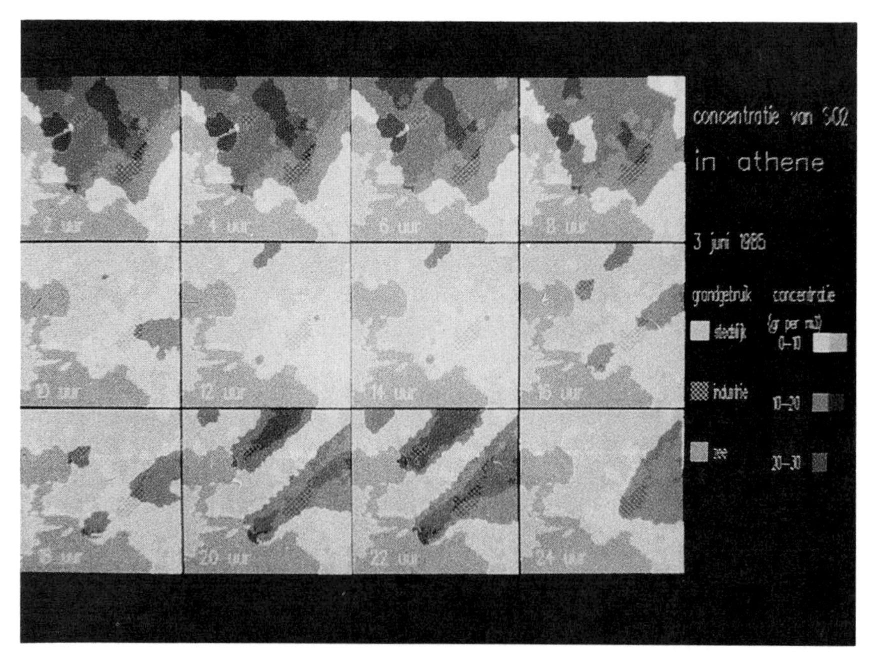

FIG. 13.9. Air pollutant concentration over the GAA (static series of small multiples).

Concentration "Cloud" Over Area

This is an alternative to the previous solutions, for a fast preview of changes in air quality. The display is similar to the previous function, but simpler: a body of pollution (a cloud) is displayed for visualizing the concentrations above a certain level of pollution (the different concentrations are not distinguished). It is useful for a quick preview as the detailed animation requires considerable time to generate a display.

City Expansion

Athens and its port (Piraeus) have expanded during the post-war years (1945–1975). Three options are offered for visualizing the change in the occupation of urban space through time: (a) the new urban area has the same colour as the existing one — the expansion is visualized by the growing area; (b) the new urban area has another colour indicating the latest growth of the urban activities; and (c) during the animation the new urban area is coloured with a value that depends on its age, while the same happens to all of the previously acquired areas, so that the result of the animation is a thematic map of the age of the urban neighbourhoods. With all options it is possible to interact and play the animation backwards, forwards or stop it at a certain frame. Time is indicated by a space bar.

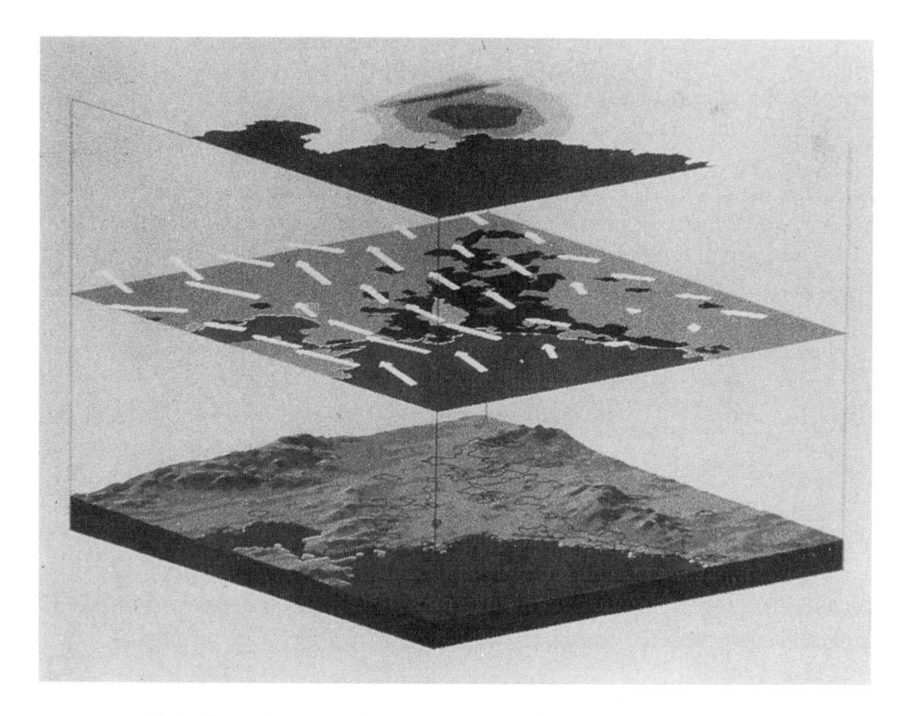

FIG. 13.10. Three-dimensional overlays (terrain, land use and wind, pollution).

Population Migration

Moving arrows display the routes of regional migration in Greece. Time is indicated by a digital clock and some interaction with the display is possible (stop and then proceed forward).

Visualization Functions for Composite Maps

Two-dimensional Overlays

Combined display (two-dimensional) of various thematic data files is possible for the purposes of visual correlation of the elements involved in air pollution.

Three-dimensional Overlays

Various thematic data files can be visualized in three dimensions. Various options are offered to the user, such as to view the terrain only, or the terrain and land use combined in a three-dimensional map, to view the three-dimensional terrain with one or two superimposed layers (two-dimensional, displayed in perspective) containing thematic information such as pollutant concentrations, and wind field (in this type of visualization, animation of one of the layers was tried: although this was practically possible the resulting animation was slow because of the complexity of

FIG. 13.11. Three-dimensional overlay with statistical surface (pollutant concentrations).

the image). Other options offered to the user are to display multiple views of the terrain from different viewpoints in different windows on the screen and to visualize the mixing layer and the inversion layer and combine them with the features of the terrain in a three-dimensional display. The user can select the files to be viewed and rotate, zoom and scale the display before the final visualization (Fig. 13.10).

Three-dimensional Overlay with Statistical Surface

This is a visualization of a three-dimensional (solid) statistical surface for displaying a selected continuous phenomenon (e.g. pollutant concentrations) over a two-dimensional layer of the topography of the GAA viewed in perspective (Fig. 13.11).

Visualization Versus Communication

During the development of some of the modules presented here, namely those with animation possibilities, it was considered worthwhile to test the performance of some types of animated map displays to evaluate how this relatively new cartographic product is perceived by the user. The final result of the test (performed for only certain types of animated maps) indicated that the correctness of answers is not influenced by the sort of the map (i.e. static or animated), but the users tend to retrieve information more quickly from animated map displays (Koussoulakou and Kraak 1992). Nevertheless, a limitation had to be imposed on the animated maps to perform this communication test: the full interactive capabilities of the map displays were not used at all (so that the test subjects would not be confused). This means that with the full use of its options an animated, interactive cartographic product might perform even better than the test indicated; it also indicates that for a product created for exploratory map use rather than the mere presentation of a theme or phenomenon, the communication criteria are limiting for the product's evaluation. In this sense communication is clearly a subcomponent of cartographic visualization, as mentioned elsewhere in this book.

Possible Future Developments

The work described here was carried out in the framework of a PhD project at the Faculty of Geodesy of the Technical University of Delft, The Netherlands. Topographic, meteorological and emission data from the GAA were initially used for simulation model runs, which provided the (simulated) pollutant concentrations (actual, measured, concentrations were also available). The (numerical) results of these model runs were of interest to researchers and technical staff of the Athens Pollution Office. The cartographic/visualization modules which were further developed, however, were not used in Athens mainly due to a lack of funds and the necessary equipment. The work was restricted to a research environment and has never been implemented in a "production" environment (which might have revealed more interrelationships between the various elements — atmospheric, topographic, polluting — each playing a part).

The visualization modules for urban air pollution monitoring and control presented here are by no means exhaustive. They only give the most important elements involved and some ways (both traditional and less traditional) to visualize them globally.

Cartographic visualization of the theme can be further developed for the elements involved and to invent new ways for the visual exploration of data related to air pollution. For instance, what could be developed further is the visualization of statistical quantities such as the correlation coefficients between measured and simulated data sets of concentrations. Visual data exploration on a local scale, which is of importance for the pollution authority, would require a more detailed visual data exploration of the emission inventory (i.e. the detailed recording of emissions from all kinds of sources). Another possible development is the detailed exploration of the compliance status of the industries for pollution control purposes.

Other further developments concern user interaction: apart from user interfaces for manipulating map views, map symbols or the process of the animations, as described above, it is also useful to develop user interfaces for direct data interaction, i.e. access to the database by graphical methods.

Another interesting possibility would be to experiment with alternative methods of cartographic visualization of the so-called "communication maps" (usually maps used by planners: overloaded maps with multiple themes) by means of display and interaction with the multiple variables related to air pollution.

The visualization modules presented here attempt to give an overview of the possibilities for cartographic visualization offered by the urban air pollution problem, rather than to treat a specific element of the phenomenon in depth. More detailed visualizations can be developed: the complexity of the phenomenon is a challenge for experimentation and research in a cartographic visualization environment. This complexity, on the other hand, makes cartographic visualization the most suitable means for understanding the phenomenon.

The development of cartographic visualization methods and techniques requires experimentations, which will require time. Nevertheless, the exploration of visualization possibilities is the best way of dealing with data, suggesting that from data to information, the shortest path is visualization.

References

Abel, D. J., S. K. Yap, R. Ackland, M. A. Cameron, F. D. Smith and G. Walker (1992) "Environmental decision support system project: an exploration of alternative architectures for geographical information systems", *International Journal of Geographical Information System*, Vol. 6, No. 3, pp. 193–204.

DiBiase, D. (1990) "Visualization in the earth sciences", *Earth and Mineral Sciences*, (Bulletin of the College of Earth and Mineral Sciences, Penn State University), Vol. 59, No. 2, pp. 13–18.

DiBiase, D., A. M. MacEachren, J. Krygier, C. Reeves and A. Brenner (1991) "Animated cartographic visualization in earth system science", *Proceedings 15th ICA Conference*, Bournemouth, pp. 223–232.

Emmott, A. (1992) "Scientific visualization — a market mosaic", *Computer Graphics World*, July, pp. 29–39.
Eurocarto 7 (1988) "Environmental applications of digital mapping", *Proceedings Eurocarto 7*, Enschede.
Gantz, J. (1992) "Scientific visualization — a market mozaic", *Computer Graphics World*, July, pp. 27–28.
Giddings, H. V. (1983) "A graphics-oriented computer system to support environmental decision making" in Teicholz, E. and B. J. L. Berry (eds.), *Computer Graphics and Environmental Planning*, Prentice Hall, Englewood Cliffs.
Grotjan, R. (1982) "Computer animation of three-dimensional time-varying meteorological fields", *Processing and Display of Three Dimensional Data, SPIE*, Vol. 367, pp. 107–108.
Hayes, S. R. (1979) "A technique for plume visualization in power plant siting", *Journal of the Air Pollution Control Association*, Vol. 29, No. 8, pp. 840–843.
ICA (1980) *Examples of Environmental Maps*, Instituto Geografico Nacional, Madrid.
IEEE, (1984) "System monitors hazardous materials dispersion", *Computer Graphics and Applications*, January, pp. 72–73.
Kadmon, N. (1983) "Photographic, polyfocal and polar-diagrammatic mapping of atmospheric pollution", *The Cartographic Journal*, Vol. 20, No. 2, pp. 121–126.
Koeman, C. (1971) "The principle of communication in Cartography", *International Yearbook of Cartography*, pp. 169–176.
Koussoulakou, A. (1990) "Computer-assisted cartography for monitoring spatio-temporal aspects of urban air pollution", *PhD Thesis*, TU Delft, Delft University Press.
Koussoulakou, A. and M.-J. Kraak (1992) "Spatio-temporal maps and cartographic communication", *The Cartographic Journal*, Vol. 29, pp. 101–108.
Kruse, F. A., A. B. Lefkoff, J. W. Boardman, K. B. Heidebrecht, A. T. Shapiro, P. J. Barloon and A. F. H. Goetz (1993) "The spectral image processing system (SIPS) — Interactive visualisation and analysis of imaging spectrometer data", *Remote Sensing of the Environment*, Vol. 44, pp. 145–163.
Lang, L. (1992) "GIS comes to life", *Computer Graphics World*, October, pp. 27–36.
Lang, L. and V. Speed (1992) "Environmental consciousness", *Computer Graphics World*, August, pp. 57–70.
Lavin, S. J. and R. S. Cerveny (1987) "Unit-vector density mapping", *The Cartographic Journal*, Vol. 24, pp. 131–141.
Liepnik, M. R. (1993) "GIS and 3D modelling fight subsurface contamination at federal site", *GIS World*, May, pp. 38–41.
MacEachren, A. and M. Monmonier (1992) "Introduction", *Cartography and Geographic Information Systems*, Vol. 19, No. 4.
Ormeling, F.-J. (1989) "Environmental mapping in transition", in *Proceedings of the Seminar Teaching Cartography for Environmental Information Management*, Enschede, pp. 11–26.
Papathomas, T .V., J. A. Schiavone and B. Julesz (1988) "Applications of computer graphics to the visualization of meteorological data", *Computer Graphics*, Vol. 22, No. 4, pp. 327–334.
Pape, H. (1973) "Urban Cartography — town planning", *International Yearbook of Cartography*, pp. 191–199.
Pape H. (1977) "Karten ueber Luftverschmutzung und Umweltbelastung", *Kartografische Nachrichten*, Vol. 2, pp. 50–53.
Pape H. (1980) "Planning restriction area maps in town Planning", *Nachrichten aus dem karten und Vermessungswesen*, Vol. II, No. 38, pp. 61–67.
Pfitzer, G. (1990) "Forecast graphics", *Computer Graphics World*, January, pp. 80–84.
Rouleau B. (1984) "Theory of cartographic expression and design", in *Basic Cartography for Students and Technicians*, Vol. 1, ICA, pp. 81–111.
Stenberg, B. (1993) "Military and civilian GIS work in tandem", *GIS Europe*, April, pp. 16–17.
Stern, A. C., R. W. Boubel, D. B. Turner and D. L. Fox (1984) *Fundamentals of Air Pollution 2nd edn*, Academic Press, New York.
Tufte, E. (1983) *The Visual Display of Quantitative Information*, Graphics Press.
Vent-Schmidt, V. (1980) "Analytische und synthetische Klimakarten", *Kartografische Nachrichten*, Vol. 4, pp. 137–143.
Wolfe, R. H. Jr and C. N. Liu (1988) "Interactive visualization of 3D seismic data: A volumetric method", *IEEE Computer Graphics and Applications*, pp. 24–30.

CHAPTER 14

Interactive Modelling Environment for Three-dimensional Maps: Functionality and Interface Issues

MENNO-JAN KRAAK

Department of Geo-information

Delft University of Technology

The Netherlands

Introduction

Today's world of spatial data handling offers users of the disciplines involved tremendous opportunities. Developments in hardware and software allow for the introduction of models to solve problems that cross boundaries of individual disciplines. The problems are solved by complex spatial analysis operations and the results are visualized with sophisticated cartographic tools. However, when large data sets are involved other techniques are necessary to uncover unknown spatial patterns. Before using the data researchers would like to explore them. It is in this context of exploration that scientific visualization has the most to offer. It allows the user to visualize these data, while having direct interactive control over the image on display and the model behind the graphics (McCormick *et al.* 1987; MacEachren and Monmonier 1992). Often the data have a three-dimensional character, although the temporal component can be involved as well.

This chapter discusses the characteristics of a prototype of a cartographic modelling system that can handle three-dimensional data linked to a digital tèrrain model. The purpose of the prototype is to allow the user to relate two-dimensional maps with the corresponding terrain surface to make them discover and understand relations between the spatial data sets. The main question addressed here, and which is of interest to three-dimensional cartography in general is "What kind of environment is needed to be able to visualize and manipulate these three-dimensional data". The title of this chapter contains some relevant keywords to answer this question. The data are visualized in three-dimensional maps in an interactive environment that allows the user to query and model. The prototype does not incorporate photo-realistic images, although the technology would allow this. Experiments in this respect are known (Sheppard 1989; Bishop 1992), and virtual reality is around the corner (Pittman 1992). This could even lead to a level of realism where users' tactile and aural senses are affected as well. Such realism could be useful in planning and architecture environments, where fantasy and reality are sometimes deliberately mixed. Experiments with three-dimensional maps are known as well; Moellering (1980), for instance, created a videotape to show the results of real-time dynamic mapping software. The software could handle small three-dimensional cartographic objects, but did not allow interaction and query.

What is "three-dimensions"? The answer will be influenced by how one approaches the question. Does the question refer only to the image or to the database representation? When the spatial data include a z-value they would be classified without doubt as three-dimensional. In the prototype's database the terrain is represented in x, y and z coordinates, so they are three-dimensional. The visualization by the prototype, whether on screen or paper, is perceptually three dimensional as well, because the images contain stimuli that make the user perceive their contents as three dimensional. Some would argue this, however, because the third dimension is not tangible (i.e. only the surface of the three-dimensional terrain can be seen). As a result these images are often described as two and a half dimensional only (see Chapter 15 for related ideas).

The prototype's interactive spatial manipulations are executed in three-dimensional space. This approach provides the user with the opportunity to experiment with the full three-dimensional map data set. Such experimentation can result in interesting non-orthogonal three-dimensional cartographic representations, which give the viewer the opportunity to look at the data from unusual viewpoints. An application with extended digital terrain models is discussed to illustrate the need for an interactive modelling environment for three-dimensional maps.

To be an effective analysis tool, an interactive terrain modelling system requires a set of specific independently accessible functions. The accessibility of these functions will depend on the quality of the user interface. This chapter, therefore, includes two main sections. After providing a brief overview of the prototype modelling environment, the specific functions implemented in the prototype are described. Typical applications of the modelling environment by cartographers (for the design of finished maps) and by earth scientists (for terrain analysis) are used to illustrate the need for each function and how it might be used. This section is followed by a discussion of the graphical user interface developed to make the interactive modelling environment more accessible to the researcher.

Environment

The core of the prototype described here to illustrate the functionality and interface issues is a special environment built for a process called cartographic terrain modelling (CTM) (Kraak 1993a). The method of CTM offers the user an interactive three-dimensional environment to manipulate map layers in relation to the terrain surface. The *C(artographic)* in CTM stands for visualization and query. Maps in CTM present the terrain and terrain-related data. As the maps are models of reality, they try to synthesize reality from a certain perspective; they do not intend to represent the terrain as realistically as possible. *T(errain)* refers to the triangle or tetrahedron-based digital terrain model. A digital terrain model in this paper means a digital three-dimensional representation of the terrain surface and of selected zero-, one-, two- and three-dimensional spatial objects that relate to the surface. Because the triangles and tetrahedrons are the basic spatial units, CTM represents a vector-based approach to three-dimensional space. The *M(odelling)* stands for the option to integrate any other terrain related data with the terrain model.

During the CTM process, it is assumed that the spatial analysis operations necessary to solve specific problems have been executed. The user can retrieve map layers containing the basic data or results of spatial analysis and combine them with the terrain model of the study area. The CTM interface offers the user a set of visualization utilities to manipulate the map in three-dimensional space (e.g. to find an appropriate orientation of the map). Cartographic design functions are available as well. They include choosing the appropriate symbology based on an object's attributes and cartographic rules. Modelling functions allow the incorporation of map layers into the terrain model. The final display utilities activate query functions and options for dynamic data exploration (real-time movement through the terrain model). Figure 14.1 illustrates the CTM process.

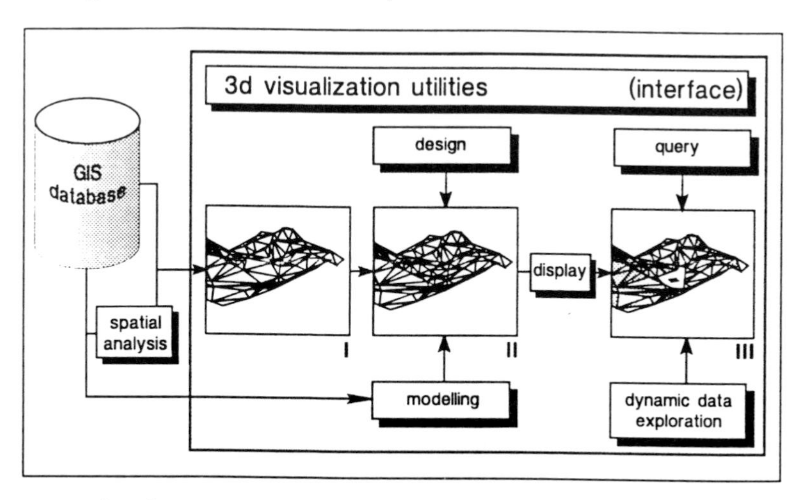

FIG. 14.1. The cartographic terrain modelling (CTM) environment with its main functional categories.

Functionality

The nature of the six main functional categories illustrated in Fig. 14.1 are detailed below. Although described separately, they are integrated in practice. Some of the functions, such as the three-dimensional visualization utilities, are needed irrespective of the type of user or application. Others, such as the cartographic design utilities or the dynamic data exploration utilities are of more interest to specific users, such as cartographers in the first or earth scientists in the second case.

Three-dimensional Visualization Utilities

These utilities are concerned with the user's view of the map. They include geometric map transformations such as rotation, scaling, translation and zooming to position the map in three-dimensional space with respect to the map's purpose and the phenomena mapped. Geometric manipulations are necessary because some elements will disappear behind others when a three-dimensional image is presented on a flat screen. The hiding of key features might lead to misinterpretation or limit the curiosity of the researcher. To avoid this, a proper position should be found, using the possibility to interactively rotate the map around each of the *x*-, *y*- and *z*-axes separately. An important additional feature is the option to scale the map along the *z*-axis to find an appropriate vertical exaggeration. When displaying the Alps the maximum height of less than 5 km would vanish if compared on equal scales with the geographic extension of the mountain ridge. Figure 14.2 shows an example of rotation and scaling along the *z*-axis. During the mapping process it should also be possible to use expedients, such as a stereoscope, to view the map in "real" three dimensions. The three-dimensional manipulation utilities should all be available in an interactive real-time mode (see Fig. 14.6). They are basic facilities to a three-dimensional environment and of help to any user, irrespective of their application. Cartographers can look for the best single perspective from which to view a map they intend to produce, whereas a geologist might continually reposition the map to obtain an optimal view of several individual phenomena.

Cartographic Design Utilities

From a communicative perspective the main design functions should include options to choose proper symbology (Robinson *et al.* in press). Among the choices to be made include the definition of colours, line sizes and fonts, and the positioning of a legend, orientation arrow and scale bar. Information about the orientation of the non-orthogonal map with respect to the more familiar two-dimensional view is of great importance (see the section on the interface issues).

The design process is influenced by three-dimensional perception rules (see Gibson 1979; Marr 1982). This means that next to the use of graphical variables to represent the character of geographic objects, pictorial depth cues, such as shading, texture, perspective and colour should be applied to create a three-dimensonal illusion. The relative importance of each of the depth cues depends on the degree of realism of the final image. An interesting study by Wanger *et al.* (1992) demonstrates this. In their research they addressed questions such as "what visual information do we need to correctly interpret spatial relations in images" and "what

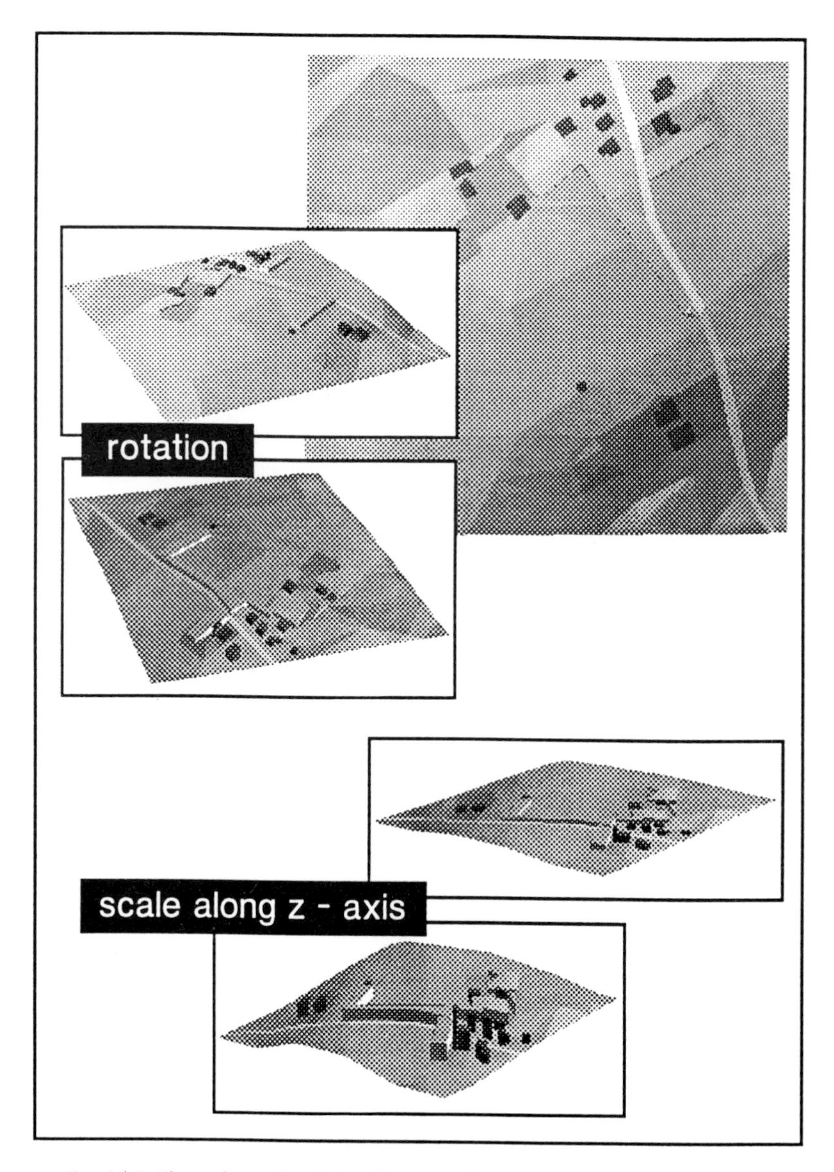

FIG. 14.2. Three-dimensional visualization utilities: an example of rotation and scaling along the *z*-axis.

is the relative importance of each of the depth cues in regard to metric judgements of spatial relations". From their experiments with relatively simple geometric objects it became clear that, to perform positioning tasks, perspective and shadow were of great importance. For orienting tasks, the same cues are relevant, but motion is also very important. Earlier research by Overbeeke and Stratmann (1988) points towards the same conclusion. Experiments in the CTM environment indeed show that being able to rotate the three-dimensional map at will, in combination with the design options, improves understanding of the map's orientation.

A question to be answered for the kind of cartographic environment described is whether the properties of the graphical variables are still effective when used together with the pictorial depth cues. Both are used because of their perceptual properties, the first to display spatial distributions and the second to create a three-dimensional impression. How does this combination function in a three-dimensional map? Assuming they indeed can be combined, do they strengthen or neutralize each other? Let us consider an example. In cartography the perceptual properties of value are used "to express information with an ordered or selective level of organization". The best known application of value in cartography is the choropleth map. In this map differences in the intensities of a phenomenon are displayed by differences in value. Variations in value are also used in three-dimensional images to create an impression of depth. However, the terminology is different. Value is called shading, as its effect is obtained by using a light source that lights the object and results in shadow and shading. Questions that arise are: "What is the effect of the combined use of the graphical variable value and the depth cue shading?" "How does the use of an expedient such as a stereoscope influence the map reading task?" "Is the map user distracted from the 'choropleth data' by the shading of the map?" Research shows that value and shading can be combined. Further details and answers to similar questions are given by Kraak (1988).

Other basic operations that might influence the design, such as coordinate transformation, selection, classification and generalization of the data, are assumed to be executed by external software. More details on the design of three-dimensional maps in general can be found in Kraak (1993b). These utilities are of primary interest to the cartographer.

Cartographic Modelling Utilities

Modelling, or more precisely cartographic modelling, can be seen as manipulating maps or map layers. Tomlin (1990) describes it as a geographic data processing method. In his book, cartographic modelling is executed in the raster mode. Ervin (1993) extrapolates this approach toward a three-dimensional environment with his program Emaps, which is able to create pseudo three-dimensional landscapes based on "planning parameters". Its maps are created to show how the landscape might look in the future using symbolic graphics. Cartographic modelling in the protoype described here involves abstracting the real-world terrain. The utilities included provide the link with a geographical information system (GIS) database (or other source of attribute data) and allow the cartographer or researcher to retrieve other map data and combine them partly or as a whole with the basic three-dimensional map data already displayed. As an example, Fig. 14.3 illustrates the snow fields of Mount Kilimanjaro combined with the elevation model. A geologist could, however, have chosen the geological map instead to look for relations between relief and the mountain's geological composition. In Fig. 14.2 not only land use boundaries have been incorporated, but additional height information as well. Houses and woodlands are represented by two z-values for each x, y-value. As the maps in the CTM environment are built from triangles and tetrahedrons, terrain-related data have to be processed before their incorporation into the terrain model can be realized. To incorporate the data instead of draping them allows the user to query all data in a later phase (Kraak 1993a).

The above can only be realized because of the data structure of the triangulated irregular network based terrain model. It adheres to the Delaunay triangulation criteria. Midtbö (1992) offers an extensive discussion on this triangulation method. The basic elevation data are triangulated directly. In the next step a map layer (for instance a road network or a land use cover) is incorporated in the triangular network, using the "mid-point-division algorithm" (Kraak and Verbree 1992). This addition is necessary as the Delaunay triangulation cannot guarantee the proper inclusion of the data from the map layer. The "mid-point division algorithm" is applied to make sure that the chains representing the map layer's objects can be found in the Delaunay network. The basic principle of the algorithm is to check if

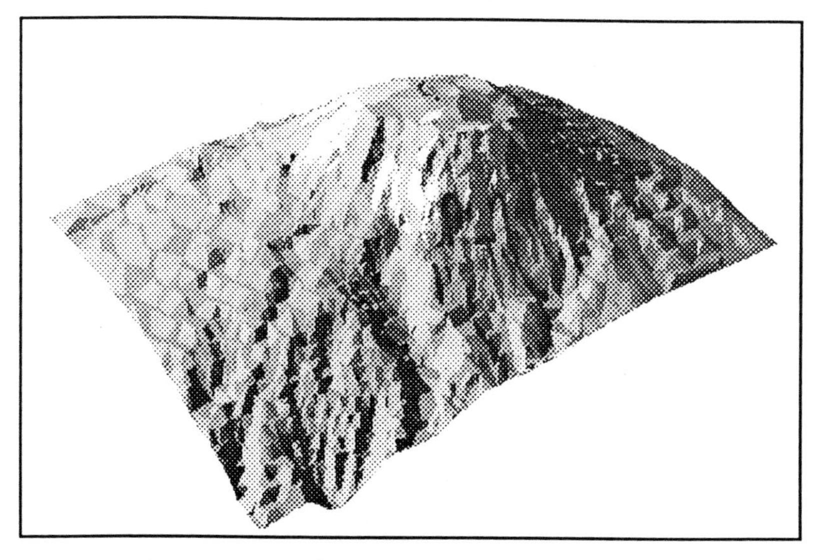

FIG. 14.3. Example of a CTM map product: snow-capped Kilimanjaro in Tanzania.

a particular chain segment can be included as a triangle side in the network. If not, it is split and the new point added to the network. The situation is checked again, and if necessary the split is repeated, and so on. The object topology (a reference to the map layer's objects) linked to the chains is used to classify the separate triangles. This data structure makes it possible to query, model and visualize the triangular network on the object level. The basic triangulation results in several tables, which contain vertex coordinates and triangle descriptions (e.g. triangle vertexes and neighbours). These tables are connected by pointers and allow the algorithm to construct the triangles and search the triangular network. The tables are, in combination with map chain data, used to generate additional tables that contain triangle attributes and chain segment topology.

Final Display Utilities

The display facilities should help the user to produce the final map using known cartographic techniques as well as appropriate computer graphics techniques, both depending on the output medium (for instance the screen or a PostScript printer). From the computer graphics world complete rendering processes can be implemented. Rendering techniques do not only take care of hidden surface removal, texturing and shading, but also include complete atmospheric models for realistic images (Kennie and MacLaren 1988). These last methods should only be applied if, from a cartographic point of view, they enhance the main task of the map: the clarification of spatial patterns. This set of utilities allows the cartographer to give the map a final touch, whereas the researcher might be satisfied with a more basic graphic display.

Query Utilities

When the final display utilities result in an image on the screen the three-dimensional visualization utilities are still active. Users can influence the position of the map. More importantly, the map can be interrogated. The queries can be at a low graphical level or a high application level. Answers to the low level queries are generated by the graphics language (PHIGS^{+}) and the data structure. The queries provide reference to coordinates, line types, line thickness, colours used and reference to the data structure's building blocks such as triangle's internal number, area, perimeter or tetrahedron's volume or neighbours. A non-cartographic user of CTM will be more interested in answers to high level queries. Basic terrain data such as slope and aspect are available. Triangle or tetrahedron attributes give access to the regions they belong to. The incorporated theme can be queried as well. Key data on the regions is provided (class, area, perimeter, volume etc.), and there is a direct link to a GIS database. The need to query is strongly linked to dynamic data exploration. Users can access the map with a three-dimensional cursor as shown in Fig. 14.4. The cursor also allows users to open extra windows where part of the map can be enlarged, or answers displayed. The cartographer can use these

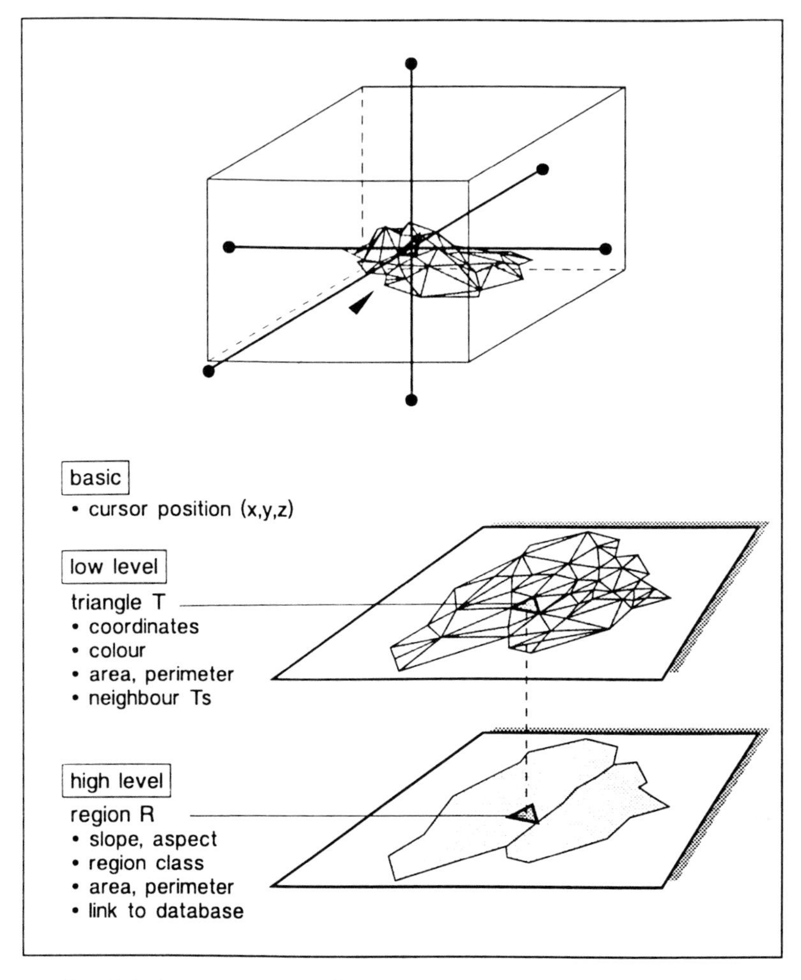

FIG. 14.4. Querying a map in the CTM environment: the three-dimensional cursor to access the map for low level and/or high level questions.

functions in a feedback process to create a better map, whereas a researcher can probably increase his or her understanding of the mapped phenomena.

Dynamic Data Exploration Utilities

As explained, CTM allows the user to link a theme (e.g. geology or land use) to the three-dimensional topography, represented by a digital terrain model. To increase the understanding of the mapped phenomena or provide tools for exploration it is possible to simulate a trip through the terrain. The landscape is visualized in a variation of what is called a flyby map. It is generated by displaying subsequent views of the terrain with a gradually changing observer's viewpoint (movement). However, the difference with the approach discussed here and flyby maps previously described (Moellering 1991; DiBiase *et al.* 1992) is that previous flybys do not allow interactive use. Animations of flybys, such as those create by the Jet Propulsion Laboratory and the USGS, have taken considerable computing time to create and follow predefined routes. In a data exploration environment interaction is a necessity. The user should be able to define the route and change directions whenever the information seen in the image makes him or her want to do so. It should also be possible to pause and look backwards, sidewards or take a full 360° turn. The main difference between movie-like flybys and the prototype discussed here is that the latter offers an environment with options for interactive positional change and for attribute change (see Fig. 14.5). The system allows, at any point during the travel, the possibility to switch the contents of map layers (partly) on or off and query the database.

To allow this dynamic functionality, a subdata structure is derived from that given above. Three tables are added to improve speed and allow interaction. As soon as a route section (chain) is identified and the direction of movement is known, the tables in the subdata structure allow the program to act immediately in response to the users' choices of direction when arriving at a crossing. For each node it is known which other chains start in the node, and which chain segments connect, and the subsequent perspective image can be shown.

In dynamic data exploration mode all CTM functionality is cumulated. Let us now look at some examples. To prepare stops for a field trip a geomorphologist could combine the terrain model with a relevant soil map, and while moving through the terrain recognize potential areas for severe erosion. Similarly, a biologist could switch on vegetation maps from different sampling dates and make temporal comparisons as well as querying the map to obtain information about specific species. The three-dimensional CTM environment should make interpretation much easier.

The functionality of the prototype terrain modelling environment has been described in six separate categories. Again, it should be stressed that all utilities interact and can be accessed in any order.

Interface

The functionality described in the previous section is available in the CTM environment. Before explaining how these capabilities are presented to the user, some remarks on the basic hard- and software are needed. One of the basic requirements is a high-end graphics workstation with specific peripherals. The desire to manipulate large fully shaded three-dimensional models requires a fast machine with a 24-bit graphics screen to allow for on-the-fly hidden-surface/hidden-line operations. Special peripherals are needed to improve the interaction with the program. To rotate a model by typing orientation parameters on the command line is less convenient than turning some dials. Access via the

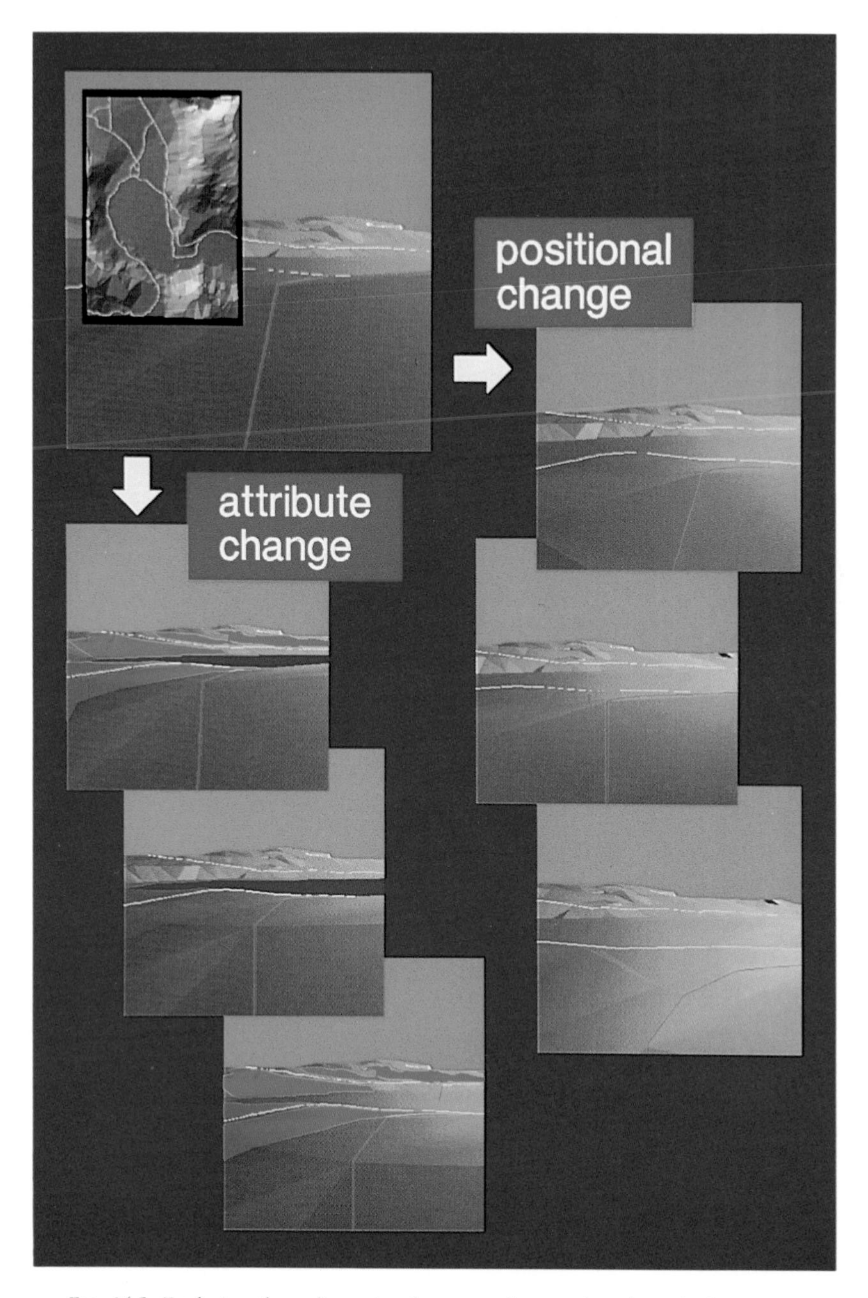

FIG. 14.5. Exploring three-dimensional topography: moving through the terrain while switching maps layers on or off. The three images below "positional change" show subsequent positions during the movement. The images below "attribute change" show examples of different terrain attributes. The attributes can be changed during the movement and can represent different themes or temporal versions of a single theme.

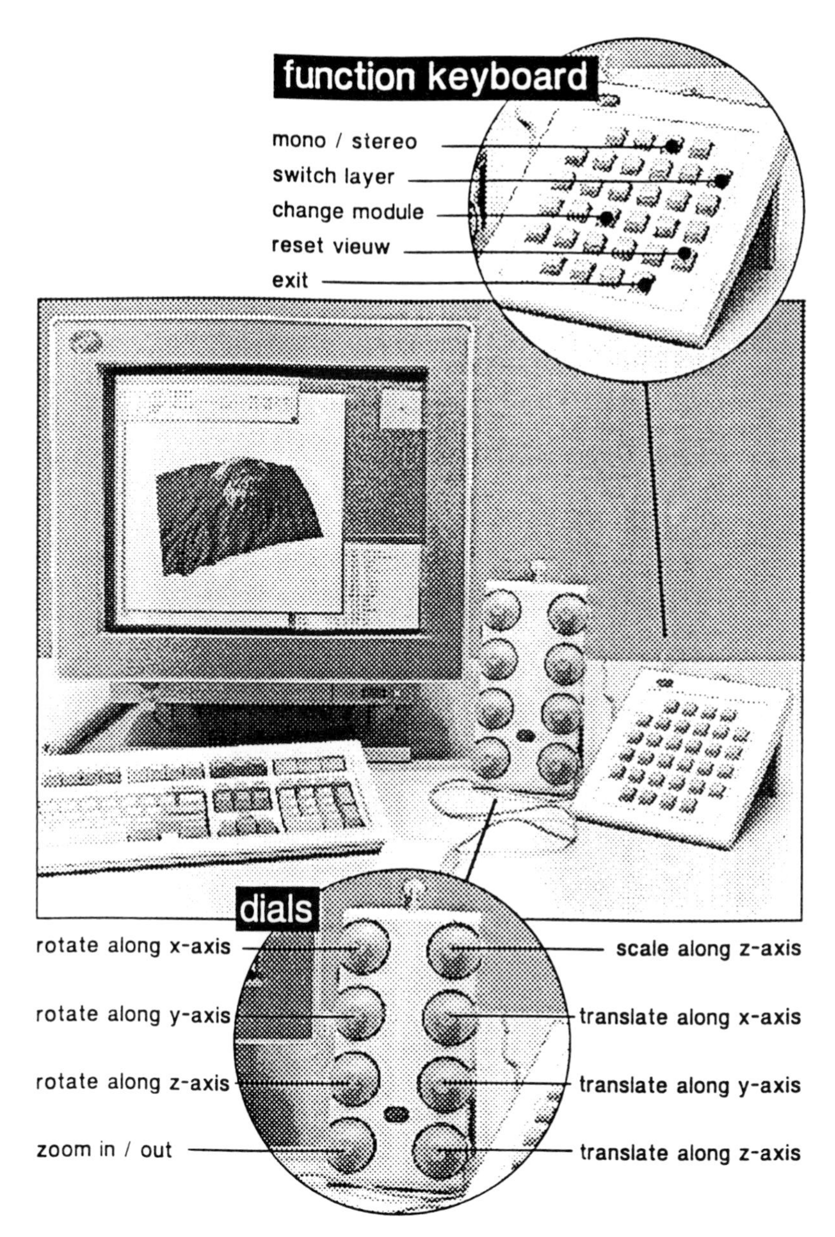

FIG. 14.6. CTM's hardware environment: a high-end graphics workstation with special peripherals.

command line is only useful when the user knows the exact orientation needed. Figure 14.6 shows two of the more interesting peripherals, the dials and a function keyboard. Both can be programmed to invoke an action after turning the dials or pushing a button. In Fig. 14.6 the functions linked to each of them can be seen. For instance, rotation is linked to the dials and a switch between mono and stereo display to the keyboard. A mouse activates the screen's graphical user interface (GUI). The general concepts of a GUI can be found in Apple (1987) and GUI in relation to GIS is described by Raper (1991).

The software runs in an X-windows environment that allows multiple windows to be active. Off the shelf software could not be used because it cannot yet fulfil the functional requirements. The software has been written with graphics routines from the extended Programmer's Hierarchical Interactive Graphical Standard (PHIGS$^+$). PHIGS$^+$ is an API programming interface (subroutine library) used in the development of graphics applications, such as mechanical design, robotics simulation and three-dimensional cartography. The applications, where the ISO standard (ISO 1988, 1989; Blake 1992) is used, all require a dynamic and interactive user interface and often deal with the visualization of three-dimensional objects. How movement is simulated will be described below. PHIGS$^+$ has the default capability to interface to the peripherals mentioned above.

Three aspects of the interface are of particular cartographic importance. These are orientation, legend and map access (query). With an unlimited freedom to position the map in three-dimensional space, it is necessary to inform the user of the map's orientation. In a three-dimensional environment the user is likely to become lost because of a "strange" position of the map, which might invoke visual illusions. In a two-dimensional map the problem is solved with a north arrow and a scale bar. In three-dimensional maps the use of an expedient such as a stereoscope can be of great help. In the CTM environment there are two situations to address in respect to this orientation problem. First, when the whole model is visible (see Figs 14.3 and 14.6), and second when only part of the model is visible (see Fig. 14.5). In both CTM situations an extra window that displays the more familiar two-dimensional map as well is helpful. In the second situation the movement is along the terrain surface. The two-dimensional map can be displayed in the plane of the screen with north to the top. A clear arrow symbol pointing towards the direction of the movement and following the positional change will help the user to keep track of his/her whereabouts. This method, however, only provides information on the position in the horizontal plane. A window with a profile of the route, including the arrow symbol, can be displayed additionally. It is the first situation, while the whole model is visible, where users are most likely to lose their feeling for orientation (the ambiguity of the so-called neckar's cube effect). As a solution the generalized two-dimensional map of the terrain could be positioned in a sphere. It will follow movements in the compass directions, and when the horizontal plane tilts the orientation map follows the movement as well.

In relation to the map's orientation a direct link between symbols in the map and those in the legend is necessary.

From a cartographic point of view a legend is essential. However, screens are relatively small. Alternatives to traditional legends include the use of a pop-up legend or simply allowing the user to query each map element separately at all times. With a pop-up legend the symbols in the legend should have the same orientation and other characteristics as those in the main map, as identification will be very difficult when symbols in the legend are of a different shape or are seen under different light conditions (shades, colours, etc.). This brings us to map access. In the CTM environment the map elements are accessible up to the level of individual triangles and tetrahedrons. With the mouse the user can point to the map element and retrieve information. A pop-up menu lets the user decide the level of information (see Fig. 14.4).

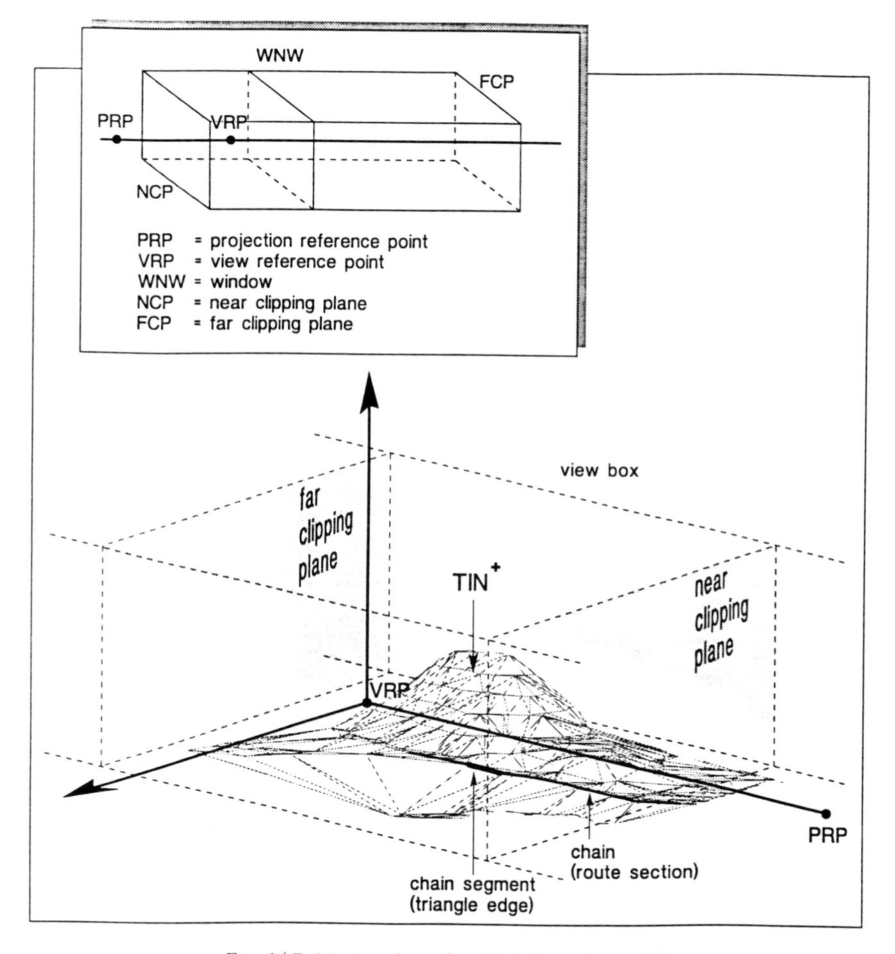

FIG. 14.7. Moving through a digital terrain model.

To understand how movement is simulated with the PHIGS$^+$, it is necessary to understand its view environment. In the three-dimensional environment, a model, such as a digital terrain model, is defined in world coordinates (for instance, the National Grid) and is positioned in a virtual box. This box is defined by clipping boundaries in a viewing coordinate system. The inset in Fig. 14.7 illustrates this approach. The origin of the viewing coordinate system is the view reference point (VRP). The front and back of the box are defined by the near and far clipping planes (NCP and FCP). The sides of the box are defined by the window (WNW). Only those parts of the model which are within the clipping boundaries (view box) can be made visible to the user. The Kilimanjaro depiction in Fig. 14.3, for example, fits in the view box as a whole, whereas the landscape in Fig. 14.5 is clipped on all sides to simulate the position of the user within the landscape. The view seen by the user is described by a view transformation matrix, to which the VRP and the projection reference point (PRP) are important parameters. The vector between the VRP and the PRP defines the line of sight. When the user perceives a change in the orientation of the model it is the result of a change in the view transformation matrix; in other words, the view box is manipulated by translation and rotation and the model itself is not modified.

To give the flyby map its dynamic character, movement is simulated by a constant change of position and orientation of the view box. Basic movement is achieved by moving the view box along the route of travel. This is accomplished by keeping the line of sight (view box) parallel with a chain segment of the current position. If one moves to the next chain segment the orientation of the view box is adapted. The near clipping plane is positioned between the user and the current position, and as a result hides the terrain which has been travelled just before.

Conclusions

Cartographic terrain modelling is about the visualization of three-dimensional "thematic" topography. In its application it can be used to create maps whose primary goal is visual communication, or create maps that are used to explore the data represented. A minimum functionality has been described as well as some basic elements of the user interface needed. Depending on the type of use, a different kind of cartography is needed. When CTM is used to create maps, traditional cartographic rules can be applied. When used to explore data, map design can be very different. An example is their use in an exploratory environment where researchers are likely to work with the original and unclassified data. Traditional classified data presentation will not work; neither will the visualization with all colours offered by the hardware in use. Even in this exploratory situation, however, maps are intended to provide insight into spatial relations. New cartographic tools have to be found. The challenge for the cartographer is to develop these new tools. Probably these new tools and rules are not as restrictive as traditional cartographic grammar, but neither are they as free as the technology allows.

References

Apple computer (1987) *Human Interface Guidelines*, Addison Wesley, Amsterdam.

Bishop, I. (1992) "Visualization in the natural environment: a look forward", *Landscape and Urban Planning*, Vol. 21, pp. 289–291.

Blake, J. W. (1992) *PHIGS and PHIGS+, an Introduction to Three-dimensional Computer Graphics*, Academic Press, London.

DiBiase, D., A. M. MacEachren, J. B. Krygier and C. Reeves, (1992) "Animation and the role of map design in scientific visualization", *Cartography and Geographic Information Systems*, Vol. 19, No. 4, pp. 201–214.

Ervin, S. M. (1993) "Landscape visualization with Emaps", *IEEE Computer Graphics and Applications*, No. 3, pp. 28–33.

Gibson, J. J. (1979) *The Ecological Approach to Visual Perception*, Houghton Mifflin, Boston.

ISO (1998) "Information processing systems — Computer Graphics — Programmer's Hierarchical Interactive Graphics System (PHIGS)", *ISO 9592*, ISO Central Secretariat, Geneva.

ISO (1989) "PHIGS PLUS", *ISO/IEC JTC1/SC24/WG2 N18* (March), ISO Central Secretariat, Geneva.

Kennie, T. J. M. and R. A. McLaren (1988) "Landscape visualisation", *Photogrammetric Record*, Vol. 72, No. 12, pp. 711–741.

Kraak, M. J. (1988) "Computer-asssisted cartographical three-dimensional imaging techniques", Delft University Press, Delft.

Kraak, M. J. (1993a) "Cartographic terrain modelling in a three-dimensional gis environment", *Cartography and Geographic Information Systems*, Vol. 20, No. 1, pp. 13–18.

Kraak, M. J. (1993b) "Three-dimensional map design", *Cartographic Journal*, Vol. 30, No. 2.

Kraak, M. J. and E. Verbree (1992) "Tetrahedrons and animated maps in 2d and 3d space", *Proceedings, International Symposium on Spatial Data Handling*, Charleston, Vol. 2, pp. 63–71.

Marr, D. (1982) *Vision*, Freeman, San Francisco.

MacEachren, A. M. and M. Monmonier (1992) "Introduction", *Cartography and Geographic Information Systems*, Vol. 19, No. 4, pp. 197–200.

McCormick, B. H., T. A. DeFanti and M. D. Brown (eds.) (1987) "Visualization in scientific computing", *Computer Graphics*, Vol. 21, No. 6 (complete issue).

Midtbö, T. (1992) "Spatial modelling by Delaunay networks of two and three dimensions", *Thesis*, University of Technology, Trondheim.

Moellering, H. (1980) "The real-time animation of three-dimensional maps", *The American Cartographer*, Vol. 7, No. 1, pp. 67–75.

Moellering, H. (1991) "Stereoscopic display and manipulation of larger 3-d cartographic objects", *Proceedings 15th ICA Conference*, Bournemouth, Vol. 1, pp. 122–129.

Overbeeke, C. J. and M. H. Strattman (1988) "Space trough movement", *Thesis*, Faculty of Industrial Design, Delft.

Pittman, K. (1992) "A laboratory for the visualization of virtual environments", *Landscape and Urban Planning*, Vol. 21, pp. 327–331.

Raper, J. F. (1991) "User interfaces" in Masser, I. and M. Blakemore (eds.), *Handling Geographical Information*, Longman, London, pp. 102–114.

Robinson, A. H., J. L. Morrison, P. C. Muehrcke, S. C. Guptill and A. J. Kimberling (in press) *Elements of Cartography*, 6th edn, Wiley, New York.

Sheppard, S. (1989) *Visual Simulation: a User's Guide for Architects, Engineers and Planners*, Van Nostrand Rheinhold, New York.

Wager, L. R., J. A. Ferwerda and D. P. Greenburg (1992) "Perceiving spatial relationships in computer-generated images", *IEEE Computer Graphics and Applications*, No. 5, pp. 44–55.

Tomlin, C. D. (1990) *Geographic Information Systems and Cartographic Modelling*, Prentice Hall, Englewood Cliffs.

CHAPTER 15

Multivariate Display of Geographic Data: Applications in Earth System Science

DAVID DIBIASE, CATHERINE REEVES, ALAN M. MACEACHREN, MARTIN VON WYSS, JOHN B. KRYGIER, JAMES L. SLOAN and MARK C. DETWEILER*

Department of Geography
The Pennsylvania State University
University Park, PA 16802, USA

**Department of Psychology*

Introduction

This chapter discusses methods for displaying three or more geographically referenced data variables. In particular, we are concerned with methods appropriate for exploratory analyses in Earth system science. Earth system science is an integrative, multidisciplinary approach to understanding "the entire Earth system on a global scale by describing how its component parts and interactions have evolved, how they function, and how they may be expected to continue to evolve on all time scales" (Earth System Sciences Committee 1988). Numerical modeling is an important method of formalizing knowledge of the Earth system and for attempting to predict natural and human-induced changes in its behavior. The volume of quantitative data produced by model simulations and Earth-observing satellites increases the value of effective graphic techniques for identifying potentially meaningful features and anomalies. Because the Earth system comprises many interrelated phenomena, the ability to display multiple data variables in formats that foster comparison is especially desirable.

The initial goal of the chapter is to acquaint readers with the variety of multivariate display methods published in the areas of statistics, computer graphics and cartography. In so doing we will set a context for describing an interface we have developed for displaying up to four paleoclimate data sets produced by a global climate model at Penn State's Earth System Science Center. Key features of the interface include: (1) display of multivariate data in computationally efficient two-dimensional map formats; (2) optional display of a single map on which up to three variables are superimposed or four juxtaposed maps depicting up to four variables; and (3) the ability to quickly revise data variable selections, to focus on subsets of the data and to experiment with different data classification schemes and combinations of point, line and area symbols. In this approach we have attempted to incorporate the three functional characteristics that differentiate cartographic visualization from cartographic communication: the interface is tailored to the needs of a specific set of users; its intended use is more to foster discovery than to present conclusions; and it is an interactive environment that encourages users to experiment with different combinations of data and graphic symbols (MacEachren this volume). We conclude with a brief discussion of the problem of evaluating the effectiveness of this and other multivariate exploratory methods.

Four Choices in Designing Multivariate Displays

Attention to multivariate display among cartographers has seldom exceeded the design of bivariate choropleth maps (for example, Olson 1981; Eyton 1984). In fact, the limited effectiveness of bivariate maps as vehicles of communication reported in those studies may have discouraged the innovation of displays of higher dimensionality. As discussed elsewhere in this book, however, cartographic visualization implies emphasis on the construction of knowledge more than on its dissemination. Complex multivariate displays are likely to be more effective for geographic analysis than for communication because analysts have access to specialized tools and expertise not available to the general public. Statisticians and computer graphics specialists have outpaced cartographers in innovative multivariate display methods because they have paid greater attention to the needs of analysts. Some cartographers have recently begun to adapt these innovations for geographic applications.

To organize our review of major published multivariate display methods we pose four choices about how to "map" data to a display that analysts address when adopting any particular method. The order in which we present the choices is arbitrary. One choice is between cartographic and data spaces. Though we have limited our purview to geographically referenced data, these may be expressed in graphic spaces whose axes represent measurement scales corresponding to either distance or quantities of phenomena distributed over the Earth's surface. As either two- or three-dimensional graphic spaces may be constructed, a second choice between two- and three-dimensional displays follows. Third, analysts may also choose between single and multiple views. Finally, assuming the availability of suitable computing facilities, a fourth choice between static and dynamic displays also comes into play. In the following four sections we describe examples of outcomes of these choices and discuss the decision criteria involved in selecting multivariate displays for Earth system science visualization. We also identify several opportunities for further research to evaluate the effectiveness of the various approaches.

Cartographic Versus Data Spaces

Maps and graphs both express relations among two or more data sets by positioning observations within spaces defined by two or more measurement scales or "axes". The axes of most graphs represent ranges of values measured for a particular variable, such as temperature or precipitation. By "data space" we mean a graphic field defined by relations between two or more such measurement scales. Maps may be conceived as a special case of graphs in which axes represent distances in "cartographic space", such as longitude, latitude and perhaps elevation relative to some datum. The problem of displaying three or more quantitative variables is exacerbated when the axes of the display are reserved for representing distance, as they are on most maps.

Data Space Representations

One example of a data space representation is the geographic biplot (Fig. 15.1). Extending the work of Gabriel (1971), Monmonier (1991) describes the geographic

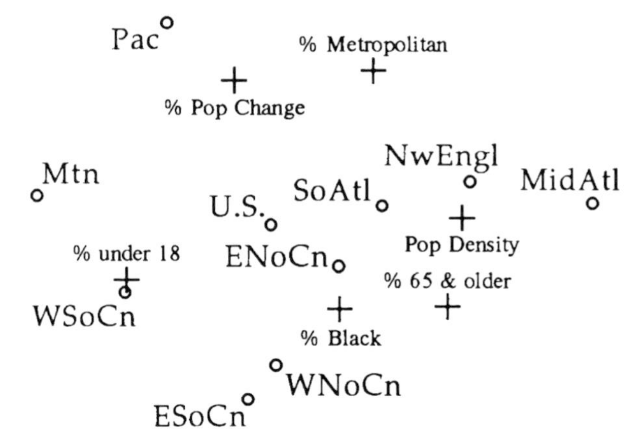

FIG. 15.1. Geographic biplot expressing relationships among nine Census divisions of the USA and six demographic indicators for 1988 (Monmonier 1991a). Principal components analysis collapses the six variables into two orthogonal axes that account for 71% of the total variation. The axes (not shown) originate at the center of the display. Census divisions are plotted using component scores as coordinates. Positions of the demographic variables are determined by normalized eigenvectors of the two principal components for each variable. The proximity of place-points represents similarity with respect to the six demographic variables. The proximity of points that stand for the demographic variables represents geographic similarities among the variables.

biplot as "a two-dimensional graphic on which a set of place points represents enumeration areas or sample locations and a set of measurement points represents different indicators or different years for a single measure." The biplot's axes are usually derived from principal components analysis. Attributes are plotted by eigenvector coefficients, places by component scores. "Relative proximity on the biplot reflects relative similarity among places and variables." The biplot is most effective when two principal components account for a large proportion of the variation in a multivariate data set. When three axes are required to explain a satisfactory fraction of variation, three-dimensional biplots can be constructed using eigenvector coefficient and component score triplets (Gabriel *et al.* 1986).

For Bertin, all geographic data analysis begins with the construction of a "double entry data table" with columns standing for enumeration areas and rows representing thematic data. Instead of numerals, each cell in the table is coded with graphic marks that vary in size or darkness to signify quantities. Reordering the rows to bring out patterns in high and low quantities constitutes a graphic analogue to cluster analysis that Bertin calls "matrix analysis". One impressive example reveals regional patterns in the cost of 70 basic commodities among 31 cities (Bertin 1981). Bertin goes on to demonstrate how matrix analysis can be used as a precursor to the classification of thematic maps.

As the human perceptual system is limited to envisioning three spatial dimensions, relations among more than three variables must be projected to two- or three-dimensional spaces for exploratory analysis. The more variables an analyst wishes to consider together, the more projections of the data are possible. Friedman and Tukey's (1974) PRIM-9 data analysis system included an algorithm called "projection pursuit" that suggested "optimal" two-dimensional projections for multivariate data based on measures of point spread, local cluster density or a combination of both. As in cartography — where interest in "optimal" data classification has been augmented by concern for observing the effects on pattern stability of alternative class assignments — projection pursuit was succeeded in statistics by the "grand tour" (Asimov 1985). The aim of the grand tour is to reveal structure in multidimensional data sets by presenting analysts with an animated sequence of many two-dimensional projections, as if passing a two-dimensional viewing plane through a point cloud in multidimensional space. Monmonier (1990) again has adapted this concept for use by geographers with the "atlas tour", a programmed sequence of maps and other information graphics intended to introduce viewers to geographic subjects from multiple perspectives. Beyond the grand tour, other statisticians have more recently developed interactive "guided tours" which enable analysts to define transformations between selected data projections. For example, Hurley's (1988) demonstration of the Data Viewer program included interpolated, real-time transformations between geographic and attribute planes for nine criteria of Rand McNally's "places rated" index (Fig. 15.2).

Cartographic Representations

For most geographers and geoscientists, however, latitude and longitude represent not variables, but a standard frame of reference. In the private realm of exploratory analysis as well as in the public realm of policy debate, the map format is insuperable whenever the geographic implications of data distributions are pertinent. The extent to which viewing geographic data in cartographic spaces is an ingrained custom in geography is evident in how leading cartographic commentators have disregarded Bertin's most powerful graphic variable "position" for mapping (for example, McCleary 1983; Morrison 1984; Muehrcke and Muehrcke 1992). Exceptions to this predilection include thematic transformations of

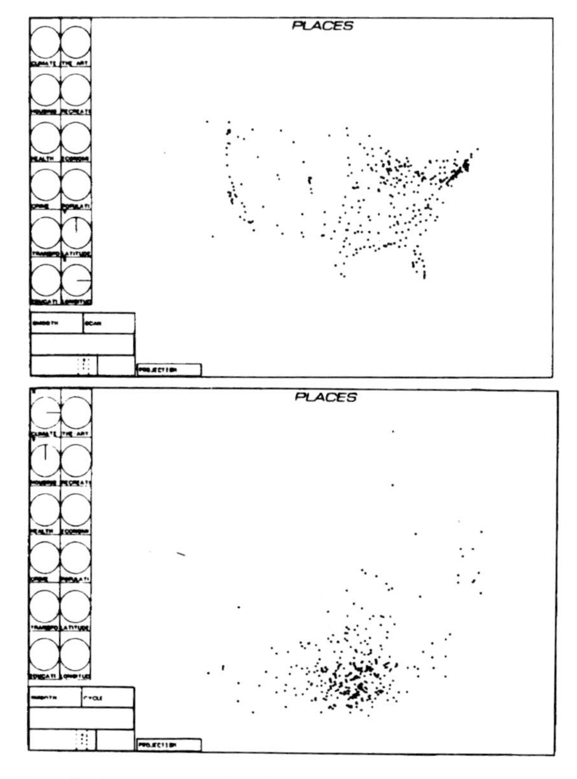

FIG. 15.2. Two displays generated with the Data Viewer (Hurley 1988) showing bivariate distributions selected from nine "livability" criteria for 329 US cities. The upper display plots locations of the cities in cartographic space. The lower display depicts the cities in a data space defined by climate versus housing criteria. Users of the Data Viewer can select variables and specify geometric relations among variables by manipulating dials stacked to the left of the display. A "guided tour" can be created by generating a smooth sequence of intermediate displays.

cartographic spaces such as logarithmic projections (Hägerstrand 1957) and cartograms (Haro 1968), which have been used occasionally and effectively in human geography, but rarely for multivariate display.

When a position is reserved for representing geographic spatial relations, other symbol characteristics must be relied upon to depict multiple data variables. Cartophiles' fascination with topographic and thematic maps is in a large part due to the ingenious graphic symbols cartographers have devised for combining many categories of information. The 1 : 24,000 scale topographic map series of the United States Geological Survey, for example, depicts 17 categories of features of the natural and built landscapes with 138 distinct symbols (Thompson 1982). Masterworks of multivariate geographic information display can be found in thematic atlases such as the *Historical Atlas of Canada*. One distinguished example from Volume I (Harris and Matthews 1987) is Plate 69 – "Native Canada ca 1820" — on which trading posts and missions, settlement populations, territorial boundaries, commerce and migration flows, predominant linguistic families and native economies are superimposed. Transfer of the design expertise evident in traditional cartographic products to interactive visualization systems is difficult because the design objectives and display media involved differ substantially.

Two- Versus Three-dimensional Displays

Most display media for computer-based analysis available to geoscientists and analysts in other disciplines provide two-dimensional images. Primary among these are raster CRTs and paper or film imaging devices. Three-dimensional illusions can be created by employing psychological or physiological depth cues (Kraak 1988). Though simulated three dimensions offer higher information-carrying capacity than two dimensions, adding a dimension to the depiction also introduces additional perceptual and computational costs. For example, in oblique perspective three-dimensional terrain images, the scale varies continuously and aspects of the surfaces tend to be hidden from view. The relative advantages of two versus three dimensions depend mainly on the dimensionality of the phenomena to be visualized (and their surrogates, the data variables) and the spatial relationships among the variables (phenomena may be measured or modeled at one or several different layers of the atmosphere, for instance). Also pertinent are the hardware and software available to the analyst, the cognitive tasks to be performed and his or her aesthetic preferences.

Two-dimensional Approaches

One class of two-dimensional approaches exploits "glyphs" — compound point symbols such as faces, trees, castles and the "icons" used in the Exvis system developed by Grinstein and co-workers at the University of Lowell (Smith *et al.* 1991). An Exvis display consists of a dense two-dimensional matrix of strings of

Fig. 15.3. "Iconographic" display created with the Exvis system (Smith *et al.* 1991) showing a composite of five bands of a remotely sensed image centered on the southern peninsula of Ontario, bounded by Lakes Huron, Erie and Ontario. Pixel values for the five images are expressed in the relative angles of five line segments joined end to end.

short line segments (Fig. 15.3). The orientation of each line segment relative to those connected to it represents a quantity for one of several variables at a given location. The system provides user control for the size of each icon, line segment length, line segment orientation limits and displacement of icons to reduce artifactual patterns. Remarkably differentiable patterns can emerge as textural differences in properly "focused" Exvis displays. Employing the analogy of a carpet to describe these displays, Grinstein and Smith (1990) remark that "one can obtain specific kinds of information about the length and materials of individual fibers, but one also receives ... an impression of the overall texture of the carpet".

For analysts willing to suspend the orthogonality of longitude and latitude, relations among many geographic variables can be viewed at once using parallel coordinate diagrams (Bolorforoush and Wegman 1988). In this method each variable is assigned a parallel linear scale (Fig. 15.4). Corresponding cases on each scale are connected by straight lines. Positive and negative correlations, clusters and

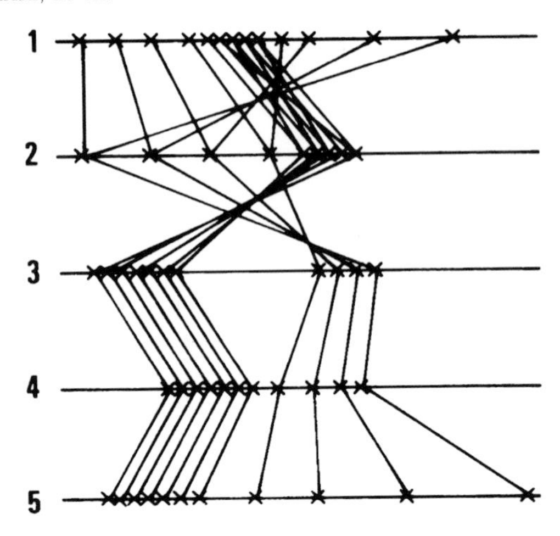

FIG. 15.4. Hypothetical example of a five-variable distribution expressed in parallel coordinates (Bolorforoush and Wegman 1988). Patterns of lines connecting observations across parallel axes reveal multidimensional correlations, clusters and modes. Non-crossing connecting lines indicate positive correlations; cross-overs indicate negative correlations. Three-dimensional clustering is exhibited among variables 3, 4 and 5. A multidimensional mode ("the location of the most intense concentration of probability") is represented in the band of connecting lines beginning near the middle of axis 1 and finishing on the left-hand side of axis 5.

modal cases among many variables can be revealed in patterns of parallelism, convergence and central tendency among the connecting lines. When the number of cases is large, the density of overlapping connecting lines can be interpolated and displayed as isolines (Miller and Wegman 1991). Coordinate scales must often be normalized or transformed in other ways to produce meaningful patterns. Similar to Bertin's matrix analysis, the order of scales affects the appearance of patterns among variables. We have not yet seen or attempted to construct a parallel coordinate diagram for geographically referenced data, but we expect that the method might be useful for displaying the gridded data produced by general circulation models used by Earth system scientists.

Cartographers and geographers have also demonstrated display methods for multivariate quantitative data on two-dimensional maps. Most efforts by cartographers to date have been directed to maps intended for presentation rather than exploratory use. Exceptions include Bertin's (1981) "trichromatic procedure" for superimposing three point symbol maps of quantitative data colored with the three printers' primaries (cyan, magenta and yellow), whose combination results in an informative range of hues and lightnesses. Dorling (1992) produced a time series of population cartograms of England superimposed with arrows symbolizing the mixture of alliances to three political parties (by hue and saturation) and the magnitude and direction of voting swings (by length and orientation) in more than 600 constituencies in Parliamentary elections.

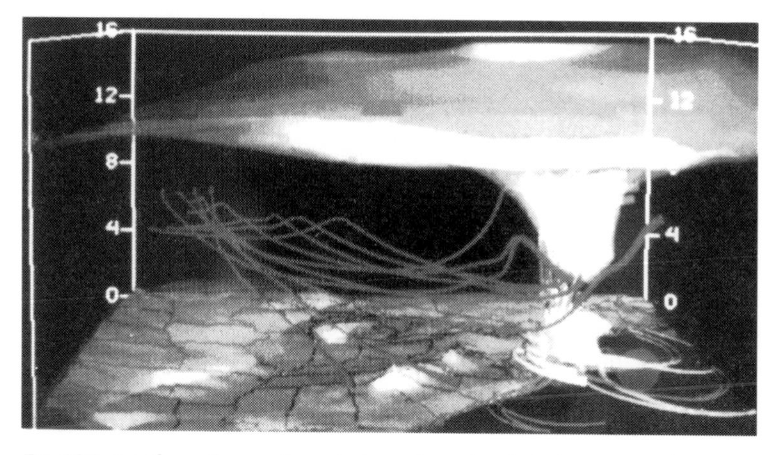

FIG. 15.5. One frame from a dynamic three-dimensional display generated with the McIDAS system (Hibbard *et al.*, cited in Keller and Keller 1993). Streamer symbols represent converging high and low vorticity flow contributing to a mid-latitude cyclone.

Three-dimensional Approaches

Three-dimensional representations are invaluable for portraying volumetric phenomena. Describing the animated stereoscopic display terminals developed for use with the McIDAS weather display system (Fig. 15.5), Hibbard (1986) remarks that "they create an illusion of a moving three-dimensional model of the atmosphere so vivid that you feel you can reach into the display and touch it". Interactive three-dimensional displays are also desirable for investigating forms of relationships among non-geographic variables in three-dimensional scatterplots (Becker *et al.* 1988). For map-based displays, however, three-dimensional does not necessarily simplify the problem of effectively superimposing three or more data distributions.

In a review article on meteorological visualization, Papathomas *et al.* (1988) note that two approaches to multivariate three-dimensional visualization have been attempted: "One is to portray each variable by a different attribute [graphic variable] ... another is to assign different transparency indices to the various surfaces that represent the variables". The success of the latter approach has been limited. As Hibbard (1986) observed, "when viewing a transparent surface and an underlying opaque surface simultaneously, the eye is quite good at separating the shade variations of each surface and deducing their shapes. However, this breaks down when there are several transparent layers". The former method has been exploited successfully in a revealing simulation of the cyclogenesis of a severe storm by Wilhelmson and co-workers colleagues at the National Center for Supercomputing Applications at the University of Illinois (Wilhelmson *et al.* 1990). Tracers, streamers, buoyant balls and translucent shading are artfully combined to reveal storm dynamics. Display of high resolution three-dimensional imagery can be costly, however; the seven and a half minute simulation required the equivalent of one person's full-time effort for more than a year and 200 hours rendering time on a Silicon Graphics Iris workstation.

Two- or Three-dimensional?

Tufte (1983) argues that the dimensionality of a graphic representation should not exceed the dimensionality of the data. Evidence supporting Tufte's claim is found in a study by Barfield and Robless (1989), who asked novice and experienced business managers to solve a management problem using bivariate data presented in either two- or three-dimensional histograms. Subjects using two-dimensional graphs formulated solutions 28% more quickly than those using three-dimensional graphs and reported feeling more confident about their answers. The accuracy of responses associated with three-dimensional graphs was not significantly better than with two-dimensional graphs, though response times associated with three-dimensional graphs were greater. Results of an experiment by Marin (1993) suggest that viewers asked to interpret vegetation patterns mapped on a shaded terrain surface perform no better with oblique three-dimensional representations than with two-dimensional planimetric displays. Kraak (1988) also reported no significant difference in response times associated with two- (choropleth) versus three-dimensional (prism) univariate thematic maps. Considering the additional computing performance required for three-dimensional rendering, two-dimensional representations ought to be preferred for two-dimensional data.

The phenomena of greatest interest to Earth system scientists are volumetric, such as the circulation of the atmosphere and the oceans. However, the products of numerical model simulations are often analyzed for specific surfaces. Tufte's recommendation to match the dimensionality of the graphic to the dimensionality of the data becomes ambiguous when phenomena of interest are undulating surfaces that are neither volumetric nor planar, such as surface temperature or 500 millibar pressure heights. Although statistical surfaces are represented numerically by lists of coordinate triplets, these may be interpretated as representing two spatial dimensions (x, y) plus an attribute (z) rather than three spatial dimensions. Planimetric two- and simulated three-dimensional visualization methods have both been used for such data (some workers, including Raper (1989), subdivide simulated three-dimensional displays into "2.5-D" perspective surface representations and three-dimensional solid models). Which approach is more appropriate for surface data? Setting aside technical requirements and aesthetic preferences, we suggest two conflicting criteria on which the choice might be based: comparability (the ease of noticing spatial correspondence among data sets) and distinguishability (the ease of distinguishing between data sets). By depicting related statistical surfaces as separate overlapping layers, simulated three-dimensional displays may maximize distinguishability at the expense of comparability. When several statistical surfaces are superimposed in a planimetric two-dimensional depiction, both comparability and distinguishability depend on the quality of graphic symbolization. In the case study reported here we present our attempt to maximize both criteria by representing three data distributions with point, line and area symbols on two-dimensional maps.

Single Versus Multiple Displays

Diverse goals motivate exploratory analyses. In a survey of 90 published studies involving visualization, Wehrend and Lewis (1990) identify 11 categories of visualization operations ("identify, locate, distinguish, categorize, clusters, distribution, rank, compare, within and between relations, associate, correlate"). Analysts engage in a variety of perceptual tasks when they use graphics to explore data. Carr *et al.* (1986) observe that "in general, we do not know how to produce plots that are optimal for any particular perceptual task". As a data display may be called upon by a variety of users to serve several tasks, search for a single optimal display method is likely to be futile. In the context of cartographic visualization, MacEachren and Ganter (1990) argue forcefully that "*there is no optimal map!*" (authors' italics). Monmonier (1991b) has argued that single displays created to communicate data relationships to the public may even be unethical.

Graphical user interfaces now make it easy to produce multiple data displays for exploratory analysis. Two strategies are constant formats (for example, a sequence of maps of the same extent and scale showing changes in time series data) and complementary formats (such as a scatterplot linked to a map).

Constant Format Multiple Views

Tufte (1990) refers to ordered series of information graphics displayed in a constant format as "small multiples", and asserts that "for a wide range of problems in data presentation, small multiples are the best solution". Bertin (1981), who generalizes the variety of visualization tasks as two types of questions ("What is at a given place?" and "Where is a given characteristic?"), declares that "it is not possible, with complex information, to represent several characteristics in a comprehensive way on a single map while simultaneously providing a visual answer to our two types of question". Only comparison of several juxtaposed univariate maps can answer both types, he argues.

Visual comparison of juxtaposed, constant format maps can be an error-prone perceptual task, however (Steinke and Lloyd 1983). In a study of map pattern comparison abilities by experienced and inexperienced map users, Muehrcke (1973) demonstrated that comparison errors associated with five different map symbol types all increased as a function of pattern complexity. Though experienced map users fared better than inexperienced ones, the accuracy of their comparisons declined at a similar rate. Subjects' performance with isoline maps — the most common symbolization type in climatological graphics — suffered the most with increased complexity. Even so, their impressive overall success rate (81 errors of a possible 750 — 89% correct) bolsters claims that small multiples can be effective.

Complementary Format Multiple Views

Monmonier has been a leader in introducing multiple view, complementary format display techniques for geographic analysis. In 1989 he described the concept of "geographic brushing" as an extension of scatterplot brushing (Becker *et al.* 1988). Geographic brushing of trivariate distributions involves the linked display of scatterplot matrices or three-dimensional scatterplots and a map (Fig. 15.6). Locations of selected data points in the scatterplot are highlighted on the accompanying map. Conversely, map selections are highlighted in the scatterplot.

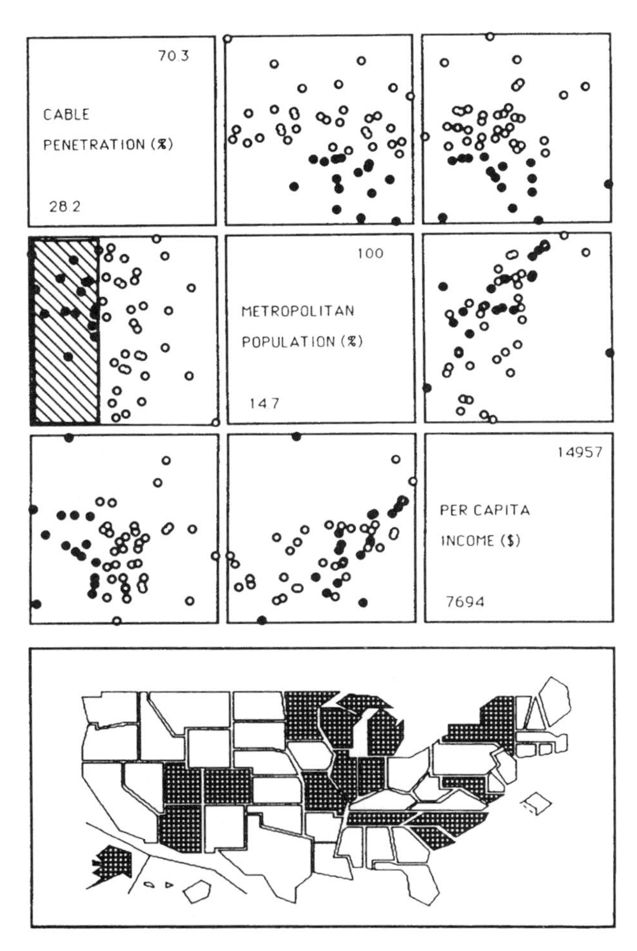

FIG. 15.6. Geographic brushing (Monmonier 1989) is an interactive, complementary format multiple view visualization technique. Cases selected in one cell of the scatterplot matrix are highlighted automatically in other cells. Corresponding geographic areas are simultaneously highlighted on the map. Conversely, geographic areas may be selected, highlighting corresponding cases in the scatterplot.

The geographic biplot (described above) also serves as a useful complement to map-based displays. The efficacy of complementary displays relative to constant format small multiples and single superimposed displays is a matter of conjecture, ripe for empirical testing.

Single View Superimpositions

Single displays on which three or more variables are superimposed will not be made obsolete by multiple view techniques, however, in spite of Bertin's proclamations and empirical test results by others. Synoptic climatologists, for example, rely on superimpositions to characterize interactions of atmospheric phenomena that give rise to severe weather events. A perhaps extreme example is a series of seven-variable thematic maps representing a "composite mean analysis for a synoptic flow type that is both favorable for the formation of a capping inversion and severe thunderstorms over the southern Great Plains in early spring" published by Lanicci and Warner (1991) (Fig. 15.7).

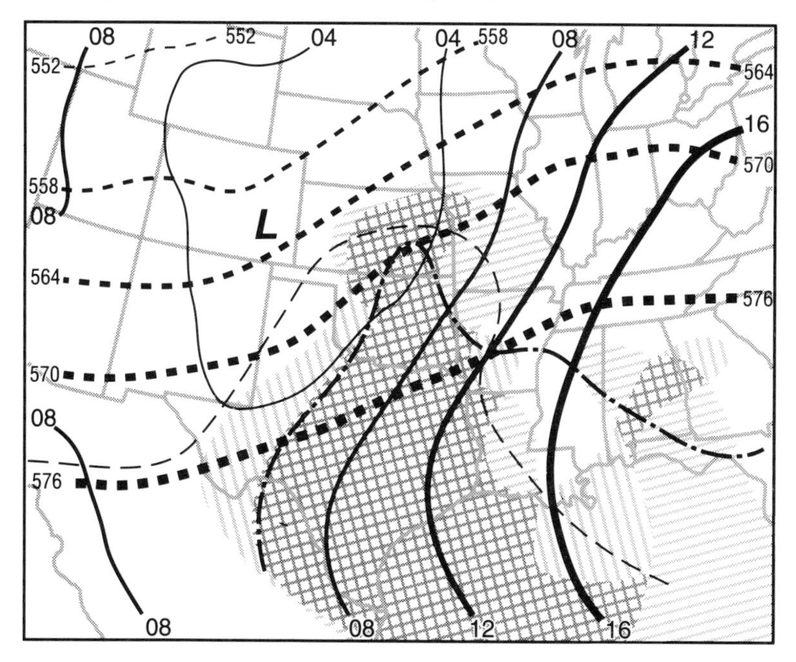

FIG. 15.7. Static, single-view superimposition constructed for a synoptic climatological analysis illustrating the use of contrasting symbols (Lanicci and Warner 1991). Seven atmospheric variables are represented, including mean sea-level isobars (solid weighted isolines), mean 500 millibar heights (dashed weighted isolines), mean surface 55°F dewpoint (dot-dashed line), 700 millibar 6°C isobar (thin dashed line), areas characterized by greater than 50% frequency of unstable buoyancy (near-vertical shaded hatches), the presence of capping inversion with well mixed layer above (near-horizontal shaded hatches) and areas where buoyantly unstable air is overlain by capping inversion (shaded cross hatches).

Analysts who require superimpositions face a secondary choice in designing multivariate displays: whether to employ composite symbols or contrasting symbols. The icons used in Exvis displays (Fig. 15.3) are composite symbols composed of several connected line symbols whose relative orientations signify data quantities. The Lanicci and Warner example (Fig. 15.7) uses contrasting line symbols to help discriminate the several variables depicted. The contrasting symbol strategy was suggested by Bertin (1981) and has also been employed by Crawfis and Allison (1991) for their "scientific visualization synthesizer". Composite symbols seem to provide clearer views of global patterns among variables, whereas contrasting symbols apparently allow easier inspection of individual distributions. Here is another opportunity for experimental evaluation.

Static Versus Dynamic Displays

A recurring theme in the published work we have consulted is the advantage of dynamic graphical methods over their static counterparts. By "static" displays we mean those whose form is essentially atemporal; although time is required to perceive them, the graphic elements that constitute their appearance are unchanging. Conversely, change is intrinsic to "dynamic" displays. Change implies the passage of time. The incorporation of time into data displays provides at least three modes of expression not available in static representations: interaction, animation and sonification (DiBiase *et al.* 1992). Interaction requires time for an analyst to invoke change in a representation. Animation requires time to express changing positions and/or attributes of graphic elements in a sequence of representations. Sonification — the representation of data with sound — requires time to express data-driven changes in pitch, volume, timbre and other sonic elements. Several of these modes are often combined in practice.

Interaction

For most analysts, the term "dynamic graphics" connotes interaction: the "direct manipulation of graphical elements on a computer screen and virtually instantaneous change of elements" (Becker *et al.* 1987). Both fast hardware and an interface that promotes the efficient performance of intended tasks are required for fruitful interaction. Two of the most common interaction operations are selection and transformation.

Highly interactive systems allow users to select easily among lists of data variables to be displayed. A study by McGuinness *et al.* (1993) of experienced and inexperienced analysts asked to draw conclusions from a nine-variable data set demonstrates that although expert analysts tend to consider fewer data variables in combination at any one time, they will also wish to view and review a greater number of combinations than novices (see the chapter by McGuiness for details). Whether such selections are more conveniently accomplished by mouse clicks on menus or by key-entering text on command lines depends on the previous experience and cognitive style of the user (Foley *et al.* 1984).

Becker *et al.* (1987) of AT&T's Bell Laboratories describe an interactive data analysis system that features several other selection tasks, including the selection of individual cases or groups of cases within a three-dimensional scatterplot or a two-dimensional scatterplot matrix (brushing). Cases highlighted in a scatterplot are identified by highlighted names in an accompanying list. Deleted selections may be recalled later. Multiple selections are highlighted by distinguishable symbols. An interactive system for geographic brushing incorporating some of these features was introduced by Tang (1992).

Meaningful relations among multiple data variables are more likely to be revealed if the analyst can vary his or her perspective on the data. Interactive rotation is required to explore multivariate distributions displayed as three-dimensional point clouds and translucent volumes. As most data represented in cartographic spaces are generalized for display (for example, continuous statistical surfaces are typically represented by isolines, whose number and interval are variable), the ability to change generalization parameters also provides alternative views. Convenient access to coordinate transformations such as multidimensional scaling, projection pursuit and guided tours also improve analysts' chances of revealing structure in multivariate data. Young and Rheingans (1990) demonstrated a real-time guided tour through a six-dimensional data space derived from principal components analysis of rates of seven types of crime for the 50 states in the USA. Cubes representing the states were "rocked" back and forth between two three-dimensional biplots formed by the first three and last three principal component coefficients, respectively. Clusters in the six-dimensional space were revealed by similar patterns of movement among the cubes through intermediate frames interpolated between the two biplots.

Animation

Though the success of their visualization depended on "patterns of movement", Young and Rheingans would object strongly if we were to describe it as animation. In the computer graphics literature, animation implies "trivial" display of pre-computed scenes, as opposed to "real-time graphics" in which each scene is computed and displayed from scratch. From an interaction perspective the latter is obviously desirable, but exploitation of the illusion of movement to depict change is fundamental in both approaches.

The most straightforward way to use animation for the display of three or more variables is to view a loop of univariate maps. This method has been called "flickering" (Felger and Astheimer 1991) and has been implemented for exploratory geographical analysis by Tang (1992). Flickering has also been suggested as a way to relate data with metadata (MacEachren *et al.* 1993). Animation can be used to inspect complex two- or three-dimensional multivariate portrayals through zooms (scale changes) and pans (horizontal viewpoint changes). Depending on the hardware power, software efficiency and the desired rate of change, zooms and pans can be accomplished either with pre-computed scenes or in real time. Dorling (1992) found these techniques for "animating space" most useful in exploring a population cartogram of England superimposed with trivariate hue and saturation symbols signifying proportions of three occupation categories for over 10,000 wards.

Dorling also experimented with "animating time" for the multivariate election data described earlier, but in general found time series animations unsatisfactory due to the fleeting nature of visual memory. Nonetheless, time series continue to be the most common subject of scientific animations.

In another paper (DiBiase *et al.* 1992) we discuss an additional way of using animation to express change, for which we borrow Tukey's term "re-expression". Re-expression implies transformation of one or more measurement scales in the hope of revealing hidden structure in the data. The order of scenes in an animation (whether produced in advance or in real-time) can be treated as a measurement scale. One way to re-express an animation is to transform a chronologically-ordered sequence into an attribute-ordered sequence. In combination with a time-line display that shows the temporal order of re-expressed scenes, we expect that this sort of representation may occasionally be helpful in identifying regional patterns and discontinuities in the simulated behavior of atmospheric and oceanic phenomena.

Sonification

Sound can serve as an informative complementary stimulus for exploratory analysis. One of the more fruitful applications to date is the Exvis system (Grinstein and Smith 1990), in which each graphic icon is associated with a unique synthetic "voice". A stereophonic auditory display is created by spreading the voices' apparent positions from the left ear to the right ear in an order corresponding to the range of some data variable. Data-driven reverbation is also used to create a proportional depth effect. Krygier (elsewhere in this book) provides a fuller treatment of this and other experiments with this intriguing display technique.

Summary

Of the four choices posed in the preceding sections, three can be given straightforward recommendations. For exploratory analysis of multivariate Earth science data, cartographic representations are desirable, but not to the exclusion of graphs. Multiple displays — either in constant formats (small multiples) or complementary formats (linked maps and graphs) — are preferable to single "optimal" views. Dynamic displays — particularly those that promote experimentation with alternative data variable combinations and transformations of viewpoint, generalization and measurement scales — are essential for exploratory analyses. The choice of planimetric two-dimensional versus simulated three-dimensional displays should be based on the principle of matching the dimensionality of the display with that of the data, though that recommendation is ambiguous for data representing statistical surfaces that are neither volumetric nor planar.

Case Study: A Map Interface for Exploring Multivariate Paleoclimate Data

The project we describe here is being carried out in the Deasy GeoGraphics Laboratory, a cartographic design studio affiliated with the Department of Geography that serves the Department, the College of Earth and Mineral Sciences and The Pennsylvania State University. One of the leading clients of the laboratory over the past five years has been Penn State's Earth System Science Center (ESSC), a consortium of 23 geoscientists, meteorologists and geographers who study the global water cycle, biogeochemical cycles, the earth's history and human impacts on the earth system. The ESSC is one of 29 interdisciplinary teams participating in NASA's Earth Observing System (EOS) research initiative.

In response to an emerging demand for animated and interactive multimedia presentations, Deasy GeoGraphics has adopted a new emphasis on software development. When asked how the labotatory might apply its cartographic expertise in a development project that would benefit ESSC analysts, ESSC's director replied that he had long wished to be able to simultaneously display multiple spatial "fields" on a single map base. Regions of climate sensitivity, for example, are identified by the spatial correspondence of a positive hydrological balance, a large variation in precipitation minus evaporation and a large variation in temperature. Model dynamics might be effectively assessed by studying trivariate maps of precipitation, evaporation and runoff.

Typically, ESSC researchers rely on the juxtaposed comparison of laser-printed univariate isoline maps to assess the spatial correspondence of model-produced atmospheric and oceanographic distributions. ESSC staff describe how the director performed a synoptic analysis of Cretaceous precipitation, upwelling and winter storms for an award-winning paper (Barron *et al.* 1988) by overlaying paper maps on his office window.

Overlay analysis is a generic GIS problem. More specifically, however, the problem we have addressed is how to effectively display multiple spatial distributions for ESSC researchers' exploratory analyses. As the above discussion should demonstrate, multivariate display is not amenable to a generic solution.

Paleoclimate Data

Our display interface is designed specifically for data produced by numerical models such as the National Center for Atmospheric Research Community Climate Model (CCM). The CCM is a three-dimensional climate model whose resolution is 7.5° longitude by 4.5° latitude, yielding a regular grid of 1920 predictions. At ESSC, the Genesis version of the CCM runs on a Cray Y-MP2 supercomputer. A typical application of the model is to simulate three to five climatic phenomena for nine to 12 atmospheric levels for the Cretaceous (100 million years before present). Once initial model parameters are set, the model "spins up" for a period of about 12 simulation years. After spin up, two daily averages for each of the 30-odd "fields" are recorded for a simulation period of about three years after the model reaches equilibrium. The product of the simulation is a digital "history volume", which consumes about two gigabytes of disk space or tape in binary form. Analysts subsequently use a separate software module called the CCM Processor to produce "history save tapes" in which the model-produced data may be aggregated to monthly, seasonal or yearly averages. One kind of history save tape is a "horizontal slice tape," an IEEE-format binary file containing aggregate univariate data predicted for a spherical shell that parallels the Earth's surface. The data are quantitative, gridded representations of continuous atmospheric phenomena.

The ESSC computing facility has developed a set of procedures collectively known as "SLCProc" for plotting slice tape data. SLCProc utilities are based on the graphics package of the National Center for Atmospheric Research (NCAR Graphics). NCAR Graphics provides a variety of command line-driven two-dimensional vector tools for isoline and unit-vector mapping. It is widely used in the geosciences (over 1000 installations) because the package is inexpensive and is distributed with a Fortran source code that facilitates the development of customized utilities such as SLCProc.

We have christened our map interface "SLCViewer" to denote its role in the ESSC software toolbox as an interactive, mouse-driven utility for viewing multiple horizontal slice tapes. We developed SLCViewer using IMSL/IDL (Visual Numerics, Houston, TX, USA) on a Sun Sparc2 workstation under UNIX (see Slocum *et al.* this volume, for more about IDL and other visualization systems). IDL provides a library of two- and three-dimensional plotting, mapping and image processing procedures, a toolkit for creating custom interface tools ("widgets") and an extensible, Fortran-like programming language for developing specialized visualization software. Our project is currently in the "proof of concept" stage. Procedures developed in IDL may be implemented later in other visualization or GIS applications if warranted.

The SLCViewer Interface(s)

SLCViewer offers analysts a choice of two interfaces: a single map or four juxtaposed maps. After launching IDL, then entering "slcview" to compile a main call program, the SLCViewer base widget appears. The base widget provides two buttons named "One" and "Four" by which users select an interface. Figure 15.8 includes both a complete screen showing the single map interface (with several SLCViewer widgets) and the multiple map interface.

SLCViewer enables analysts to plot horizontal slice tape data as gray scale gridded images, as isolines and as point symbols. Users select a plotting symbol type by clicking one of three buttons provided beside each map interface. After selecting a symbol type, a file selection widget appears, allowing the user to navigate through folders, then select and read a file. The file name subsequently appears in both the map interface (under the corresponding symbol selection button) and in the plot widget that replaces the file selection widget.

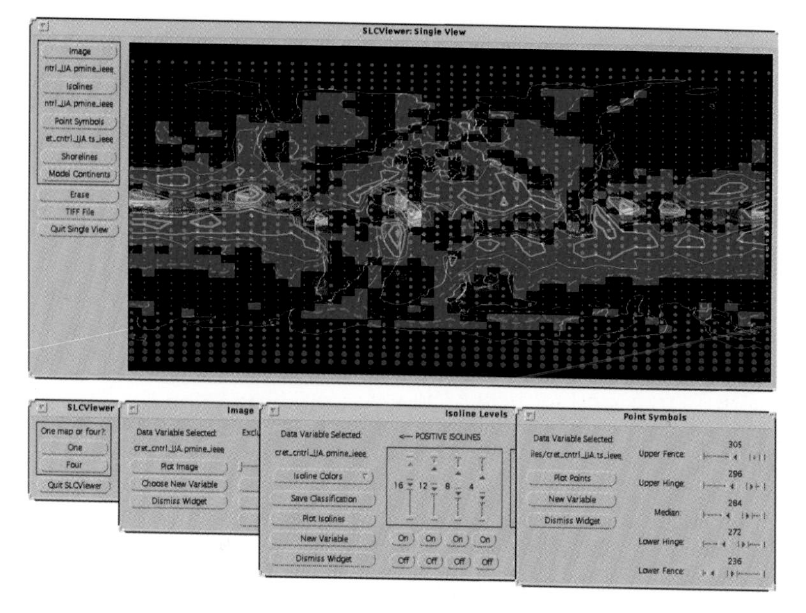

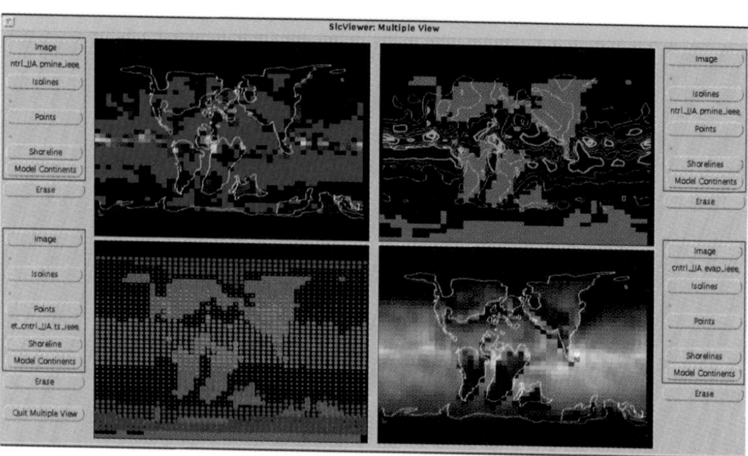

FIG. 15.8. The SLCViewer interfaces. Upper display shows the single view window and several interface "widgets". Superimposed in the single view are three model-produced climatological data variables for a Mid-Cretaceous summer, including areas of positive hydrological balance (gray scale image), precipitation minus evaporation (colored weighted isolines) and surface temperature (point symbols). Mid-Cretaceous shorelines appear in white. The lower display shows four data variables expressed as small multiples.

Images

The image-plotting widget includes buttons to plot the selected data, to recall the file selection widget and read a new file, and to dismiss itself from view. When the "Plot Image" button is clicked, the selected data are mapped onto a 128-step gray scale and plotted on a cylindrical equidistant projection as 1920 7.5° by 4.5° tinted cells, where the lightest cells signify the highest data values.

Focusing on Subsets of the Data

Our work has been influenced by two ESSC scientists who have met with us on several occasions to observe our progress and offer suggestions. Early on they outlined the sequence of exploratory tasks they expected to perform with the interface. The sequence involves the initial inspection of univariate maps of complete distributions, followed by the focused observation of selected subsets of distributions (a process they referred to as "exclusion", called "focusing" in published papers on exploratory data analysis), then superimposition of several complete or subsetted variables. SLCViewer supports the focusing process by providing a slider bar in the image plotting widget by which users can restrict displays to values greater than or less than a specified threshold value. Excluded values in data sets are symbolized with dark blue grid cells.

Isolines

Clicking the "Isolines" button in a map interface first calls up the file selection widget. After a file is read, the file selection widget is replaced by an isoline plotting widget. In addition to buttons for plotting isolines, reading a new file and dismissing the widget, the isolines widget provides nine slider bars by which users can control the value of each isoline. A maximum of four isolines are provided for positive values, one for zero and four more for negative values. SLCViewer calculates equal-interval default slider values separately for positive and negative subsets of the data range.

SLCViewer uses the "weighted" method of isoline symbolization, by which isolines increase in width as the values they represent diverge from zero (DiBiase *et al.* 1994). By default, positive isolines are plotted in orange, negative isolines in cyan and the zero isoline in yellow dashes. Other color schemes may be selected by clicking a pop-up menu called "Isoline Colors". Any individual isoline can be turned on or off by clicking buttons located beneath the sliders.

Points

In addition to the standard plot, new file and dismiss buttons, the point-plotting widget includes five sliders that provide control over data classification. By default, selected data are categorized by the same robust, median-based logic devised for constructing box graphs (McGill *et al.* 1978). The middle two categories of points — divided at the median and comprising 50% of all observations — are symbolized by small, desaturated red and blue dots. The default upper and lower class limits of the next two categories are the upper and lower "fences", beyond which values can be considered outliers. Values outside the middle 50% of values but within the fences are symbolized by larger desaturated red and blue dots. Outliers are symbolized by still larger, fully saturated red and blue dots. We expect that the box plot classification scheme will be particularly useful for identifying anomalies in newly generated slice tapes.

Multiple Views

SLCViewer provides the optional display of a single map on which up to three variables are superimposed or the display of four maps among which up to four variables are juxtaposed. In an interactive exploratory environment the choice between single and multiple displays need not be exclusive — SLCViewer allows analysts to have it both ways. The single map interface facilitates superimposition of three variables (or more, if the analyst is willing to deal with multiple isoline displays). The four-map interface provides small multiples.

Two-dimensional Displays

SLCViewer displays multivariate paleoclimate data in two-dimensional, planimetrically correct map formats. Although the CCM is a three-dimensional model of the atmosphere, the horizontal slice tape data produced by the CCM Processor represent predicted values for discrete spherical surfaces in the atmosphere. As each spherical surface is parallel to the Earth's surface, all surfaces are parallel to one another. When the surfaces chosen for the combined display represent phenomena at the same level of the atmosphere (as they often do), they are coincident. When these conditions are met, superimposed two-dimensional map displays are faithful representations of multivariate paleoclimate data. The main practical benefit of the two-dimensional approach is that substantially less computing power is required for interactive performance, meaning that analysts need not have access to very high performance workstations to use the SLCViewer effectively.

Interactivity

SLCViewer is a dynamic environment in which analysts may select data variables to be displayed and express each variable with a distinctive set of point, line or area symbols. The ability to revise data variable combinations quickly, to focus on subsets of the data and to try different symbolization and classification strategies could increase the efficiency of exploratory analyses of model data at ESSC. By easing the exploratory process we hope that SLCViewer may even foster new insights. The ESSC scientists who have consulted with us in SLCViewer's development expect that the interface will also be useful as a teaching tool in geoscience classes.

Symbol Dimensions and Graphic Variables

As discussed earlier, to effectively superimpose three or more variables on two-dimensional maps, the distributions must be both comparable and distinguishable. In his usual unequivocal style, Bertin (1975) argues that "the only instant distinction which the eye can make is between the three orders: point, line and area". The trivariate map strategy we adopted involves contrasting the dimensionality of symbols (point, line and area) and enhancing each symbol type with the graphic variables that are most potent for expressing quantitative data (size for lines and points, lightness for area symbols). This approach contrasts with the iconographic approach used for Exvis, another two-dimensional method in which all variables are represented as compound line symbols that vary in size, orientation and color. Though Exvis displays excel in revealing overall patterns among multiple variables, the interpretation of patterns in the distribution of individual variables seems difficult. We have not yet begun to evaluate formally the effectiveness of our strategy, however.

Evaluating Effectiveness

Two crucial tests of the effectiveness of an exploratory technique are whether it is used at all and if so, whether its use profits the analyst. If comparable methods exist, these can be compared in controlled studies. Such studies are valuable as visualization is expensive both in terms of time and capital outlay. Benefits ought to be documented. The initial development of SLCViewer continues even as this chapter takes shape — the interface has not yet been used by ESSC scientists for actual research. Our plan is to promote SLCViewer within ESSC and, assuming it is adopted, closely monitor the results.

Conclusion

With long experience in the visual display of quantitative information, cartographers have much to contribute to the practice of scientific visualization. We have only occasionally encountered references to cartographic research, however. In an earlier publication, some of us called on cartographers to engage their experience and creativity in applied visualization projects and to observe carefully what scientists find useful (DiBiase *et al.* 1992). In this project we have aimed to practice what we preach. Our efforts have been guided by the needs and expertise of scientists among whom visualization is a routine but inefficient activity. Our contribution is a prototype exploratory multivariate data analysis tool for geographically referenced data that maintains the planimetric integrity and computational efficiency of two-dimensional maps, offers analysts a choice of single or multiple views and provides user control over data variable, data classification and graphic symbol selections. We cannot yet gauge the value of this contribution.

Acknowledgements

The comments of an anonymous reviewer were helpful in preparing this chapter. Thanks to Bill Peterson, Jim Leous and Jeff Beuchler of the ESSC Computing Facility who provided the workstation, advice on IDL and access to the slice tape data. We also appreciate the interest of ESSC researchers Karen Bice and Peter Fawcett and ESSC Director Eric Barron's unflagging support.

References

Asimov, D. (1985) "The grand tour: a tool for viewing multidimensional data", *SIAM Journal on Scientific and Statistical Computing*, Vol. 6, No. 1, pp. 128–143.

Barfield, W. and R. Robless (1989) "The effects of two- and three-dimensional graphics on the problem-solving performance of experienced and novice decision makers", *Behaviour and Information Technology*, Vol. 8, No. 5, pp. 369–385.

Barron, E. J., L. Cirbus-Sloan, E. Kruijs, W. H. Peterson, R. A. Shinn and J. L. Sloan II (1988) "Implications of Cretaceous climate for patterns of sedimentation", *Annual Convention of the American Association of Petroleum Geologists*, Houston.

Becker, R. A., W. S. Cleveland and A. R. Wilks (1987) "Dynamic graphics for data analysis", *Statistical Science*, Vol. 2, pp. 355–395, reprinted in Cleveland W. S. and M. E. McGill (1988) *Dynamic Graphics for Statistics*, Wadsworth, Belmont, pp. 1–50.

Becker, R. A., W. S. Cleveland and G. Weil (1988) "The use of brushing and rotation for data analysis" in Cleveland, W. S. and M. E. McGill (eds.), *Dynamic Graphics for Statistics*, Wadsworth, Belmont, pp. 247–275.

Bertin, J. (1975) "Visual perception and cartographic transcription" in E. S. Bas (ed.), *Proceedings, Seminar on Regional Planning and Cartography*, Enschede, The Netherlands.

Bertin, J. (1981) *Graphics and Graphic Information Processing*, deGruyter, New York.

Bolorforoush, M. and E. J. Wegman (1988) "On some graphical representations of multivariate data" in Wegman, E. J., D. T. Gantz and J. J. Miller (eds.), *Computing Science and Statisitics, Proceedings of the 20th Symposium on the Interface*, Alexandria, VA, pp. 121–126.

Carr, D. B., W. L. Nicholson, R. J. Littlefield and D. L. Hall (1986) "Interactive color display methods for multivariate data" in Wegman, E. J. and DePriest (eds.), *Statistical Image Processing and Graphics*, Marcel Dekker, New York, pp. 215–249.

Crawfis, R. A. and M. J. Allison (1991) "A scientific visualization synthesizer" in Mielson, G. M. and L. Rosenblum, (eds.), *Proceedings Visualization '91*, Los Alamitos, IEEE Computer Society Press, pp. 262–267.

DiBiase, D., A. M. MacEachren, J. B. Krygier and C. Reeves (1992) "Animation and the role of map design in scientific visualization", *Cartography and Geographic Information Systems*, Vol. 19, No. 4, pp. 201–214.

DiBiase, D., J. L. Sloan II and T. Paradis (1994) "Weighted isolines: an alternative method for depicting statistical surfaces", *The Professional Geographer*, Vol. 46, No. 2, pp. 218–228.

Dorling, D. (1992). "Stretching space and splicing time: from cartographic animation to interactive visualization", *Cartography and Geographic Information Systems*, Vol. 19, No. 4, pp. 215–227.

Earth System Sciences Committee, NASA Advisory Council (1988) *Earth System Science: a Closer View*, The National Aeronautics and Space Administration, Washington.

Eyton, J. R. (1984) "Complementary color, two-variable maps", *Annals of the Association of American Geographers*, Vol. 74, No. 3, pp. 477–490.

Felger, W. and P. Astheimer (1991) "Visualization and comparison of simulation results in computational fluid dynamics" in Farrell, E. J. (ed.), *Extracting Meaning from Complex Data: Processing, Display, Interaction II, Proceedings SPIE,* 1459, pp. 222–231.

Foley, J. D, V. L. Wallace and P. Chan (1984) "The human factors of computer graphics interaction techniques", *IEEE Computer Graphics and Applications,* November, pp. 13–48.

Friedman, J. H. and J. W. Tukey (1974) "A projection pursuit algorithm for exploratory data analysis", *IEEE Transactions on Computers,* Vol. C-23, pp. 881–890.

Gabriel, K. R. (1971) "The biplot graphic display of matrices with application to principal-component analysis", *Biometrika,* Vol. 58, pp. 453–467.

Gabriel, K. R., A. Basu, C. L. Odoroff and T. M. Therneau (1986) "Interactive color graphic display of data by three-dimensional biplots" in Boardman, T. J. (ed.), *Computer Science and Statistics — Proceedings of the 18th Symposium of the Interface,* American Statistical Association, Washington, pp. 175–178.

Grinstein, G. and S. Smith (1990) "The perceptualization of scientific data" in Farrell, E. J. (ed.), *Extracting Meaning from Complex Data: Processing, Display, Interaction, Proceedings SPIE,* 1259, pp. 164–175.

Hägerstrand, T. (1957) "Migration and area" in *Migration in Sweden, Lund Studies in Geography, Series B, Human Geography,* Vol. 13, Royal University of Lund, Department of Geography, Lund, Sweden pp. 27–158.

Haro, A. S. (1968) "Area cartogram of the SMSA population of the United States", *Annals of the Association of American Geographers,* Vol. 58, No. 3, pp. 452–460.

Harris, R. C. and G. J. Matthews (1987) *Historical Atlas of Canada,* Vol. I, *From the Beginning to 1800,* University of Toronto Press, Toronto.

Hibbard, W. L. (1986) "Computer-generated imagery for 4-D meteorological data", *Bulletin of the American Meteorological Society,* Vol. 67, No. 11, pp. 1362–1369.

Hurley, C. (1988) "A demonstration of the data viewer" in E. J. Wegman, D. T. Gantz and J. J. Miller (eds.), *Computing Science and Statisitics, Proceedings of the 20th Symposium on the Interface,* Alexandria, VA, pp. 108–114.

Keller, P. R. and M. Keller (1993) *Visual Cues,* IEEE Computer Society Press, Los Alamitos.

Kraak, M. J. (1988) *Computer-assisted Cartographical Three-dimensional Imaging Techniques,* Delft University Press, Delft.

Lanicci, J. M. and T. T. Warner (1991) "A synoptic climatology of the elevated mixed-layer inversion over the southern plains in spring. Part II: the life cycle of the lid", *Weather and Forecasting,* Vol. 6, No. 2, pp. 198–213.

MacEachren, A. M. and J. H. Ganter (1990) "A pattern identification approach to cartographic visualization", *Cartographica,* Vol. 27, No. 2, pp. 64–81.

MacEachren, A. M., D. Howard, M. von Wyss, D. Askov and T. Taormino (1993) "Visualizing the health of Chesapeake Bay: an uncertain endeavor", *GIS/LIS Proceedings,* pp. 449–458.

Marin, R. (1993) "Evaluating three-dimensional digital terrain maps for visual interpretation of thematic data", Unpublished Masters Thesis, Department of Geography, San Diego State University.

McCleary, G. F. (1983) "An effective graphic 'vocabulary'", *Computer Graphics and Applications,* March/April, pp. 46–53.

McGill, R., J. W. Tukey and W. A. Larsen (1978) "Variations of box plots", *The American Statistician,* Vol. 32, No. 1, pp. 12–16.

McGuiness, C., A. van Werswch, and P. Stringer, (1993) "User differences in a GIS environment: a protocol study", *16th International Cartographic Conference,* Cologne, pp. 478–485.

Miller, J. J. and E. J. Wegman (1991). "Construction of line densities for parallel coordinate plots" in Buja, A. and P. A. Tukey, (eds.), *Computing and Graphics in Statistics,* Springer, New York.

Monmonier, M. (1989) "Geographic brushing: enhancing exploratory analysis of the scatterplot matrix", *Geographical Analysis,* Vol. 21, No. 1, pp. 81–84.

Monmonier, M. (1990) "Strategies for the visualization of geographic time-series data", *Cartographica,* Vol. 21, No. 1, pp. 30–45.

Monmonier, M. (1991a) "On the design and application of biplots in geographic visualization", *Journal of the Pennsylvania Academy of Science,* Vol. 65, No. 1, pp. 40–47.

Monmonier, M. (1991b) "Ethics and map design: confronting the one-map solution", *Cartographic Perspectives,* Vol. 10, pp. 3–7.

Morrison, J. L. (1984) "Applied cartographic communication: map symbolization for atlases", *New Insights in Cartographic Communication, Monograph 31, Cartographica*, Vol. 21, pp. 44–84.

Muerhcke, P. C. (1973) "Visual pattern comparison in map reading", *Proceedings of the Annual Meeting of the Association of American Geographers*, Atlanta, GA, pp. 190–194.

Muehrcke, P. C. and J. O. Muehrcke (1992) *Map Use: Reading, Analysis, Interpretation*, 3rd edn, JP Press, Madison.

Olson, J. M. (1981) "Spectrally-encoded two-variable maps", *Annals of the Association of American Geographers*, Vol. 71, No. 2, pp. 259–276.

Papathomas, T. V., J. A. Schiavone and B. Julesz (1988) "Applications of computer graphics to the visualization of meteorological data", *Computer Graphics*, Vol. 22, No. 4, pp. 327–334.

Raper, J. F. (1989) "The 3-dimensional geoscientific mapping and modelling system: a conceptual design" in Raper, J. F. (ed.), *Three Dimensional Applications in Geographic Information Systems*, Taylor and Francis, London.

Smith, S., G. Grinstein and R. Pickett (1991) "Global geometric, sound and color controls for iconographic displays of scientific data" in Farrell, E. J. (ed.), *Extracting Meaning from Complex Data: Processing, Display, Interaction II, Proceedings SPIE*, 1459, pp. 192–206.

Steinke, T. R. and R. E. Lloyd (1983) "Judging the similarity of choropleth map images", *Cartographica*, Vol. 20, pp. 35–42.

Tang, Q. (1992) "From description to analysis: an electronic atlas for spatial data exploration", *ASPRS/ACSM/RT '92 Technical Papers*, Vol. 3, *GIS and Cartography*, pp. 455–463.

Thompson, M. M. (1982) *Maps for America*, 2nd edn, United States Geological Survey, Boulder.

Tufte, E. R. (1983) *The Visual Display of Quantitative Information*, Graphics Press, Cheshire.

Tufte, E. R. (1990) *Envisioning Information*, Graphics Press, Cheshire.

Wehrend, S. and C. Lewis (1990) "A problem-oriented classification of visualization techniques" in Kaufman, A. (ed.), *Proceedings of the First IEEE Conference on Visualization — Visualization '90*. IEEE Computer Society Press, Los Alamitos, pp. 139–143.

Wilhelmson, R. B., B. F. Jewett, C. Shaw, L. J. Wicker, M. Arrott, C. B. Bushell, M. Bauk, J. Thingvold and J. B. Yost (1990) "A study of a numerically modeled storm", *The International Journal of Supercomputing Applications*, Vol. 4, No. 2, pp. 20–36.

Young, F. W. and P. Rheingans (1990) "Dynamic statistical graphics techniques for exploring the structure of multivariate data" in Farrell, E. J. (ed.), *Extracting Meaning from Complex Data: Processing, Display, Interaction, Proceedings SPIE*, 1259, pp. 164–175.

CHAPTER 16

Visualization of Data Quality

FRANS J. M. VAN DER WEL, ROB M. HOOTSMANS
and FERJAN ORMELING

Faculty of Geographical Sciences —
Cartography Section, Utrecht University
P.O. Box 80.115, 3508 TC Utrecht
The Netherlands

Introduction

Cartographers assume that the maps they produce are used in decision-making processes: map users decide about the values or nature of control point attributes, about slopes or shortest routes or about the suitability of areas for specific purposes, on the basis of the (electronic or paper) maps consulted for these purposes. For proper decisions, the map itself might not be sufficient, but may evoke a need for additional data (e.g. about data quality) on the basis of the mapped data. To avoid lengthy research, this data quality should be visualized.

In the following sections, we focus on five aspects of the visualization of data quality (1) the stage in the information process at which visualization of data quality is needed; (2) the type of map use intended; (3) the nature of the data quality parameters; (4) their level of measurement; and (5) the graphic variables used to represent quality. The digital tools available nowadays for rendering data quality are discussed consecutively, followed by an extended example. However, the chapter starts with developments before the digital revolution.

Pre-digital Data Quality Visualization Techniques

Cartographers have always tried to produce maps with a high and homogeneous degree of accuracy. In those cases where that was impossible, they used graphic means to advise the map user about local or regional deviations from the quality standards, and textual means to indicate overall causes of data inaccuracy, valid for the whole map area. An example of the graphic means is the reliability diagram,

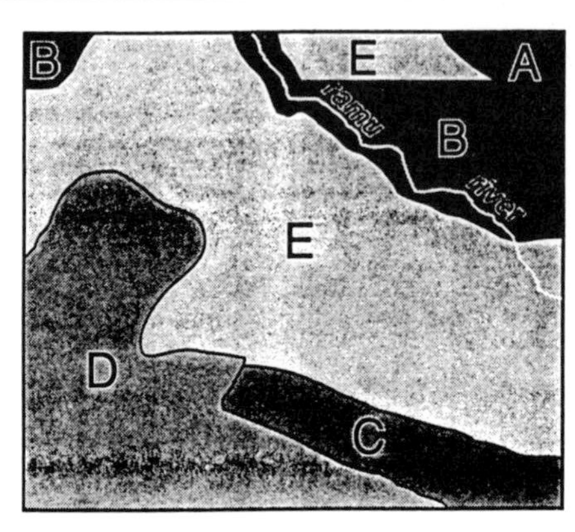

A Accurate: for shape and position some doubt regarding village positions

B Reliable: from oil company traverses and German map

C Fairly reliable: from plans of doubtful accuracy

D Unreliable: from many conflicting plans of doubtful accuracy

E Very sketchy: from sketch plans

FIG. 16.1. Example of a reliability diagram. Reproduced with permission from *Sheet Ramu, New Guinea* 1 : 253,440, LHQ (Australia) Cartographic Company, 1942.

added to map margins (Fig. 16.1). An example of the second are map accuracy standards (e.g. Thompson 1980: 104), or statements documenting the data gathering methods and/or cartographic characteristics and methods used. This could refer, for example, to the nature and the density of the sampling technique used in data gathering. Ives (1980) provides an example of the cartographic characteristics, in which the scale, projection, data resolution ("the smallest map unit represented is approximately 200 m x 200 m") and data accuracy ("map unit boundaries should be accurate within ±100 m; in areas indicated as specific soil units, this unit occupies at least 75% of the area") are documented.

As a special service to the non-expert map user, the effects of these data restrictions can be explained for specific intended applications. In Ives (1980) we find "generalized data suitable for definition of broad bands/corridors approximately 1500 m in width; generalized data should be used only at the scale at which it was produced". It is especially in environmental assessment maps that restrictions for use are indicated to safeguard the area against developers.

The use of specific thematic map series usually takes place in an environment where the nature of the data and the restrictions caused by the cartographic methods are known. The changeover to GIS implies that this link to documentary and procedural knowledge might be severed; the nature of GIS operations with their quick combination of data sets makes it improbable that users would indulge in lengthy study of documentary information. Within a GIS, a map is just one of many possible graphic representations of the same information contained in the database and subjected to the whims of a user who is not necessarily an expert. The quality of mapped data is no longer implicitly considered, as was the case for the traditional cartographer, whose skills guaranteed a concordance between the quality of the data and their visual representation. The use of quality maps is

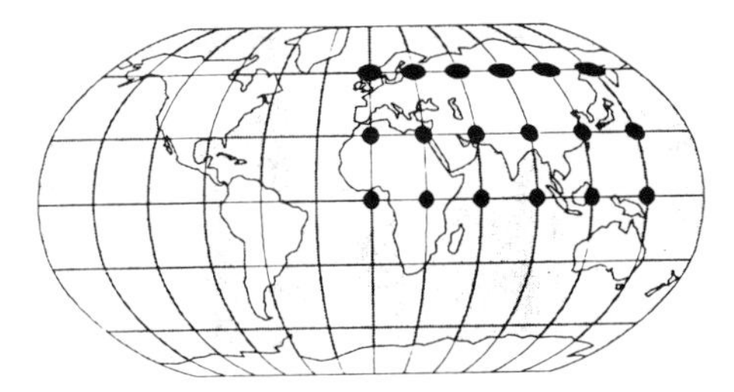

FIG. 16.2. Example of Tissot's indicatrix, used here to show the areal and angular distortion of the Robinson projection. Reproduced with permission from MacEachren *et al.* (1992).

something hardly ever referred to. In the map design literature the graphic results of data display methods are regarded as something final, not as a means to an end. However, for appropriate decision support, data quality information is still needed, both for assessing the overall accuracy and the local/regional deviations from it.

For data on the overall accuracy we should be able to immediately obtain alphanumerical accuracy information from the system (Hootsmans *et al.* 1992). To obtain information on accuracy anomalies, quality maps should be made available (Van der Wel and Hootsmans 1993), providing more detailed information than reliability diagrams [Wood (1992) is by no means impressed by the expressiveness of these diagrams]. To make decisions on the basis of spatial patterns in the data we should be aware of both the general quality restrictions and the local deviations in data quality. One way in which the decision-making process can be improved is by alternating the final decision map and its quality counterpart on the display screen.

Both the overall and the local quality indicators refer to fitness for use (Chrisman 1984) and to assess that fitness, we try to indicate the accuracy of the data, i.e. the closeness of the observations to their true value (DCDSTF 1988). In accordance with current conventions, the following dimensions of accuracy are discerned here:

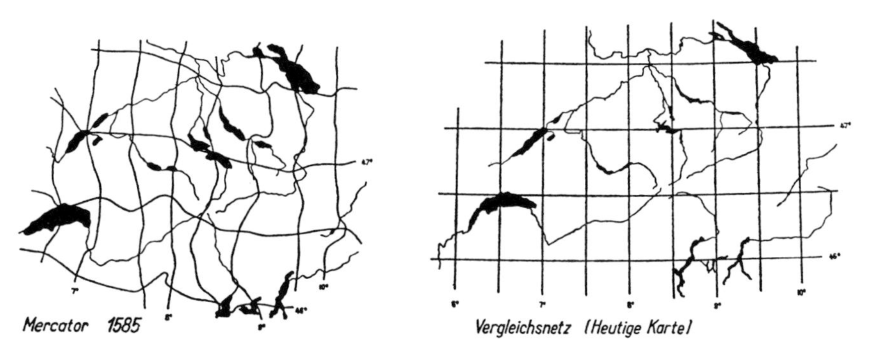

FIG. 16.3. Indication of the accuracy of Mercator's map of Switzerland (1585) by drawing in a distortion grid based on today's maps. After Imhof (1964).

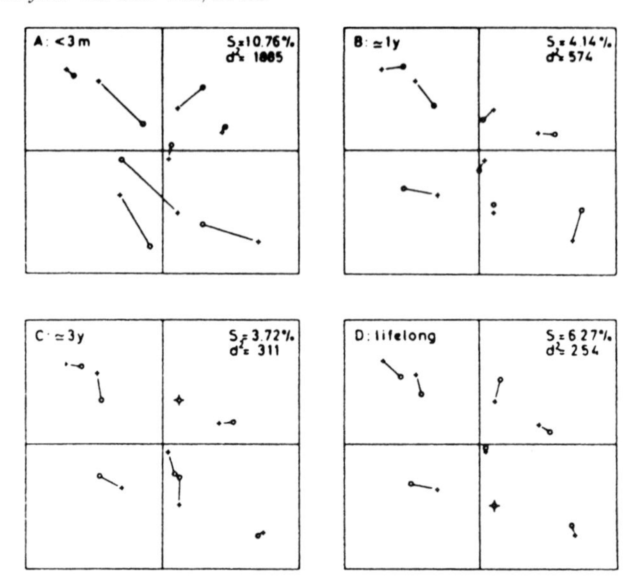

FIG. 16.4. Quality of mental maps: use of vectors to indicate direction and distance of average misrepresentation of urban elements by specific groups of city inhabitants (based on Kruskal's multidimensional scaling). A–D = duration of inhabitants' stay in city; d^2 = squared distance; s = indication of "stress" needed to represent data two-dimensionally.

lineage, positional accuracy, attribute accuracy, logical consistency and completeness (DCDSTF 1988). Logical consistency will only refer to the whole data set; positional and attribute accuracy, lineage and completeness might deviate locally, so it should be possible to present a spatial image of their variation. Overall accuracy information will be used to relativize decision data sets, whereas local variations in accuracy will be used to restrict the decision-making process to specific areas, pending the gathering of more reliable information for the remaining areas.

However, it is not only in the final decisions on spatial matters that data quality information is called for: during the processing of the data and their analysis the immediate availability of this information could inform the GIS user instantly about the usefulness of combining specific data sets, performing specific operations and generally easing their exploration.

In cartography the main emphasis in data quality visualization has been on positional accuracy. There are only a few examples of the visualization of other quality aspects (e.g. MacEachren 1985). In representing positional accuracy, the variation in line character was probably the first means used to indicate uncertainty, a dotted or tagged line referring to a less precisely known course or route. This method was joined later by Perkal's epsilon bands, representing the zone around a digitized line within which the original line will be situated with a given probability

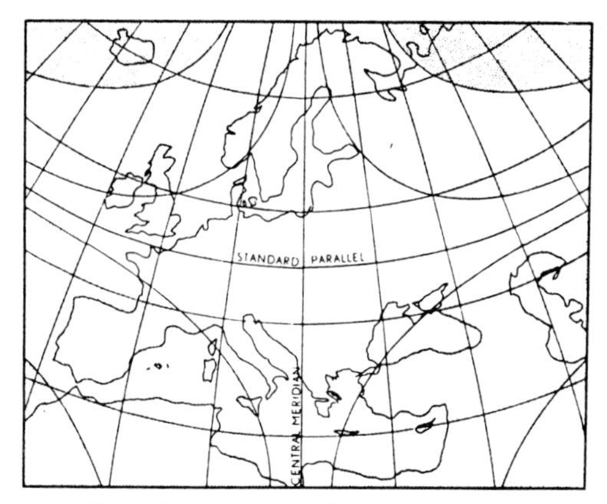

FIG. 16.5. Europe in Bonne's projection with isograms for 1° and 5° of maximum angular distortion. Reproduced from *Basic Cartography for Students and Technicians*, Volume 1, 1st Edition with the permission of the International Cartographic Association.

(Perkal 1966). Accuracy at points has been indicated by circles that vary in size (Maling 1973; Mekenkamp 1989) or by vectors, which could also show the direction of the distortion forces. The accuracy of the representation of area characteristics such as scale or conformity was indicated by Tissot's indicatrix (strictly for point locations; Fig. 16.2), or by grids. Grids are also used for indicating irregular representations due to insufficient local knowledge, for instance in analysing projections of old maps (Fig. 16.3) or mental maps. The distortion of mental maps can also be shown by vectors (Fig. 16.4).

The use of isolines, based on interpolation between point accuracy data, has emerged relatively recently. Though the earliest examples found for indicating distortion in map projections date from 1912, the technique has only gained popularity since the 1940s. On the basis of interpolation between point data, isograms can be drawn, showing either the maximum areal distortion or the maximum angular distortion (Fig. 16.5) at a point. Bregt (1991) uses isolines to visualize confidence intervals to call attention to the uncertainty that is related to the geostatistical interpolation method of kriging.

The visual impression of these techniques — most relevant for decision-making — depends upon a correct application of the graphic variables as discerned by Bertin (1983). Based on Bertin's "rules", only Tissot's indicatrix and the vector maps are able to give a correct impression of the size of the areal distortion and the size and/or nature of the angular/direction distortion. The isograms, when combined with a variation of grey tints, are able to communicate the location of the areas with high (light) and low (dark) data quality, as could be done by the reliability diagrams. The interpretation of the grid maps scores the lowest as here only form differences are shown, by which, according to Bertin, it is hardest to convey a correct overall image of distortion magnitude. However, no experimental attempts have yet been made to test the efficiency of the cartographic methods for displaying the data quality information described above.

Towards a Framework for the Visualization of Information on Data Quality

With the general acceptance of digital techniques in spatial data processing, the relevance of quality and related issues concerning accuracy and error propagation has been established, and this has induced elaborate initiatives (DCDSTF 1988; Goodchild and Gopal 1989; Heuvelink 1993). More recently, visualization has gained considerable attention (in the context of data quality) as several workers have made an appeal to its capacities to inform users about the integrity of data and to facilitate understanding of the potential of that data within a GIS (Buttenfield and Beard 1991; Chrisman 1991; MacEachren *et al.* 1992).

Visualization goes beyond the communication paradigm generally accepted in cartography, and comprises a more exploratory approach to spatial information extraction as well [e.g. MacEachren and Ganter (1990), as well as the introductory chapter in this volume by MacEachren].

The conveyance of data quality information fits into this context, as its aim is to provide users with meta-information during the information process as a whole, thus supporting decisions concerning the selection, processing, interpretation and presentation of spatial data. The digital era has made the provision of quality information inevitable and, at the same time, computers have enabled its dynamic and versatile visualization.

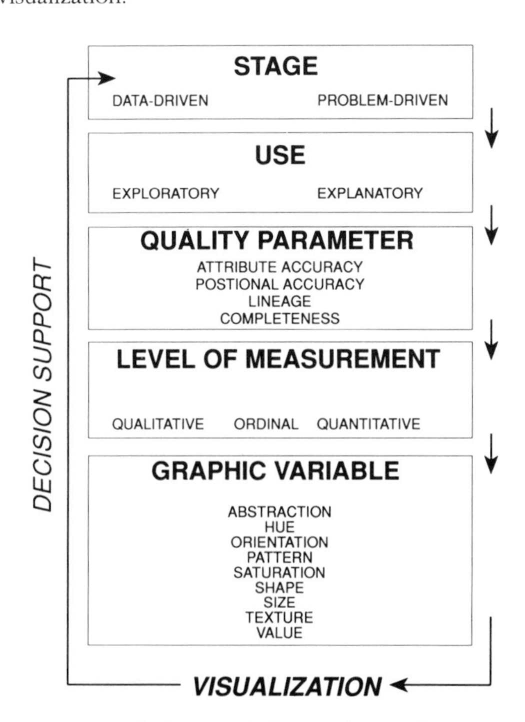

FIG. 16.6. Components of a framework for visualizing information on data quality.

The problem of visualization of data quality concentrates on the following question: which graphic variables can be used at what stage of the information extraction process to represent a particular quality parameter that refers to a certain data set, given the purpose aimed at by a user (Fig. 16.6)? A prerequisite for visualization is of course the existence of methods to derive quality statements (e.g. Chrisman 1991; Goodchild and Gopal 1989), but this is assumed to be beyond the scope of this chapter. Attention will therefore be focused on five components whose mutual relations are considered essential for the sound application of quality visualization techniques.

Stage in the Information Process

The use of visualized quality information can be described according to the life cycle of spatial data. A distinction can be made between a data-driven and problem-driven approach to the information process. The former applies to a process in which the selection of data is subjected to criteria such as availability and accessibility; quality information refers to the status of the data and their inherent uncertainty (source, processing and product error). In a decision-making environment, however, a problem-driven approach is more justified because of the assumption that not data but questions are decisive for the information (answers) to be derived. Quality information is concerned with the steps involved in decision-making, ranging from the definition of a hypothesis to the acceptance of a scenario obtained after several simulations. Obviously, this approach links with the aims of visualization. In the near future, the role of the GIS user will shift towards a more pioneering one, in which quality information provides the criteria on which decisions are based concerning the data and processing methods to be used.

Use

The definition of cartographic visualization implies an important role of map-based visualizations during the complete information process. Here, DiBiase *et al.* (1992) refer to the distinction between graphic representations to support exploratory visual thinking and explanatory visual communication, respectively. The former aims at the generation of new knowledge by examining the different visualizations of a data set, similar to the observation of the facets of a turning diamond. Parallel to the functional taxonomy of map types presented by Armstrong *et al.* (1992), who define the use of different cartographic displays for different tasks during the decision-making process, a framework for the visualization of quality information links the intended use of the graphic representation to a place in the information extraction process.

Quality Parameters

Completeness, lineage and both positional and attribute accuracy can vary within a data set and their spatial variation can be visualized in an appropriate way. A separate description and visualization of positional and attribute accuracy could give rise to discussion if there is a complex spatial relationship between these two components, as is the case for soil data. The uncertain representation of boundaries on a soil map affects the attribute accuracy at these places as well.

Level of Measurement

Quality information can be identified at all four levels: nominal lineage information (source, age), ordinal uncertainties (boundary fuzziness) and accuracy values at interval and ratio scale (quantitative). Lineage comprises descriptive meta-information, nevertheless indirectly enabling the derivation of ordinal accuracy statements. For example, a land cover map can be derived from agricultural research, verbal correspondence and old maps, and these sources imply a certain order of reliability.

The way in which level of measurement, the six graphic variables, implantation[1] and pursued perceptual effect (selection[2], association[3], order, quantity) are interrelated has been described extensively by Bertin (1983). With the definition of "new" variables, some of which are noted below, the relations have to be reconsidered and extended.

Graphic Variables

Bertin's (1983) original distinction of six graphic variables forms a useful starting point for the development of a model for the visualization of data quality information. Colour hue, orientation, shape, size, texture (grain) and value have subsequently been supplemented by other variables or their combinations, such as structure or pattern arrangement (Muehrcke and Muehrcke 1992) and colour saturation. The latter is considered "the most logical one to use for depicting uncertainty", according to MacEachren (1992). Brown and Van Elzakker (1993) also stress the importance of this variable by considering some practical guidelines for its application to represent quality information — in addition to attribute information — in a bivariate map.

To visualize the spatial variation with respect to the completeness of a data set, different levels of abstraction can be applied. Uncertain data can be represented in an abstract, highly generalized way, whereas detailed parts of a visualization correspond with more reliable data. Although the power of abstractions has been discerned by cartographers (Muehrcke 1990), the link with uncertainty level is new. Of course, to a certain extent this application of abstraction can benefit from a consideration of more general efforts, such as the distinction of the mimetic to arbitrary symbol range by Robinson and Bartz Petchenik (1976).

Still more variables can be added, some of them being at the fringe of cartography (e.g. sound, see Chapter 8). Their successful application is bound to the availability of computers and gives cartographic visualization a challenging interpretation.

	nominal	ordinal	interval	ratio	
positional accuracy	x	o	o	o	o: possible
attribute accuracy	x	o	o	o	
lineage	o	o	x	x	x: not possible
completeness	x	o	o	x	

level of measurement	nominal	ordinal	interval	ratio
graphic variable	L	L A P C	A P C	A P (9)
colour hue	+ (2)	- -/+ -/+ - (5)	+ + -/+ (8)	- -
orientation	+	- - - -	- - -	- -
shape	-/+ (3)	- - - -	- - -	- -
size	-	- + + - (6)	+ + -	+ +
texture	+ (4)	-/+ -/+ -/+ -/+	- - -	- -
value	-	+ + - +	+ + -/+	- -
abstraction	-	- - + + (7)	- - -	- -
pattern	+	- - - -	- - -	- -
colour saturation (1)	-	+ + + +	+ + -	- -

- not suitable
-/+ suitable under predefined conditions
+ suitable

L lineage
A attribute accuracy
P positional accuracy
C completeness

(1) Colour saturation can be used in combination with colour hue in bivariate maps.

(2) By using colour hue, selective perception is maintained for a considerable amount of classes, for all types of implantation.

(3) The first example in figure 9 shows a compilation graph, summarizing the data collection methods used in a particular area by means of point symbols, indicating field work (drill symbol), aerial photographs (plane symbol) and images (satellite symbol).

(4) Texture provides order as well.

(5) Colour hues don't impose order except for the case that the colours are ordered according to their values. The reason for proposing this variable at an ordinal level of measurement lies in its ability to support the interpretation of meta-information. The "traffic light principle" at an ordinal level is in fact an interpretation of the underlying meta-information; accuracy values have been translated into safety classes. The reader associates green, orange and red with increasing danger. The effectiveness of this principle depends largely on such issues as cultural background and therefore the term subjective association is justified.

(6) For point and line features.

(7) Different levels of abstraction correspond with different levels of completeness.

(8) At an interval level of measurement it seems difficult to defend the use of different colour hues. Besides forcing order, the "distance" between classes must be assessed. The depiction of temperature (zones) on a map could benefit from the associations red-hot and blue-cold. Combined with red values and blue values such a map could be informative, but it is very difficult to derive such information as "the difference between the dark blue zone and the light red zone equals the difference between the light blue zone and the dark red zone". Only if a dichotomy has been defined, based on association (red-hot/blue-cold), interpretation at interval level is enabled, but only within the two temperature ranges. For meta-information, an extension of the "traffic light principle" seems obvious. Bertin (1981) has a completely different point of view.

(9) Visualization itself is questionable for line and area features at a ratio scale of measurement. For area features, size could be used to create an anamorphosis, but this is considered too complicated. Bertin's ideas on using size and number of repeated symbols within areas as a way to depict ratio-level information result in complex visualizations (see the graduated sizes in a regular pattern, Bertin, 1983). Its effectiveness to depict meta-information is judged as low (-/+). Therefore, with the exception of point symbols, preference is given to visualization at another level of measurement.

FIG. 16.7. An extended framework for the visualization of quality information. Graphic variables are linked with quality parameters at different levels of measurement.

Understanding the way in which the above components are interrelated is a prerequisite for the formulation of guidelines for the sound visualization of data quality. Buttenfield (1991) has made an attempt to construct a framework for visualizing cartographic metadata by proposing a matrix in which the relation between the quality components and data types is established through differing graphic variables. From this, it appears that differences in size, shape, value and colour saturation are often applied in generally accepted design strategies to represent the — spatially — varying quality. The four geographical dimensions of space and time can be used separately or combined with the graphic variables to attain special effects (see next section).

Figure 16.7 gives an extended outline for the visualization of quality information, based on the components that have been distinguished in the above. A starting point is the definition of a continuum ranging from general to specific quality information, corresponding to the level on which the information itself is available and the variety of needs of different users. A further distinction has to be made with respect to the purpose and use of the visualization and its place in the information process. This distinction touches upon the exploratory, confirmatory, scientific functions of cartographic visualization in the highly interactive, private realm of a pioneering user, and the informative, arranging and decision-supporting properties of cartographic communication in the public realm of "end users", as discussed by DiBiase (see MacEachren *et al.* 1992).

The next step is to decide what kind of quality component has to be conveyed and whether or not it is suited to visualization. For instance, during exploratory analysis a user may benefit more from a quality report if, for a limited number of classes, only a global estimate of classification accuracy is available or if the lineage information lacks any spatial reference and instead refers to the data set as a whole. Regarding the continuum mentioned above, positional accuracy can be specified according to the level of spatial aggregation, whereas the indication of attribute accuracy is dependent on the level of thematic aggregation.

Before deciding on the graphic variable to be used for the visualization, the level of measurement of the quality component must be assessed. Qualitative lineage information revealing indirect, intuitive observations on the reliability of a data set ("area between A and B mapped by means of photo-interpretation, area between C and D visited during fieldwork") can be visualized by means of different patterns or colour hues. For lineage information revealing a kind of order ("the northern areas are sampled before 1990, the central part has been visited between 1990 and 1992 and the southern areas are only recently mapped"), the use of value is justified (with lighter values for better information).

The graphic variables that can be used theoretically can be restricted by specifying the visual relation between data and quality (after MacEachren 1992):

- Combined
  Bivariate map (data and quality represented by one visualization)
- Separated
  Static map pair (data and quality represented separately but at the same time)
  Succeeding map pair (first data, then quality)
  Alternating map pair or map sequence (data, quality, data, quality,...)

Although the construction of map pairs is technically simple, they make a quick comparison difficult and cumbersome. If, on the other hand, not the eyes but the map pair is subjected to movement by toggling, the retina will record both data and quality information and induce a combined sensation. The selected time interval is important because a fixed observation of slowly interchanging images can cause annoying effects resulting from phenomena such as chromatic adaptation. Monmonier (1992) shows the usefulness of this technique of alternating maps with respect to the exploration of geographic correlation between two variables.

A bivariate map offers the most compact view, but its complex interpretation key could discourage a majority of inexperienced map readers. Nevertheless, colour hue combined with either value or saturation are promising combinations for the visualization of nominal data and their uncertainties, respectively (see the chapter by Brewer for suggestions on colour schemes for bivariate maps). MacEachren (1992) draws attention to the potential of combining sequential presentations and bivariate maps in a dynamic visualization environment.

The framework that results from the overview of several interrelated dimensions is, of course, not complete. Undoubtedly, different distinctions can be made (between accuracy and precision, between error and accuracy), but this does not affect the validity of the presented relations. In an extended example, the framework will be further elucidated, although some of the relations remain difficult to illustrate as a consequence of the limitations of the static paper map and the immaturity of appropriate techniques. It is, however, expected that with computer technology still evolving, the feasibility of these visualizations will soon become within reach.

Digital Tools, New Techniques

Computers facilitate and sometimes even enable the generation of sophisticated and spectacular effects, which are often erroneously considered graphic variables themselves. These effects, which are aimed at the visualization of data quality, are still evolving. In addition to the increasing capability of computers (from both a hardware and software point of view) cartographers can benefit from experiences gained in other disciplines, such as image processing. The lion's share of the effects given below has been noted by cartographers and it is expected that this development will result in even more convincing visualization tools to represent quality in the near future.

Focus

This phenomenon is valuable for the visualization of quality information because it determines the contrast between objects and thus the sharpness of boundaries (acuteness). The effect is known as fading, blurring or fuzziness, caused by spatially differing (grey) values (Fig. 16.8a). When related to the position of an object, MacEachren (1992) refers to it as contour crispness, whereas fill clarity is suggested for the thematic counterpart. If applied to large parts of a dataset, a more general impression of uncertainty is obtained (MacEachren (1992) calls this "fog").

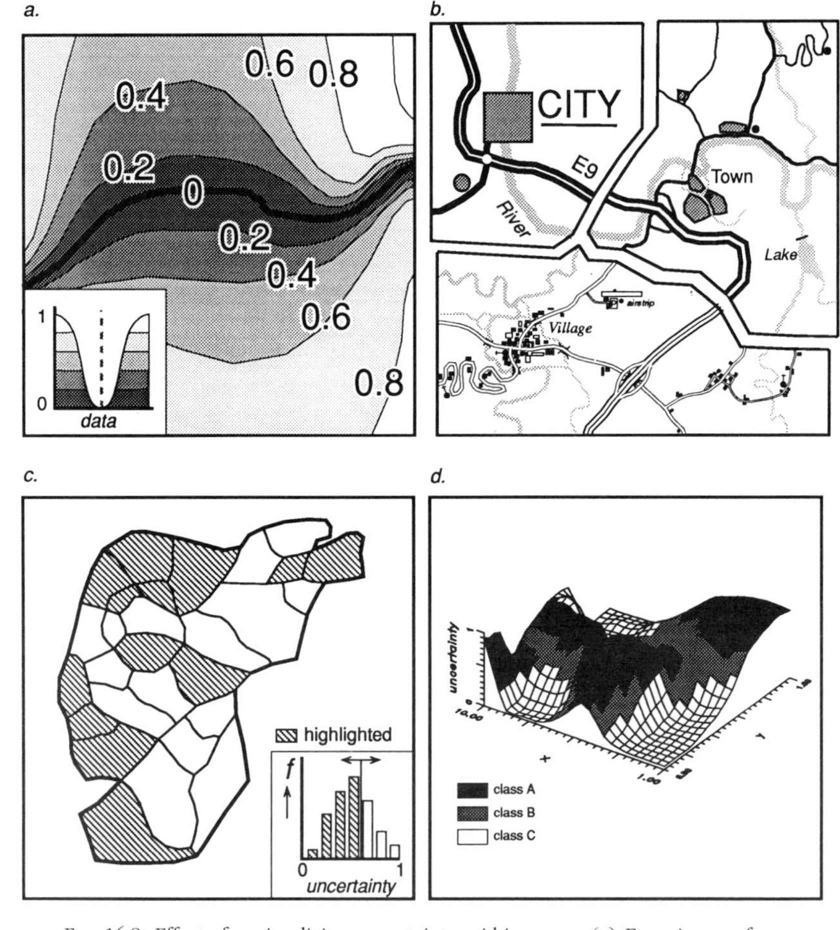

FIG. 16.8. Effects for visualizing uncertainty within maps. (a) Focusing on fuzzy boundary locations; (b) levels of detail available for zooming; (c) slicing of uncertainty levels; and (d) three-dimensional draping of thematic information over uncertainty surface.

Zooming

The amount of certainty assumed to be present in a data set can be related to the extent to which details are shown (Fig. 16.8b). Quality can be merely represented by visualizing objects from different, imaginary viewpoints. If quality falls greatly short of expectations, a bird's-eye view can be used to give a general impression of a particular spatial pattern without providing the resolution necessary for an unambiguous identification. For a more advantageous quality impression, the limited but detailed field of vision of a close-up seems appropriate.

The idea of using different resolutions as proposed by MacEachren (1992) can be considered another way to obtain the zooming effect.

Slicing

The definition of different threshold values causes a gradual shift in the spatial pattern of a data set. Adapting the (thematic) class boundaries to the doubts that attend their definition results in a series of possible classifications (Fig. 16.8c). This variation is visualized by growing and shrinking regions that correspond with the dispersion of particular classes. Mitigating class definitions could result in the gradual domination of certain colours, representing different classes, like an ink blot. Animated display can support the subsequent comparison of all possibilities (e.g. as a movie). An example of slicing is given by Monmonier (1993), as he examines the stability of a covariance pattern in a bivariate map.

Three Dimensions

If numerical (un)certainty values are visualized as heights, the surface of a three-dimensional representation reveals spikes at very (un)certain locations. Thematic information can be draped over the uncertainty landscape by using different colours to avoid interfering patterns (Fig. 16.8d).

An interesting extension of this idea would be to include some of the functions of the terrain analysis system of Kraak (see his chapter in this book). This would enable the interactive control of the exaggeration along the uncertainty axis, depending on the desired dominance.

Shading

On a two-dimensional display (e.g. paper maps), the impression of "uncertainty relief" can be obtained by shading. In addition to the cumbersome creation, the main disadvantage of this effect consists of the occurrence of grey and black shadow spots. They are required to achieve the impression of a third dimension, but do not necessarily relate to uncertainty, and as such confuse a user/map reader.

Blinking

In a more dynamic approach, uncertainty can be related to the display time of thematic information on a screen. Fisher (1992; in press) illustrates this by giving an example of a data set for which the uncertainty of class assignment is inversely related to the duration of their display on a computer screen. Twinkling spots indicate a lack of certainty and the different interchanging colours represent the alternative classes for that particular location. A different approach, using frequencies, is presented in the error animation method of Fisher (1993).

Sequence

Another application of animation can be obtained from successive visualizations and the definition of a time interval. Above, reference has been made to the usefulness of time for the evaluation of sliced images. However, explicit quality information is not always required to inform a user about the reliability of a data set. Muehrcke (1990) stresses the usefulness of a sequence of alternative map designs to assess map stability, a measure of the reliability of the cartographic representation.

A slightly differing approach to map sequence can be obtained by visualizing a development, e.g. the different realizations that represent a series of distortions corresponding to the spatial variation present in one particular data set (Goodchild *et al.* 1992). In addition to superimposing these map layers, they can be interchanged quickly and result in a movie revealing uncertainty patterns.

Colour Transformation

Known from image processing practice, this effect is not related to the geographical space but is undoubtedly worth mentioning. Colours from the RGB (red, green, blue) system are converted into their IHS components (intensity, hue, saturation). Differences in hue are used to represent the nominal information, whereas quality information (e.g. uncertainties) can replace the intensity and/or saturation values. After converting the adapted IHS values to RGB again, the quality information has been embedded within the thematic information. This transformation is only performed to include quality information in the RGB image, and is not meant as a preparatory printing procedure. The resulting images can be considered bivariate maps and are impressive, as Middelkoop (1990) shows. Nevertheless, the inherent complexity will probably discourage the inexperienced GIS user/map reader. This image analysis approach to colour transformation is just one possible method to generate such bivariate thematic maps.

Dazzling

Using different patterns can cause unpleasant effects reminicent of the "dazzle painting" technique applied extensively during World War I. The confusing interplay of different line structures served as a means to protect objects against potential attackers. The improper application of the graphic variables texture (combined with orientation), pattern and colour hue (possibly as an overlay) can be embraced to accomplish an impression of uncertainty. Confusing and interfering images prevent a user from seeing all the details and, instead, balance attractiveness of the representation and the quality of the underlying data.

Non-visual techniques

Some of the proposed additions to the list of graphic variables of Bertin can be better referred to as non-visual sensorial variables, which can be combined with visual variables to obtain effects that excite a certain sensation. The best cartographic example of the application of a sensorial variable is probably the tactile map, in which not vision but feeling is stimulated. Sound offers interesting possibilities to represent data as well, as is shown in the chapter by Krygier elsewhere in this book. If applied to information about data quality, sound has the advantage of preserving the graphic variables for conveying other data. The possibilities to "map" metadata by means of sound variables has been touched upon by Fisher (1994). For example, the application of sound as an alarm, with uncertainty directly proportional to tone, results in a strident noise that notifies a user that his cursor is entering a less reliable part of the data set. Though it is clear that the developments are promising, perceptual research experiments are essential.

Given the developments in the area of virtual reality and hypermedia, it is not difficult to imagine the usefulness of the associative properties of smell, flavour and feeling, or the intensity of their effect, with respect to quality. This is not science fiction, although the need for using penetrating, offensive smells to warn a user of a highly uncertain object is beset with questions; the exploration of visual tools seems far more urgent at this moment.

An Extended Example

During the planning of a proposed route of a railway, many decisions have to be made. The following example (Fig. 16.9) is prompted by an ongoing discussion about the construction of a railway track for a high speed train linking Amsterdam with Brussels. In a densely populated area such as the Netherlands, an affluent society and its nature reserves are likely to be hostile to such plans. Therefore, the impact of decisions is considerable. The importance of data quality as a means to minimize the risk of wrong or misleading information is clear. For reasons of convenience, the example deals with fictitious data sets.

In this case, the very common integration procedure of data layer selection and overlay is used to demonstrate sensitivity to variations in input data. Figure 16.9 gives an overview of possible visualizations of quality information during the processing stages. The fictitious data layers are presented in cartographic products, as these form the visual interface between the database and user.

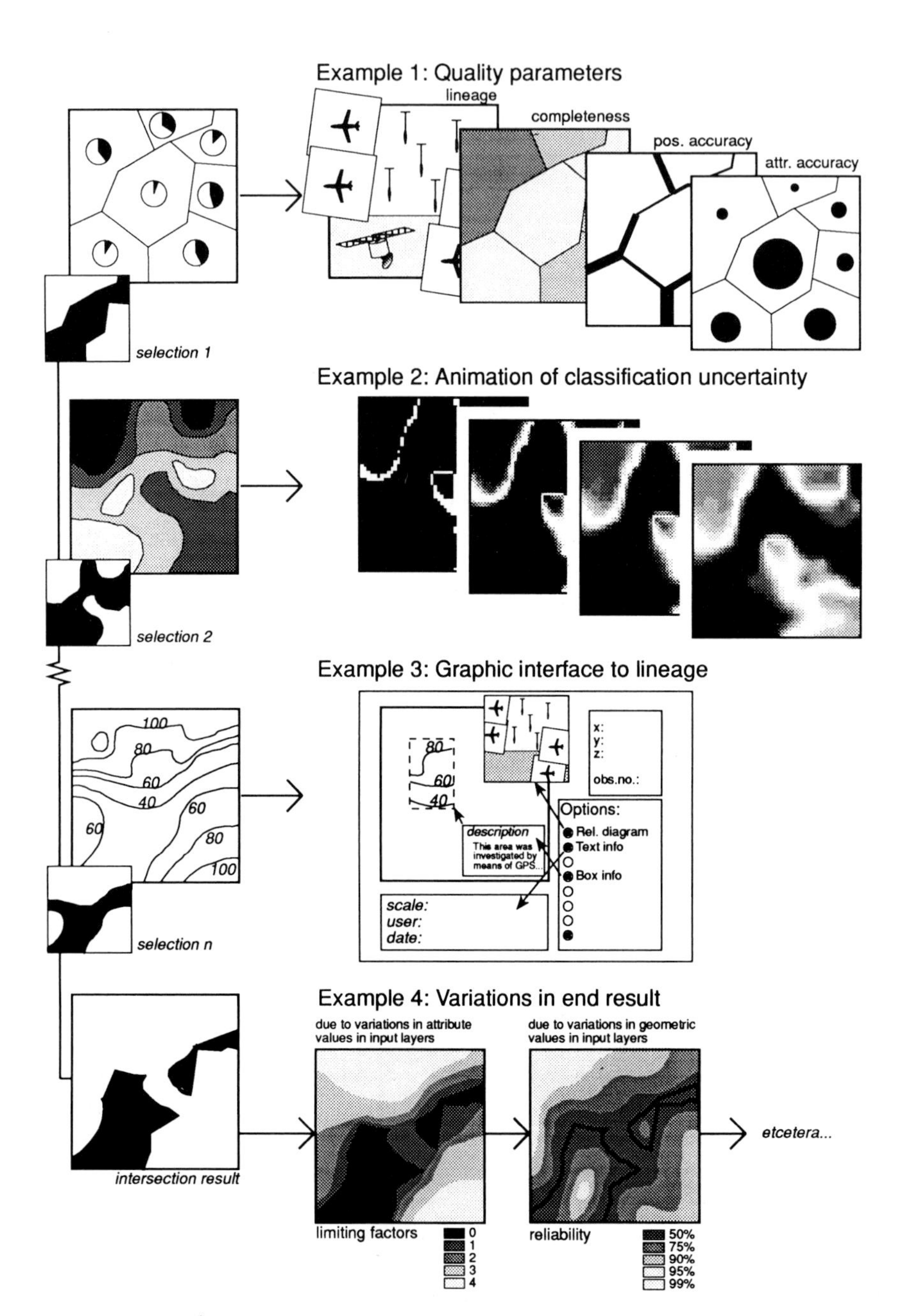

FIG. 16.9. An extended example illustrating the extra value of several visualizations of quality information.

As proposed in Fig. 16.7, each quality parameter at its specific measurement level can be linked to an optimal graphic variable (example 1 in Fig 16.9). For example, nominal lineage information on compilation sources is linked to shape; ordinal information on completeness is linked to value; interval information on positional accuracy is linked to the size of line objects; ratio information on attribute accuracy is linked to the size of point objects.

The uncertainty related to either geographical or class boundaries can be formalized through probability or fuzzy set theory (depending on the continuous or discrete nature of the source data). Reliability thresholds cause different static visualizations of quality information. Continuous animation of several thresholds between a certain minimum and maximum value (example 2 in Fig. 16.9) demonstrate the spatial variability in boundary uncertainty.

Graphic interfaces enable high user friendliness. Clear option menus and cursor handling on computer screens assist non-expert users in the easy exploration of quality information (example 3 in Fig 16.9). For example, lineage information in its many different forms can be visualized on-screen through the compilation of various windows depicting reliability diagrams, descriptive texts symbols, text on acquisition date, scale, etc.

The conventional overlaying of the selected areas of the various data layers will result in a Boolean image with suitable and non-suitable areas. However, taking into account the available quality information (example 4 in Fig. 16.9) will eventually lead to numerous other visualization possibilities. For instance, if all but one criteria are met it can be useful to know which criterion is limiting; or the user may be interested in the visualization of boundary uncertainty in the overlay result. Depending on his or her objectives, the user can make a selection of suitable visualizations revealing the extent to which uncertainty affects the "decision space".

Concluding Remarks

The extra value of quality information has been demonstrated by the above example. The development of a framework for the visualization of quality information goes beyond the boundaries of traditional cartography. Computers have both enabled and necessitated the advent of new techniques to communicate meta-information in a decision-making environment. It is obvious that the same quality information can be visualized in several ways. At this moment, only general statements can be made about the effectiveness of these visualizations because it is still too early to expect sufficient feedback from the user community. The judgement of visualizations in general requires appropriate evaluation methods, as noted by MacEachren and Monmonier (1992).

In addition to cartographic validity, a number of criteria have to be taken into account in an explicit way, such as attractiveness, readability and usefulness, or the acceptance of this kind of visualization is doomed to failure. Cartography can benefit from efforts made in other disciplines such as computer science, mathematics and perceptual psychology. Once suitable techniques have been devised for the visualization of data quality, the cartographer must still convince the vendors of commercial GIS systems of the usefulness of visualization techniques for handling quality information.

References

Armstrong, M. P., P. J. Densham, P. Lolonis and G. Rushton (1992) "Cartographic displays to support locational decision making", *Cartography and Geographic Information Systems*, Vo. 19, No. 3, pp. 154–164.

Bertin, J. (1981) *Graphics and graphic information processing* (translated by W. J. Berg and P. Scott) Walter de Gruyter, Berlin, New York.

Bertin, J. (1983) *Semiology of Graphics* (translated by W. J. Berg) University of Wisconsin Press, Madison.

Bregt, A. K. (1991) "Mapping uncertainty in spatial data", *Proceedings of the Second European Conference on Geographical Information Systems*, Brussels, Vol. 1, pp. 149–154.

Brown, A. and C. P. J. M. van Elzakker (1993) "The use of colour in the cartographic representation of information quality generated by a GIS", *Proceedings, 16th International Cartographic Conference,* Cologne, Vol. 2, pp. 707–720.

Buttenfield, B. P. (1991) "Visualizing cartographic metadata", *NCGIA Research Initiative 7: Visualization of Spatial Data Quality. Scientific Report for the Specialist Meeting (Castine, ME). NCGIA Technical Paper 91–26.*

Buttenfield, B. P. and M. K. Beard (1991) "Visualizing the quality of spatial information", *Auto-Carto 10, Technical Papers 1991 ACSM-ASPRS Annual Convention*, Baltimore, Vol. 6, pp. 423–427.

Chrisman, N. R. (1984) "The role of quality information in the long-term functioning of a Geographical Information System", *Cartographica, Vol.* 21, pp. 79–87.

Chrisman, N. R. (1991) "The error component in spatial data" in Maguire, D. J., M. F. Goodchild and D. W. Rhind, (eds.), *Geographical Information Systems. Principles and Applications.* Vol. 1, Longman Scientific and Technical, Harlow.

DiBiase, D., A. M. MacEachren, J. B. Krygier and C. Reeves (1992) "Animation and the role of map design in scientific visualization", *Cartography and Geographic Information Systems*, Vol. 19, No. 4, pp. 201–214, 265–266.

Digital Cartographic Data Standards Task Forces (DCDSTF) (1988) "The proposed standard for digital cartographic data", *The American Cartographer*, Vol. 15, No. 1 (complete issue).

Fisher, P. F. (1992) "Randomisation and sound for the visualisation of uncertain spatial information", *Working paper for the AGI Visualization Workshop*, Loughborough. .

Fisher, P. F. (1993) "Visualizing uncertainty in soil maps by animation", *Cartographica*, Vol. 30, No. 2/3, pp. 20–27.

Fisher, P. F. (1994) "Hearing the reliability in classified remotely sensed images", *Cartography and GIS*, Vol. 21, No. 1, pp. 31–36.

Fisher, P. F. "Visualization of the reliability in classified remotely sensed images", *Photogrammetric Engineering and Remote Sensing*, in press.

Goodchild, M. F. and S. Gopal (1989) *The Accuracy of Spatial Data Bases*, Taylor and Francis, London.

Goodchild, M. F., Sun Guoqing and Shiren Yang (1992) "Development and test of an error model for categorical data", *International Journal of Geographical Information Systems*, Vol. 6, No. 2, pp. 87–104.

Heuvelink, G. B. M. (1993) "Error propagation in quantitative spatial modelling. Applications in geographical information systems", *PhD Thesis*, KNAG/Faculty of Geographical Sciences of Utrecht University, Utrecht.

Hootsmans, R. M., W. M. de Jong and F. J. M. van der Wel (1992) "Knowledge-supported generation of meta-information on handling crisp and fuzzy datasets", *Proceedings, 5th International Symposium on Spatial Data Handling,* Charleston, Vol. 2, pp. 470–479.

Imhof, E. (1964) "Beiträge zur Geschichte der topographischen Kartographie", *International Yearbook of Cartography,* Vol. 4, pp. 129–154.

Ives, L. (1980) "The utilization of maps for environmental assessment", *10th International Cartographic Conference*, Tokyo, 4 pp (mimeograph).
MacEachren, A. M. (1985) "Accuracy of thematic maps/implications of choropleth symbolization", *Cartographica*, Vol. 22, No. 1, pp. 38–58.
MacEachren, A. M. (1992) "Visualizing uncertain information", *Cartographic Perspectives*, No. 13, pp. 10–19.
MacEachren, A. M. and J. H. Ganter (1990) "A pattern identification approach to cartographic visualization", *Cartographica*, Vol. 27, No. 2, pp. 64–81.
MacEachren, A. M. and M. Monmonier (1992) "Introduction", *Cartography and Geographic Information Systems*, Vol. 19, No. 4, pp. 197–200.
MacEachren, A. M. in collaboration with B. Buttenfield, J. Campbell, D. DiBiase and M. Monmonier (1992) "Visualization", in Abler, R. F., M. G. Marcus and J. M. Olson (eds.), *Geography's Inner Worlds: Pervasive Themes in Contemporary American Geography*, Rutgers University Press, New Brunswick, pp. 99–137.
Maling, D. (1973) *Coordinate Systems and Map Projections*, George Philip and Son, London, pp. 62–67.
Mekenkamp, P. G. M. (1989) "Geometric cartography: the accuracy of old maps", *Abstracts of the XIIIth International Conference on the History of Cartography*, Amsterdam, pp 87–89.
Middelkoop, H. (1990) "Uncertainty in a GIS: a test for quantifying interpretation output", *ITC Journal*, Vol. 3, pp. 225–232.
Monmonier, M. (1992) "Authoring graphic scripts: experiences and principles", *Cartography and Geographic Information Systems*, Vol. 19, No. 4, pp. 247–260, 272.
Monmonier, M. (1993) "Navigation and narration strategies in dynamic bivariate mapping", *Proceedings, 16th International Cartographic Conference*, Cologne, Vol. 1, pp. 645–654.
Muehrcke, P. C. (1990) "Cartography and geographic information systems", *Cartography and Geographic Information Systems*, Vol. 17, No. 1, pp. 7–15.
Muehrcke, P. C. and J. O. Muehrcke (1992) *Map use. Reading, Analysis, Interpretation*, 3rd edn, J.P. Publications, Madison.
Ormeling, F. J. (1984) *Basic Cartography*, Vol. 1, ICA/Elsevier, London.
Perkal, J. (1966) "An attempt at objective generalization" in Nystuen, J. (ed.) (translated by Jackowski, W.), *Michigan Inter-University Community of Mathematical Geographers. Discussion Paper 10*, University of Michigan Department of Geography, Ann Arbor.
Robinson, A. H. and B. Bartz Petchenik (1976) *The Nature of Maps. Essays Toward Understanding Maps and Mapping*, The University of Chicago Press, Chicago.
Thompson, M. M. (1980) *Maps for America. Cartographic products of the U.S. Geological Survey and Others. A Centennial volume 1879–1979*, U.S. Geological Survey, Washington, DC.
Van der Wel, F. J. M. and R. M. Hootsmans (1993) "Visualization of quality information as an indispensable part of optimal information extraction from a GIS", *Proceedings, 16th International Cartographic Conference*, Cologne, Vol. 2, pp. 881–897.
Wood, D. (1992) *The Power of Maps* (with J. Fels), Routledge, London.

Endnotes

[1] Implantation refers to the classes of representation, i.e. point, line and area.
[2] Selection is the process of immediate recognition and isolation of all map elements of a certain class, thus perceiving the image formed by that class.
[3] Association is the process of matching and grouping corresponding map elements as distinguished by a variable.

The Future of Cartographic and Geographic Visualization

CHAPTER 17

Perspectives on Visualization and Modern Cartography

D. R. FRASER TAYLOR

Department of Geography and International Affairs

Carleton University, Ottawa, Canada, K1S 5B6

Introduction

In the first volume of the *Modern Cartography* series the argument was made that visualization was central to modern cartography (Taylor 1991). It therefore seemed appropriate to make visualization the theme of the second volume to explore more fully the issues involved. Cartography, as Wood argues in Chapter 2, helps to make visible facts and concepts that might otherwise remain hidden and visual analysis and communication are historically important elements of the discipline. Modern cartographic visualization is, however, different in both quantitative and qualitative ways as a result of recent technological changes involving computer cartography and computer graphics. In quantitative terms it is now possible to produce a wide range of different cartographic products much faster and much more cheaply. Accompanying this is an important qualitative change allowing interaction with visual displays in almost real time, which will greatly increase comprehension in a wide variety of subject areas. This is at the heart of cartographic visualization. What is emerging as a result is a dynamic cartography in which the interactive manipulation of spatial information will foster a fundamental change in the discipline. The emphasis will be on dynamic rather than static map and spatial data use. The contributions to this volume clearly demonstrate the potential and progress that have been made in this respect. The purpose of this concluding chapter is to give a perspective on the future of visualization in modern cartography, including its place in cartographic theory and practice.

MacEachren gave his interesting view of "cartography cubed" in the opening chapter and, as he indicated, his views on visualization in cartography have evolved and continue to evolve. In Volume 1 a triangular diagram was used to help

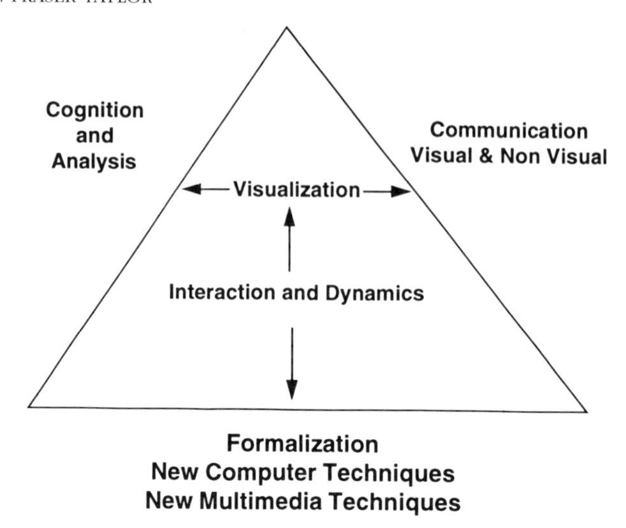

FIG. 17.1. Conceptual basis for cartography.

conceptualize thinking on cartography and this appears as Fig. 1.2 in the opening chapter. A revised version of this diagram (Fig. 17.1) is used to illustrate more recent thinking.

In the diagram visualization still appears as central to cartography although, as MacEachren rightly points out, it should not be seen as simply the equivalent of cartography itself. Visualization will never be all of cartography, but it will affect all three major aspects of cartography as shown on the diagram and will do so in an increasingly important way. These elements are: formalization or cartographic production techniques, communication and cognition and analysis.

Visualization and Cartographic Formalism

Formalization in the diagram refers to the approach in cartography which sees the main purpose of the discipline and profession in terms of cartographic production. Here, computer technology continues to dominate and a major technological development which the interest in visualization has brought to cartography has been the utilization of new multimedia techniques. Multimedia is exploding as a production tool, as Cartwright illustrates in Chapter 5. Like so many technological developments, the main developments in multimedia are not being driven by the needs of cartography, but by the emerging market for entertainment, education and training. There is also a growing interest in the corporate world in using multimedia as a marketing tool. As several chapters in this volume illustrate, multimedia has considerable potential for cartography. Developments in multimedia technology are occurring at a very rapid rate. In June 1993, for example, an interactive video CD player was introduced by Matushita and the resulting multimedia production possibilities demonstrated by 3DO Incorporated of San Mateo are very impressive indeed. In May of the same year IBM and Blockbuster Entertainment formed a new joint company called Fairway Technology Associates, whose objective is to let consumers manufacture their own disks and videos electronically. The first products will be audio compact disks and video games, but there is no reason why the same technology could not be used to create multimedia cartographic products if the demand and market are large enough. A central computerized library is linked to regional servers by high speed data links. The consumer is connected by a telephone link to a regional server at retail outlets and selects the product by inserting a blank CD in a player, paying by using a credit card which is automatically scanned.

This is simply one example of a technological convergence of computer, broadcast and telecommunications technologies and media, which many have argued will realize the ultimate promise of the information revolution which is the interactive delivery of information directly into the home.

A key element in the process is the creation of what have been called "information superhighways", which were an element in the 1992 US Presidential election campaign of Clinton and Gore. At its December 1993 meeting in Toronto, the America Dialectic Society chose "information superhighway" as its new expression for the year 1993 and in January 1994 a high level Superhighway Summit was held at the University of California, at which Vice-President Gore announced further details of the Clinton Government's policy following an investment of over US$ 2 billion announced in 1993. President Clinton again emphasized the importance of the information superhighways in his State of the Union Address on 25 January 1994. Similar highways are under construction in other countries such as Canada, where the Canadian Network for Advancement of Research, Industry and Education (CANARIE) was established in 1993 (OCRI 1993). In January 1994, the creation of an Information Highway was announced in the Throne Speech by the new Liberal Government, and the Government of the Province of New Brunswick announced the appointment to the Cabinet of a Minister for the Information Highway. Japan has invested US$250 billion to create a fibre optic network which will link every school, home and office in the country by 2015 and several European nations are also building similar networks. Gore has announced a coalition of 28 US companies, including AT&T, IBM and Microsoft, to develop policies for the new National Information Infrastructure. Millions of dollars are being invested by the private sector in preparation for the superhighway phase of the information era.

In the USA the national fibre optic network, which is the backbone of the information superhighway, is already in place and new techniques of data compression and digital switching allow delivery directly to the home TV set either on standard telephone wires or by coaxial cable. The home TV set itself is being greatly improved as a display device by the introduction of high definition television and liquid crystal display tubes.

The fibre optic network is currently used by the powerful computer information network Internet. In January 1994 it was estimated that over 20 million users were connected to about 20,000 different information network sources through Internet and the network is now being extended from the scientific and defence community, where it originated, to libraries, schools, offices and homes. This is a fundamental shift in the way individuals receive and generate information. Utilization of Internet is growing exponentially. Almost every day news is released about new technical and commercial developments in the field as well as joint ventures among computer, telephone, cable and telecommunication companies. In January 1994, for example, Bell Atlantic Corporation of Philadelphia and Tele-Communications Inc. of Denver, which agreed to merge in October 1993, announced that they will assist over 26,000 elementary and secondary schools in the areas served by the two companies to access computer and video materials free of charge. This affects about 25% of the schools in the USA.

Cartographic information is already being transmitted over the new data highways, although this is at present only a very small fraction of the information flow. The Automated Cartographic Information Center of the University of Minnesota is involved in a pilot project (NCGIA 1992) to provide library users with direct access to digital and cartographic information, and in Canada map information is part of the information available on CANARIE (OCRI 1993). Goodchild (1993) draws attention to the innovative use of Internet for geographic and cartographic research purposes by geographers in the USA.

Entertainment and games are a key driving force in a commercial sense of the new technological development and what has been called "edutainment" is a growing trend. There are already cartographic and geographic examples. The computer game Magellan, for example, which was available for around US$450 for Christmas 1993, is an interactive touchtalk globe. It is very similar to a traditional globe, but it can be used either as a computer game or as an educational tool. The globe is covered in "virtual keypads" which respond to touch and give taped audio information on each country.

Increasingly, cartographic products will appear on CD-ROM and the trend begun by the pioneering British Domesday Project (Openshaw and Mounsey 1987) will accelerate. The National Geographic Society is already marketing CD-ROMS and in a number of countries new versions of electronic atlases are emerging. In Canada the Jean Talon project (CMMM 1992) is creating modular computer-operated multimedia resources on Canada's geography and history on CD-ROM which will complement the *Electronic Atlas of Canada.*

Consideration of the impact of the technological developments facilitating cartographic visualization has been by no means exhaustive or complete, but it is clear that the new technologies are going to influence cartographic production in an increasing way and lead to the creation of new products. The new technologies, the information superhighways, interactive video and television, video on demand and virtual reality promise a revolution in communications in which cartography must play a part.

A key element in the new technologies is that they facilitate dynamism and interaction in ways which have previously not been possible. The importance of this dynamism and interaction to cartography cannot be over-emphasized. As MacEachren argues in Chapter 1, interaction is a key element in geographic and cartographic visualization and this capability has been made increasingly possible by recent technological developments.

Visualization and Cartographic Communication

Communication has already been important to cartography and the new technologies and processes connected with visualization and multimedia are leading to significant quantitative and qualitative changes. As Fig. 17.1 indicates, although visual communication is still central, additional non-visual elements, especially sound, are increasingly important. Krygier (Chapter 8) outlines the importance of sound, not only to the significant proportion of people who suffer from visual impairment, but also as an important additional communication element for the sighted. The new technologies also increase the possibilities of the use of colour in communication, which are discussed by Brewer in Chapter 7.

Cartographic visualization and multimedia techniques will affect and may revitalize the concept of cartography as a communication process. The communication model of cartography was dominant in the discipline in the 1970s but fell into disrepute. However, in the information era the need to understand and improve visual communication is increasing. At the same time the need to convert increasing volumes of data into usable information has never been greater. The map and related cartographic products are ideal media for the effective presentation and communication of information in a wide variety of subject areas.

Mention has already been made of new methods of cartographic production both in this chapter and in earlier chapters of this volume (Chapters 5–10). Indeed, the increasing flood of data demands new presentation methods and techniques.

Equally important, however, is the need to better understand the cartographic communication process. Here, the work carried out earlier in cartography on this topic may not be of great help. The new electronic products are different from the paper map and the human brain's perception of electronic images is not the same as that of traditional products.

Recent scientific research on the human visual system is leading to a new understanding of what is a much more complex system than was previously thought, as Peterson shows in Chapter 3. Researchers in the neuromechanics field have been developing detailed brain maps for visual information which reveal that different parts of the brain seem to respond to different kinds of visual input. Variables such as shape, motion, size and colour, for example, appear to be dealt with in different parts of the brain and much of the processing appears to take place outside the primary visual cortex, which was the focus of much earlier research. The significance and nature of this processing, which appears to take place in the anterior inferotemporal cortex, was not fully realized until relatively recently (Sejnowski and Churchland 1989).

Cartographic research on user reaction to the new products is relatively sparse, and as McGuinness indicates in Chapter 10, and there are many unanswered questions. The importance of the topic may lead to a revitalization of research and applications in the field of cartographic communication. McGuinness also points out, however, that attention must be paid to research which deals with concept rather than stimulus. Cartographers' interest in their own products often leads to an emphasis on the stimulus to the exclusion of other user and process variables. Psychological research (Booth 1989) strongly suggests that effective interaction between the user and the product is the key to learning and communication. If this is indeed the case then the degree of interactivity possible between the user and the emerging range of new cartographic products will be critical. In MacEachren's concept of "cartography cubed", one of the three continua is the degree of interaction. It would seem that high interaction is necessary not only for the "private and known" end of his other two continua, but also for the "public and known" end. Visualization as a form of map use requires a high degree of interaction between product and user.

Visualization and Cartographic Cognition and Analysis

The third arm of the triangle is cognition and analysis. Cartographic cognition is a unique process as it involves the use of the brain in recognizing patterns and relationships in their spatial context. Whereas this cannot be easily replicated by GIS software with its essentially linear analytical processes, it can be considerably enhanced by cartographic visualization. MacEachren, in earlier work, defined visualization as "...a human ability to develop mental images (often of relationships that have no visual form), together with the use of tools that can facilitate and augment this ability. Successful visualization tools allow our visual and cognitive processes to almost automatically focus on the patterns depicted rather on generating these patterns" (MacEachren 1991:13).

The analytical and cognitive aspects of cartographic visualization are again emphasized in the opening chapter of this volume in the concept of "cartography cubed", where visualization is defined in terms of map use distinguished by the combination of variables along three continua: private to public, unknown to known and high interaction to low interaction. MacEachren argues that it is the combination of these three variables that distinguishes visualization from other areas of cartography and this concept of cartographic visualization has been the organizing concept for this volume.

In "cartography cubed" cartographic visualization as a research and analytical tool is in one corner and communication in the other. His own emphasis is on the increased use of maps in research exploration and analysis, which he argues should draw cartographers closer to their geographic colleagues, a view which is shared by this author (Taylor 1993). It is in the cognitive and analytical field, especially as a result of the utilization of visualization, where the possibilities of a new synthesis between modern geography and modern cartography are greatest. MacEachren *et al.* (1992), following DiBiase (1990), present four stages of visualization in the geographic research sequence: exploration, confirmation, synthesis and presentation. Although these are seen as a continuum, the first two are seen as part of visual thinking and the latter two as visual communication.

The distinction between analysis and communication may not be as clear-cut as it seems. If "Visualization is foremost an act of cognition, a human ability to develop mental representations ... to identify patterns and to create or improve order" (MacEachren *et al.* 1992:101), then a highly interactive system, even at the "public" and "known" end of the other two continua in "cartography cubed" can allow a user to analyse and interpret in ways very different from the perspectives of those "communicating" the information. The nature of the technological developments described earlier will also mean that users will increasingly interact with systems on a one to one basis utilizing an individual terminal, rather than as part of a more general public audience using shared facilities.

Cartographic visualization will make an important contribution to both the communication and cognitive and analytical functions of cartography, but its importance on the analytical side is probably more significant. Rhind has argued that maps as display mechanisms are "... the main even sole future of cartography" and sees the cartographer primarily being relegated to a position he describes as "design consultant" (Rhind 1993:12). The continuing development of cartographic visualization will demonstrate the inadequacy of this very limited view of the scope of modern cartography and the abilities of cartographers. MacEachren's argument that cartographic visualization will rejuvenate cartography in the same way that GIS has rejuvenated geography is sound and visualization will be a major contributor to cartographic theory and practice.

Cartographic visualization in its modern context is an exciting extension of methods for imaginative data presentation and analysis which have been present in cartography since at least the 18th century. Visualization uses computer-based techniques and can therefore lay claim to the utilization of "scientific" methodologies. At the same time it involves the use of imagination, intuition and artistry, especially in the creation of new multimedia products and in the exploratory use of virtual reality. In conceptual terms it can therefore provide a link between different epistemological paradigms. As Abler *et al.* (1992) point out, the gap between positivism, humanism and structural/realism is growing and this is a major challenge for geographers which, as the latest debate on automated geography in the *Professional Geographer* (1993) illustrates, shows little signs of being resolved.

Social Context and Cartographic Visualization

Visualization, like all aspects of cartography, should not be divorced from its social context as the late Brian Harley has so eloquently argued (Harley 1990). In an article published after his death, co-authored with Zandvliet, the argument is made that "... content is always more important than techniques in the social history of cartography" (Harley and Zandvliet 1992:17). There are already signs that the cartographic love affair with computer techniques is being extended to the new visualization and multimedia techniques with content and social context receiving little attention.

As the new technologies explode into the market place, the influence of the vendors and the large corporate players on what is produced and what is not produced in terms of hardware and, perhaps more importantly, software, increases. The decisions of vendors also affect and limit application areas. The nature of the technological convergence described earlier is leading to new corporate mergers and alliances, which are concentrating information power in fewer and fewer hands and almost exclusively in the hands of the post-industrial nations of the developed world to the increasing exclusion of developing nations. It is also clear that the multi-million dollar investments are being made in expectation of substantial future profits. The costs of access to information may be high.

Potential control of the new information superhighways is also of concern. In January 1994, reports were appearing in the press that the US Government was advocating the use of the "Clipper-Clapstone" system as the mandatory industry standard. This would allow the Government to eavesdrop on the information highways. Reports were also circulating that the US military, which was the originator of Internet, was contemplating limiting military links to the network, which is the backbone of the new information superhighways.

Harley's writings have stimulated an interest in social and cultural cartography and the social and cultural implications of the new developments described in this book will be of concern to cartographers interested in this topic. Cartographers should not ignore the context in which cartographic information is used. There has always been a strong linkage between state power and the corporate world, which is now a major player in determining the context in which modern cartography is developing. The new developments are not of neutral value and will have significant socioeconomic consequences. Indeed, the information superhighways are seen as major generators of economic growth and renewal in the North American economy.

Conclusion

The last decade of the 20th century will witness a renaissance of cartography and cartographic visualization will be a leading factor in this process. The concept of "mapping" is already being extended into areas well beyond those with which cartography has traditionally been concerned. Hall, in his interesting book *Mapping the Next Milenium: The Discovery of New Geographies* (Hall 1992), argues that the act of mapping, when it expands beyond its usual terrestrial and scale constraints, plays a critical role in the process of scientific research and that mapping and visualization, which are by nature non-linear, help to explain what is essentially a non-linear world in new ways. He illustrates this by considering a variety of topics at a variety of scales, from the subatomic to the universal. The computer is the direct source of proliferation of data and the new ways of visualizing it. The driving force, however, is not these new computer technologies but curiosity, exploration and discovery. Cartographic visualization can, in this sense, take cartography forward to an exciting future.

References

Abler, R. F., M. G. Marcus and J. M. Olson (eds.) (1992) *Geography's Inner Worlds: Pervasive Themes in Contemporary American Geography*, Rutgers University Press, New Brunswick.

Booth, P. A. (1989) *Introduction to Human Computer Interaction*, Lawrence Erbaum Associates, Hilldale.

CMMM (1992) "The Jean–Talon Project", *The Canadian Multi–Media Magazine*, Vol. 1, No. 5, pp. 6–8.

DiBiase, D. W. (1990) "Scientific visualization in the earth sciences", *Earth and Mineral Sciences*, Vol. 59, No. 2, pp. 13–18.

Goodchild, M. F. (1993) "Ten years ahead: Dobson's automated geography in 1993", *The Professional Geographer*, Vol. 45, No. 4, pp. 444–446.

Hall, S. S. (1992) *Mapping the Next Millenium: the Discovery of New Geographies*, Random House, New York.

Harley, J. B. (1990) "Cartography, ethics and social theory", *Cartographica*, Vol. 27, No. 2, pp. 1–23

Harley, J. B. and K. Zandvliet (1992) "Art, science and power in sixteenth century Dutch cartography", *Cartographica*, Vol. 29, No. 2, pp. 10–19.

Marcus, M., J. M. Olson and R. F. Abler (1992) "Cartography as science and art" in Abler, R. F., M. G. Marcus and J. M. Olson (eds.), *Geography's Inner Worlds: Pervasive Themes in Contemporary American Geography*, Rutgers University Press, New Brunswick, pp. 99–137.

MacEachren, A. M. (1991) "Visualizing uncertain information", *Cartographic Perspectives*, No. 13, pp. 10–19.

MacEachren, A. M., in collaberation with B. P. Buttenfield, J. B. Campbell, D. W. DiBiase and M. Monmonier (1992) "Visualization" in Abler, R. F., M. G. Marcus and J. M. Olson (eds.), *Geography's Inner Worlds: Pervasive Themes in Contemporary American Geography*, Rutgers University Press, New Brunswick, pp. 99–137.

National Center for Geographic Information and Analysis (NCGIA) (1992) *NCGIA Update*, March, pp. 1–16.

Openshaw, S. and H. M. Mounsey (1991) "Geographic information systems and the BBCs Domesday interactive videodisk", *International Journal of Geographic Information Systems*, Vol. 1, No. 2, pp. 173–179.

Ottawa Carleton Research Institute (OCRI) (1993) "CANARIE and OCRINET", *OCRINET*, Vol. 9, No. 3, pp. 1–7.

The Professional Geographer (1993) "Automated geography in 1993", Vol. 45, No. 5, pp. 431–460.

Rhind, D. (1993) "Mapping for the new millenium", *Proceedings, 16th International Cartographic Conference*, Cologne, Vol. 1, pp. 3–14.

Sejnowski, T. J. and P. S. Churchland (1989) "Brain and cognition" in Posner, M. I. (ed.), *Foundations of Cognitive Science*, MIT Press, Cambridge.

Taylor, D. R. Fraser (1991) "Geographic information systems: the microcomputer and modern cartography" in *Geographic Information Systems: the Microcomputer and Modern Cartography*, Pergamon Press, Oxford, pp. 1–20.

Taylor, D. R. F. (1993) "Geography, GIS and the modern mapping sciences: convergence or divergence", *Cartographica*, Vol. 30, No. 2/3, pp. 47–53.

Wood, D. (1993) "Maps and mapmaking", *Cartographica*, Vol. 30, No. 1, pp. 1–9.

Index